Anne Håkansson, Ronald Hartung, and Ngoc Thanh Nguyen (Eds.)

Agent and Multi-agent Technology for Internet and Enterprise Systems

Studies in Computational Intelligence, Volume 289

Editor-in-Chief
Prof. Janusz Kacprzyk
Systems Research Institute
Polish Academy of Sciences
ul. Newelska 6
01-447 Warsaw
Poland
E-mail: kacprzyk@ibspan.waw.pl

Further volumes of this series can be found on our homepage: springer.com

Vol. 268. Johann M.Ph. Schumann and Yan Liu (Eds.)
Applications of Neural Networks in High Assurance Systems, 2009
ISBN 978-3-642-10689-7

Vol. 269. Francisco Fernández de de Vega and Erick Cantú-Paz (Eds.)
Parallel and Distributed Computational Intelligence, 2009
ISBN 978-3-642-10674-3

Vol. 270. Zong Woo Geem
Recent Advances In Harmony Search Algorithm, 2009
ISBN 978-3-642-04316-1

Vol. 271. Janusz Kacprzyk, Frederick E. Petry, and Adnan Yazici (Eds.)
Uncertainty Approaches for Spatial Data Modeling and Processing, 2009
ISBN 978-3-642-10662-0

Vol. 272. Carlos A. Coello Coello, Clarisse Dhaenens, and Laetitia Jourdan (Eds.)
Advances in Multi-Objective Nature Inspired Computing, 2009
ISBN 978-3-642-11217-1

Vol. 273. Fatos Xhafa, Santi Caballé, Ajith Abraham, Thanasis Daradoumis, and Angel Alejandro Juan Perez (Eds.)
Computational Intelligence for Technology Enhanced Learning, 2010
ISBN 978-3-642-11223-2

Vol. 274. Zbigniew W. Raś and Alicja Wieczorkowska (Eds.)
Advances in Music Information Retrieval, 2010
ISBN 978-3-642-11673-5

Vol. 275. Dilip Kumar Pratihar and Lakhmi C. Jain (Eds.)
Intelligent Autonomous Systems, 2010
ISBN 978-3-642-11675-9

Vol. 276. Jacek Mańdziuk
Knowledge-Free and Learning-Based Methods in Intelligent Game Playing, 2010
ISBN 978-3-642-11677-3

Vol. 277. Filippo Spagnolo and Benedetto Di Paola (Eds.)
European and Chinese Cognitive Styles and their Impact on Teaching Mathematics, 2010
ISBN 978-3-642-11679-7

Vol. 278. Radomir S. Stankovic and Jaakko Astola
From Boolean Logic to Switching Circuits and Automata, 2010
ISBN 978-3-642-11681-0

Vol. 279. Manolis Wallace, Ioannis E. Anagnostopoulos, Phivos Mylonas, and Maria Bielikova (Eds.)
Semantics in Adaptive and Personalized Services, 2010
ISBN 978-3-642-11683-4

Vol. 280. Chang Wen Chen, Zhu Li, and Shiguo Lian (Eds.)
Intelligent Multimedia Communication: Techniques and Applications, 2010
ISBN 978-3-642-11685-8

Vol. 281. Robert Babuska and Frans C.A. Groen (Eds.)
Interactive Collaborative Information Systems, 2010
ISBN 978-3-642-11687-2

Vol. 282. Husrev Taha Sencar, Sergio Velastin, Nikolaos Nikolaidis, and Shiguo Lian (Eds.)
Intelligent Multimedia Analysis for Security Applications, 2010
ISBN 978-3-642-11754-1

Vol. 283. Ngoc Thanh Nguyen, Radoslaw Katarzyniak, and Shyi-Ming Chen (Eds.)
Advances in Intelligent Information and Database Systems, 2010
ISBN 978-3-642-12089-3

Vol. 284. Juan R. González, David Alejandro Pelta, Carlos Cruz, Germán Terrazas, and Natalio Krasnogor (Eds.)
Nature Inspired Cooperative Strategies for Optimization (NICSO 2010), 2010
ISBN 978-3-642-12537-9

Vol. 285. Roberto Cipolla, Sebastiano Battiato, and Giovanni Maria Farinella (Eds.)
Computer Vision, 2010
ISBN 978-3-642-12847-9

Vol. 286. Alexander Bolshoy, Zeev (Vladimir) Volkovich, Valery Kirzhner, and Zeev Barzily
Genome Clustering, 2010
ISBN 978-3-642-12951-3

Vol. 287. Dan Schonfeld, Caifeng Shan, Dacheng Tao, and Liang Wang (Eds.)
Video Search and Mining, 2010
ISBN 978-3-642-12899-8

Vol. 288. I-Hsien Ting, Hui-Ju Wu, Tien-Hwa Ho (Eds.)
Mining and Analyzing Social Networks, 2010
ISBN 978-3-642-13421-0

Vol. 289. Anne Håkansson, Ronald Hartung, and Ngoc Thanh Nguyen (Eds.)
Agent and Multi-agent Technology for Internet and Enterprise Systems, 2010
ISBN 978-3-642-13525-5

Anne Håkansson, Ronald Hartung, and
Ngoc Thanh Nguyen (Eds.)

Agent and Multi-agent Technology for Internet and Enterprise Systems

Springer

Prof. Anne Håkansson
Department of Computer and Systems Sciences
Forum 100, SE-164 40 Kista
Sweden
E-mail: Anneh@dsv.su.se

Dr. Ronald Hartung
Division of Computer and
Information Sciences
Franklin University
Columbus Ohio 43224
USA
E-mail: hartung@franklin.edu

Prof. Ngoc Thanh Nguyen
Institute of Computer Science
Wroclaw University of Technology
Str. Wyb. Wypsianskiego 27
50-370 Wroclaw
Poland
E-mail: thanh@pwr.wroc.pl

ISBN 978-3-642-13525-5 e-ISBN 978-3-642-13526-2

DOI 10.1007/978-3-642-13526-2

Studies in Computational Intelligence ISSN 1860-949X

Library of Congress Control Number: 2010928722

Typeset & Cover Design: Scientific Publishing Services Pvt. Ltd., Chennai, India.

Printed on acid-free paper

9 8 7 6 5 4 3 2 1

springer.com

Preface

Research in multi-agent systems offers a promising technology for problems with networks, online trading and negotiations but also social structures and communication. This is a book on agent and multi-agent technology for internet and enterprise systems. It represents the extension of agent systems into an applications domain for business systems. The book is a pioneer in the combination of the fields and is based on the concept of developing a platform to share ideas and presents research in technology in the field and application to real problems. The chapters range over both applications, illustrating the possible uses of agents in an enterprise domain, and design and analytic methods, needed to provide the solid foundation required for practical systems. The organization of the chapters is arranged into several areas, however the ideas in the chapters overlap between the analytic work and the applications themselves.

The first three chapters of this book present samples of efforts of providing agents and multi-agent technology in distributed environments. It discusses handling agents in networks, and architectures, as well as, service fusion in these environments. Based on some technology in distributed systems, applications for are provided for domain knowledge, trading, information handling and network analysis.

From these chapters, the focus moves towards analytical methods for organizing and managing large agent systems. These conduct logical analyses, multi-agent learning and automatically creating hierarchies.

The following chapters of the book present formal analyses using modal languages and technical solutions to problems with large state spaces with a large set of information. These chapters also present solutions to handle cognitive architectures, mobile agents, personalized ontologies, e-commerce, multimedia and interfaces.

The contributors to the book were carefully selected authors with peer-reviewed chapters. The authors are academics, educators, and practitioners who are doing research in agents and multi-agents technology, as well as, internet and enterprise systems, to meet, improve, and overcome challenges of the internet and enterprise realm. The agent and multi-agent technology is applied in implementing real-life, exciting developments associated with the need to eliminate problems of the organizations.

The opening chapter of the book, Chapter 1, is titled Performance models for multi-agent systems and mobile agent network and authored by Mario Kusek, Kresimir Jurasovic, Ignac Lovrek, Vjekoslav Sinkovic, and Gordan Jezic.

The authors present performance models for distributed multi-agent systems and mobile agent networks applied to the Internet and enterprise systems. The concept of a mobile agent network extends the notion of a multi-agent system to a networked and distributed environment, supporting agent hosting, execution, communication and migration, as well as enabling interaction among service users and service providers. In this chapter, the authors elaborate the problem of establishing and configuring agent-based distributed systems, including agent service capability, allocation of services to agents, and assignment of agents to system nodes. The authors also discuss strategies for distributing services to agents and assigning agents to nodes, and optimization of the distribution.

The following chapter, Chapter 2, titled Service Oriented Architecture and Agents: parallels and opportunities is authored by Ronald L Hartung.

The chapter presents a synergy between Service Oriented Architecture (SOA) and multi-agent systems by comparing them and emphasizing their support for each other. In the chapter, the author introduces the reader to several theses surrounding SOA and agent technical approaches, as well as solutions using agent techniques and agent systems to support SOA. The author gives a review of the principles that underlie SOA and looks at some agent system implementations, from which, analysis of possible architectural styles for agent systems is explored. Architectural styles are similar to patterns and are a useful tool for enterprise architects and, therefore, these are easily extended into agent architectures. From some of the work in agent systems, extensions into the SOA architectures are elaborated. Conclusions of the chapter show that SOA can provide a strong framework for agent systems development and deployment.

The next chapter, Chapter 3, Service Fusion in Mobile Contexts, is authored by Jun Miao Shen, Michael O'Grady, and Gregory O'Hare

In this chapter, the authors use the agent paradigm to extend services beyond the Internet on to mobile devices. The authors reflect on some of the pertinent issues necessary for the agents' constructing a viable platform for service delivery to mobile users, including Agent Oriented Software Engineering (AOSE). Developments in AOSE and agent frameworks result in an increasing range of options for those considering harnessing the characteristics of agents for their applications and services. The discussion is illuminated by a systematic description of EasyLife – a framework that harnesses the agent paradigm to construct and deliver services to mobile users. EasyLife demonstrates the viability of the agent paradigm for integrating heterogonous information sources for service provision to mobile users. The robustness of the agent platform is verified through the use of two separate agent toolkits being harnessed to deliver EasyLife services.

The chapter that follows, Chapter 4, is titled Environmental Impact Assessment by Multi-Agent Systems and is authored by Marina V. Sokolova and Antonio Fernández-Caballero.

The authors propose a generalized multi-agent system technology (MAST) that facilitates the workflow of multi-agent system's planning and implementation in decision support systems (DSS) applied to Environmental Impact Assessment (EIA). This research has already showed promising results in the field of Environmental Impact Assessment and implementation of decision support systems, and the authors demonstrate an extension of a multi-agent decision support system to new problem domains by changing the domain ontology. The authors introduce and discuss how the functional organization and the roles, executed in the MAS, conform to the logical sequence of the data transformation flow, which includes information retrieval and fusion, data preprocessing, data mining and modeling, and simulation and decision making.

The next chapter, Chapter 5, is titled Multiagent Forex Trading System and authored by Rui Pedro Barbosa and Orlando Belo.

The authors of this chapter use multi-agent systems to implement currency automated trading systems, which is a specific case of e-commerce. The work analyzes the intelligent agent's ability to act autonomously as well as adapt and interact with the environment. The authors introduce and describe the implementation of a multi-agent system that can achieve results trading in the Forex market. The implementation includes a set of autonomous currency trading agents, using an infrastructure that consists of an ensemble of classification and regression models, a case-based reasoning system, and an expert system. The authors also demonstrate how investment diversification can be used to reduce the trading risk, and show how enabling the agents to communicate among themselves can help reduce the trading costs and the capital requirements. The system can lessen the impact of the trading costs to increase the total profit by integrating the agents in a multi-agent system in which the agents communicate with each other before opening new trades.

The Chapter that follows, Chapter 6, is titled Information coordination using Meta-Agents in Information Logistics Processes and authored by Mats Apelkrans and Anne Håkansson.

The chapter presents a work that shows the combination of multi-agent system and information logistic process (ILP) to facilitate the information workflow in business processes. The authors combine theories from multi-agent systems and Information Logistics (IL) to coordinate and deliver information at the right time and to the right place at an acceptable cost. The multi agent system contains agents and meta-agents that search, coordinate, and deliver the information. Information Logistics is, on the other hand, an information product that becomes accessible and delivered to the information receiver who can make use of the information. The agents and the meta-agents support the information flow between the manufacturing company, its customers, and its suppliers in the form of orders, requirements, and production plans, which can speed up the information exchange and optimize the production cost.

The next chapter, Chapter 7, is titled Social Network Analysis Based Approach to Trust Modeling for Autonomous Multi-agent Systems and authored by Grzegorz Kołaczek.

The chapter extends social network analysis to multi-agent systems in small-world on a scale-free network to generate trust relationships. The author presents a general concept about trust modelling and management using a network based analysis of relations between autonomous agents and an analysis method for autonomous multi-agent systems. The method is based on the assumption that multi-agent systems constitute social networks and the network specific parameters can be used to evaluate the trust levels. The author describes how metrics like centrality and clustering coefficient used in the description of these types of networks can be used to model trust relation between agents. Thus, the parameters describing small-world or scale-free networks can be also used to analyze the position and relations among agents within their societies.

The following chapter, Chapter 8, titled Logical Analysis of Multi-Agent Systems is authored by Sergey Babenyshev and Vladimir Rybakov.

The independent operation of multi-agent systems increases the difficulty of understanding and analysing the agents' operations. In this chapter, the authors discuss several interdependent theoretical tools from modern mathematical logic and computer science for formalizing the informal notion of multi-agent systems. The authors explain how the modal propositional language may be used to specify the behaviour of the processes, and formally introduce the Kripke semantics for modal languages. The authors also discuss different techniques of combining simple modal languages into complex ones, more suitable for practical purposes, and conclude with an extended example, illustrating the major points of the above-mentioned techniques.

The next chapter, Chapter 9, is titled Multi-agent systems and large state spaces and authored by Yann-Michaël De Hauwere, Peter Vrancx, and Ann Nowé.
When a multi-agent system uses reinforcement learning, the size of the state space inhibits effective learning. A solution explored by the authors of this chapter is Generalized Learning Automata (GLA). The GLA can learn generalisations over a large number of states, without needing to store large amounts of samples, perform computational intensive statistical tests, or build a model of the environment. GLA can help take the correct actions in large, unknown environments. The authors introduce a general framework for multi-agent learning, where learning happens on two separate layers, and agents learn when to observe each other. Within this framework, the authors introduce a new algorithm, called 2observe, which uses a GLA approach to distinguish between high-risk states where the agents have to take each other's presence into account and low risk states where they can act independently. The algorithm is applied to a grid-world problem since it has similarities to some real-world problems, such as autonomous robot control.

The chapter that follows, Chapter 10, is titled Automatically creating hierarchies of agents and meta-agents using clustering and authored by Anne Håkansson.

This chapter describes generating hierarchies of agents to control and master problem solving in multi-agent systems. The hierarchies are established automatically in response to the problem solving. The author presents how to create hierarchies of agents automatically from the result of searching for information. The agents, both intelligent software agents and meta-agents, are represented in a hierarchy to cluster similar agents and optimize locality. The commonality between the agents represents semantics, and constitutes information chains for solutions. The clustering can support avoiding the unsuccessful paths and makes the search efficient. In addition, the hierarchy supports reasoning at different levels in hierarchy by dividing the overall task for the agents in the hierarchy. As a test environment, the author uses the web, since it can produce a vast amount of information that needs structuring in order to be manageable and accessible. The results of the search infrastructure are automatically cast into a more manageable and understandable hierarchy.

The following chapter, Chapter 11, titled A Role-based Cognitive Architecture for Multi-Agent Teaming is authored by Jinsong Leng, Chee Peng Lim, Jiuyong Li, Dongguang Li and Lakhmi Jain.

The authors in this chapter introduce a development of a role-based architecture to enable teaming and collaboration to organise problem solving in a multi-agent system. Multi-agent teaming can be constructed in a layered architecture with individual agent behaviours and team behaviours where the architecture uses Belief, Desire, and Intension (BDI) as an argumentation framework. The role-based BDI framework facilitates optimization problems at the team level such as competitive, cooperation, and coordination problems. This BDI framework is extended on the commercial agent software development environment known as JACK Teams. The layered architecture has been used to group the agents' competitive and cooperative behaviours. The chapter includes a presentation of the use of reinforcement learning techniques to learn different behaviours through experience. These issues have been investigated and analysed using a real-time, 2D simulation environment known as Soccer-Bots.

The next chapter, Chapter 12, is titled Mobile Agents and Eventually Perfect Predicate Detection: an Intelligent General Approach to Monitoring Distributed Systems and authored by Chunbo Chu and Monica Brockmeyer.

In this chapter, the authors introduce a monitoring approach to oversee a multi-agent system using a form of prediction detection that is guaranteed to reach correct states. The authors present an application of mobile agents to solve distributed predicate detection in realistic applications. As a solution, the authors make an extension of the study of predicate detection in failure-prone systems. The extension includes applying the mobile agent technique to implement an eventually accurate failure detection sequencer in a Finite Average Response (FAR) model. In this chapter, the authors propose a general-purpose solution for predicate detection in partially synchronous distributed systems that contributes to predicate detection in large-scale distributed systems with crash failures and limited synchrony.

The chapter that follows, Chapter 13, is titled Personalized Ontology-based Recommender Systems for Multimedia Objects and authored by Krzysztof Juszczyszyn, Przemysław Kazienko and Katarzyna Musiał

The chapter demonstrates the application of ontology formation to construct personalised recommendations in the multimedia domain. The Authors propose a framework for recommendation of multimedia objects based on processing of individual ontologies. The technique compares ontologies as whole structures to assess similarity between multidimensional profiles of users and multimedia objects in the recommender system. The ontologies provide the comprehensive view of the information gathered in the multimedia sharing system. As a result, the executed recommendation process takes into account similarities calculated both between objects' and users' ontologies, which reflect the social and semantic features existing in the system. The ontologies, which are close to the current context, provide a list of suggestions presented to the user. Each user in the system possesses its own Personal Agent that performs all necessary online tasks. Personal Agents co-operate each other and enrich lists of possible recommendations. The system is developed for the Flickr multimedia sharing system.

The next chapter, Chapter 14, is titled Toward Emotional E-Commerce and authored by Nicolae Jascanu, Veronica Jascanu and Severin Bumbaru.

In this chapter, the authors apply emotional models to negotiate in E-commerce applications and introduce the main actors of the multi-agent system. The authors propose a multi-agent system for electronic commerce that integrates emotional models for each one of the three agents: the customer, the supplier, and the community. Here the authors are using a continuous approach as a model for emotions based on the seminal work of James Russell. The novelty of the work in this chapter is the mapping of the emotional model to the electronic commerce field. For this, the consumer and supplier behaviors are translated in terms of emotional models and an algorithm to process negotiation configurations. By including an emotional community agent in the negotiation to support the customer and the supplier, the work lays out a foundation for emotional based argumentation.

The last chapter of this selection, Chapter 15, is titled Web-based Affective Human-agent Interaction Generation and authored by Xia Mao and Zheng Li.

In this chapter, the authors describe the construction of perceptive user interfaces using agent technologies. This is a higher level of interface for agent applications especially in the Internet and extends the usability of agents. The authors describe the architecture of Multimodal Interaction Markup Language (MIML), a XML-based markup language for scripting the web-based intelligent interaction of virtual agents. MIML cannot only simply respond with a pre-defined utterance dependent on the user's input, but also produces dynamic responses, which are determined by the user's emotional state. Unlike existing related languages, MIML can script not only the presentations of virtual agents, but also their affective capability. The authors also describe facial expression recognition, and speech emotion recognition. They also describe emotional speech synthesis ActiveX controllers and illustrate one scenario that instantiates the affective web-based human-agent interaction scripted by MIML. With the MIML design, web-based affective interaction can be described and generated easily. By the actual web-based intelligent interaction implemented with MIML, the tutoring system shows how to mark up the interaction system that features a virtual agent with intelligent behaviour.

AnneHåkansson
Stockholms University, Kista, Sweden

Ronald Hartung
Franklin University, Columbus, Ohio, USA

Ngoc Thanh Nguyen
Wroclaw University of Technology, Wroclaw, Poland

Table of Contents

Performance Models for Multi-agent Systems and Mobile Agent Network

Mario Kusek*, Kresimir Jurasovic, Ignac Lovrek,
Vjekoslav Sinkovic, and Gordan Jezic

University of Zagreb
Faculty of Electrical Engineering and Computing
Unska 3, HR-10000, Zagreb, Croatia
Tel.: +385 1 612 98 01; Fax: +385 1 612 98 32
{mario.kusek,kresimir.jurasovic,ignac.lovrek,
vjekoslav.sinkovic,gordan.jezic}@fer.hr

Abstract. This chapter deals with performance issues in multi-agent systems and mobile agent networks applied to the Internet and enterprise systems. A mobile agent network extends the notion of a multi-agent system to the networked and distributed environment defining the Internet and enterprise systems which supports agent hosting, execution, communication and migration, and enables interaction among service users and service providers. The basic performance measures relevant to agent-based Internet applications and enterprise systems are response time and throughput. For performance evaluation purposes, a multi-agent system and a mobile agent network are described as a queuing system where an agent represents an information unit to be served. The problem of establishing and configuring agent-based distributed systems, including agent service capability, allocation of services to agents, and assignment of agents to systems nodes, is elaborated upon. Strategies for distributing services to agents and assigning agents to nodes, and their optimisation are discussed. Operational performance is analysed with network calculus using a theory of deterministic queuing. Case studies, as well as simulation-based performance evaluation results are presented.

1 Introduction

Internet applications and enterprise systems are faced with the problems associated with user mobility and the requirements for seamless service provisioning. In an environment where everything is mobile, including the users and their equipment, services should be provided on the move, whether the user is moving within the same network or while changing networks during a session. Furthermore, personalization, communication context, as well as user communities and business alliances, are becoming increasingly important to both users and service providers. The complex relationships between service users and service providers should remain or become manageable, a task in which intelligent software agents

* Corresponding author.

A. Håkansson et al. (Eds.): Agent and Multi-agent Technology, SCI 289, pp. 1–24.
springerlink.com

will play an important role. The multi-agent paradigm is intrinsically communication and cooperation-oriented and therefore suitable for the Internet and enterprise systems: agent-based services and applications that rely on agent intelligence and mobility can cope with both the number of users and the complexity of applications. The features of intelligent software agents make them applicable in enterprise systems and electronic markets (e-markets). For example, in the Trading Agent Competition Supply Chain Management (TAC SCM) game, each agent acts as an independent entity in a simulated environment. Defined game scenario provides a unique testbed for studying and prototyping SCM agents by providing a competitive environment in which independently created intelligent software agents can be simulated and tested [1].

This chapter deals with the performance issues for multi-agent systems and mobile agent network. The concept of a mobile agent network extends the notion of a multi-agent system to a networked and distributed environment, supporting agent hosting, execution, communication and migration, as well as enabling interaction among service users and service providers. The basic performance measures relevant to agent-based Internet applica- tions and enterprise systems are response time and throughput, i.e. the time required to handle a service request and the number of services completed in a time unit. For performance evaluation purposes, we describe a multi-agent system and a mobile agent network as a queuing system where each agent represents an information unit to be served.

Multi-agent technology can be evaluated from two points of view: agent-based establishment and maintenance of the Internet and enterprise systems; and agent-based services and applications in Internet and enterprise systems. Therefore, two independent problems influencing system performance are considered. First, the problem of establishing, configuring and reconfiguring agent-based distributed system is elaborated upon. Strategies for distributing service resources to systems and using agents to deploy them are discussed. Furthermore, optimisation strategies used to minimize configuration or reconfiguration set-up time, taking into account network conditions, are described. The second problem of operational performance is analysed with network calculus, using a theory of deterministic queuing based on Min-Plus algebra. The underlying idea is to regulate the flow of agents carrying services, as well as achieve service guarantees through deterministic scheduling.

The chapter is organised as follows: multi-agent systems and mobile agent networks are defined in the next section, followed by a short overview of related work in the area of agent performance research. Section 4 deals with agent-based configuration of Internet and enterprise systems and includes optimisation of service deployment strategies using a genetic algorithm. Performance analysis using network calculus is elaborated upon in the Section 5, with a case study in Section 6.

2 Multi-agent Systems and Mobile Agent Network

When applied to network-centric applications, such as distributed Internet and enterprise systems, agents represent service users or service providers and act autonomously on behalf of their owners. The functionality of an agent-based Internet or enterprise system is defined by the set of elementary services, $ES = \{es_1, es_2, \ldots es_j, \ldots, es_{nes}\}$, offered to user agents by provider agents. Agent intelligence and mobility influence their service capability with respect to functionality and performance. Depending on the complexity of the applications, agents perform the services requested of them by a user or provider either alone, or as a member of an agent team operating within a multi-agent system. Agents can compete or communicate and co-operate to serve their owners.

In order to define the system parameters required for performance evaluation, a formal model from [2] based on a mobile agent network will be used. A mobile agent network (Figure 1) is represented by the following triple: $\{A, S, N\}$ where
A – is a multi-agent system consisting of co-operating and communicating stationary and mobile agents,
S – is a set of processing nodes – servers in which the agents perform services,
N – is a network, i.e. the Internet that connects processing nodes and allows agent communication and mobility.

An agent represents a service user or service provider and performs tasks on behalf of its owner. It can communicate with other agents and migrate

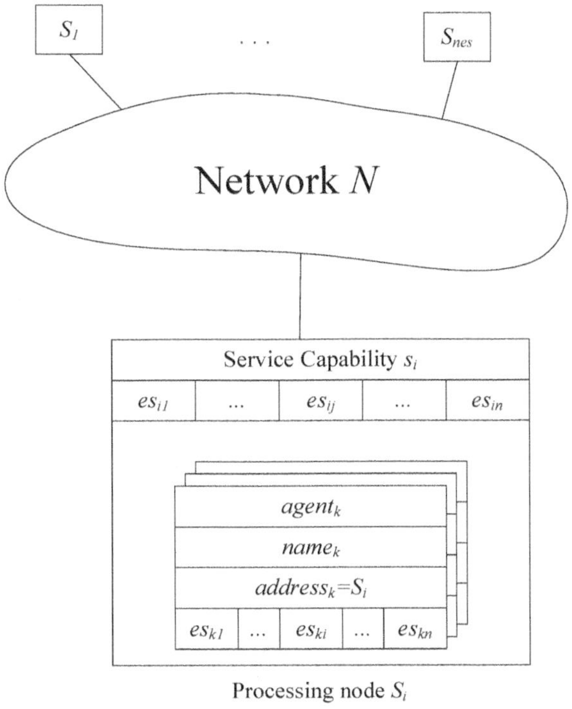

Fig. 1. Mobile Agent Network

autonomously from node to node to perform a service. An agent is defined by a triple:

$agent_k = \{name_k, address_k, service_k\}$,

where $name_k$ represents a unique agent identification, $address_k$ its location and $service_k$ functionality it provides in the form of $service_k = \{es_{k1}, es_{k2}, \ldots, es_{ki}, \ldots, es_{kn}\}$ representing a set of assigned elementary services from ES. The same elementary service can be provided by multiple agents, such as personal agents representing human users. A single agent can provide multiple elementary services , e.g, different business transactions, for its owner. Therefore, agents are considered as multiservice agents.

Each node, S_i, is characterised by the set of elementary services, $s_i = \{es_{i1}, es_{i2}, \ldots, es_{ij}, \ldots, es_{in}\}$ it supports. The same elementary service can be supported by multiple nodes, and one node can support multiple services. If $agent_k$ is hosted by node S_i or directed towards it, value $address_k = S_i$. Elementary services from the intersection $service_k \cap s_i$ can be executed at S_i.

A network N is represented by an undirected graph, $N = (S, E)$. The set of processing nodes is denoted as $S = \{S_1, S_2, \ldots, S_i, \ldots, S_{ns}\}$. E is a set of links, $E = \{e_1, e_2, \ldots, e_m, \ldots, e_{ne}\}$, where $e_m = \{S_i, S_j\}$ represents a link between nodes S_i and S_j. Figure 2a shows an example network, while Figure 2b shows its matrix representation.

From the operations point of view, each service request is decomposed into a set of elementary services that are presented by a directed acyclic operations graph $G = (T, L)$. T denotes the list of elementary services comprising a request with the address of the node where they should be executed, $\{es_i, address_i\}$. If the address is not defined es_i can be executed on an arbitrary node S_j that supports es_i, i.e. $es_i \in s_j$, while the agent decides where the elementary service will be executed according to its knowledge and intelligence. L denotes the set of directed edges that define precedence relations between elementary services. An edge is described by a quadruple $\{es_i, o_i, es_j, i_j\}$ where o_i and i_j represents

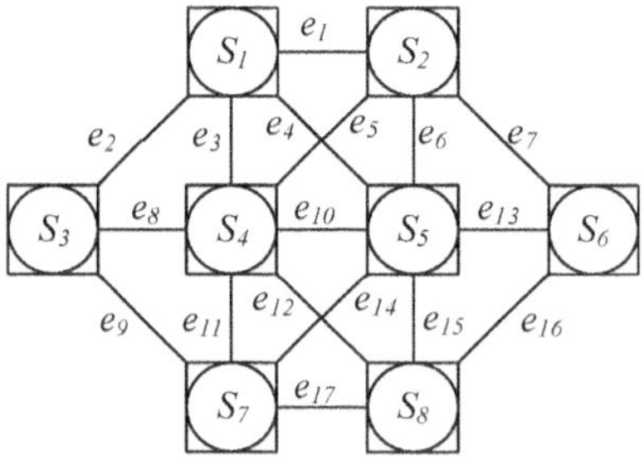

$$E = \left\{ \begin{array}{c} \{S_1, S_2\} \\ \{S_1, S_3\} \\ \{S_1, S_4\} \\ \{S_1, S_5\} \\ \{S_2, S_4\} \\ \{S_2, S_5\} \\ \{S_2, S_6\} \\ \vdots \\ \{S_7, S_8\} \end{array} \right\}$$

a) Graphical Representation b) Matrix Representation

Fig. 2. Network representation

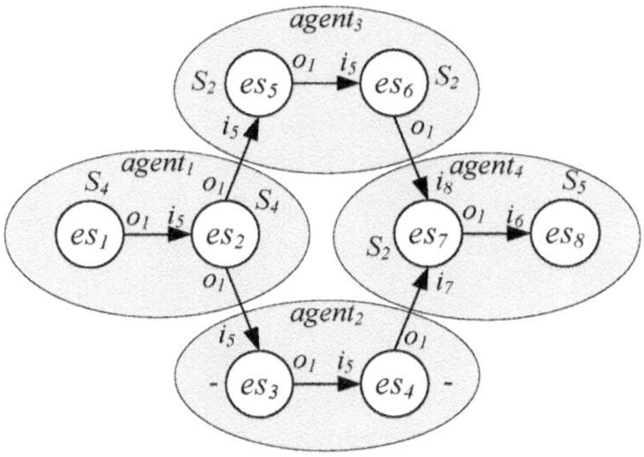

$$T = \left\{ \begin{array}{c} \{1,4\} \\ \{2,4\} \\ \{3,-\} \\ \{4,-\} \\ \{5,2\} \\ \{6,2\} \\ \{7,1\} \\ \{8,5\} \end{array} \right\} \qquad L = \left\{ \begin{array}{c} \{1,1,2,5\} \\ \{2,1,3,5\} \\ \{2,1,5,5\} \\ \{3,1,4,5\} \\ \{4,1,7,7\} \\ \{5,1,6,5\} \\ \{6,1,7,8\} \\ \{7,1,8,6\} \end{array} \right\}$$

a) Graphical Representation b) Matrix Representation

Fig. 3. The operations graph

output data of es_i and input data of es_j respectively, defining that $i_j = o_i$ and es_i precedes es_j.

An example of an operations graph is shown in Figure 3, displaying both its graphical (Figure 3a) and matrix representation (Figure 3b). Elementary service es_2 should be executed at node S_4. This is defined in the second row of T. In the third row, the dash in place of the execution node denotes that elementary service es_3 is not pre-assigned to a specific node for execution. The edge connecting output o_1 of es_1 with input i_5 of es_2 is represented by the first row of matrix L.

To execute all operations included in the graph, there is a team of agents. The process of assigning elementary services includes determining the number of agents in the team, as well as deciding which agent will execute which elementary service [3]. Consequently, intelligent agents can choose different distributions depending on their knowledge of their service execution ability, the network topology, the current network and node load, etc. Consider the following example: $agent_1$ will execute es_1 and es_2 and, thus, $service_1 = \{es_1, es_2\}$; $agent_2$ will execute the following two, i.e., $service_2 = \{es_3, es_4\}$. The next two agents, $agent_3$ and $agent_4$, are assigned services as follows: $service_3 = \{es_5, es_6\}$ and $service_4 = \{es_7, es_8\}$. It is evident that $agent_1$ will first execute es_1 and then the es_2 on the same node (i.e., S_4). However, $agent_4$ will execute es_7 on node S_1 and then migrate to S_5 where it will execute es_8. $agent_2$, on the other hand, can decide where to execute its services since no nodes are predefined.

From the performance perspective, two problems are studied:

1. service distribution, i.e. allocating service resources to processing nodes and using agens to deploy them, and
2. service execution, i.e. serving user requests.

For performance evaluation purposes, we describe a mobile agent network as a queuing system where an agent represents an information unit to be served. Network flows of agents are defined by the parameters defining agent hosting, execution, communication and migration. Parameters defining the mobile agent network as a networked distributed system are processing nodes and their capacity and network connectivity and link capacity.

3 Related Work

Agent performance research is based on different theoretical frameworks, including analytical or simulation procedures, offering various approaches to determining the basic performance measures of response time and throughput. Models are much more complicated then ones suitable for the client-server paradigm and basic network performance analysis [4], because of agent autonomy and relationships among agents that produces responsibility of a single agent or agent team for service provisioning.

Queuing theory applied to multi-agent systems and mobile agent network is based on the following modeling assumption: agents are represented as information units to be served at nodes capable of hosting and executing the agents, connected by a network providing agent communication and migration. The stochastic nature of service requests implies that agent execution, communication and migration are stochastic processes. Classical queuing theory considers stochastic processes and their average quantities in an equilibrium state, i.e. response time for a service at a node, overall service response time, agent execution time, agent migration time and overall agent response time [2,5,6].

Markov decision processes are used in multi-agent co-ordination research and team performance analysis [7,8,9], while Petri nets are studied as evaluation and verification tools [10,11]. In this paper, agent performance evaluation tools are extended to network calculus, using a theory of deterministic queuing systems [12,13], and performance assessment based on simulations. Some earlier work also applied network calculus to communicating agents [14].

4 Agent-Based Configuration of Internet and Enterprise Systems

Operations performed by agents for agent-based establishment and maintenance of Internet and enterprise systems include software delivery to system nodes, remote installation, start and stop, version handling, as well as modification of software after delivery. The latter can be used to correct faults, improve performance, adapt to a changed environment and/or improve maintainability. Such systems should be configured initially, and re-configured when corrective, perfective, adaptive or preventive actions are required.

From the formal standpoint, system nodes $S = \{S_1, S_2, \ldots, S_i, \ldots, S_{ns}\}$ should be configured in order to provide support for the set of elementary services, $ES = \{es_1, es_2, \ldots, es_j, \ldots, es_{nes}\}$. Re-configuration is required when an

ES changes (e.g., a new elementary service is introduced or an existing one is updated), an S changes (e.g., a new node is connected or an existing one is upgraded) or N changes (e.g., a change in network topology or link capacity).

Intelligent and mobile agents provide the means to establish, change and enable the evolution of Internet and enterprise systems, taking partial or full responsibility of these resource-intensive and costly tasks.

Regarding performance issues, configuration, as well as re-configuration, should minimally influence system operation, and should therefore be optimised.

4.1 Service Deployment

Agent-based frameworks for remote software deployment and maintenance are proposed in [15,16]. Service deployment must take into account the characteristics of nodes (their capacity, operating system, agent platform, installed software) and the network (its topology, link bandwidth), as well as procedures specific to the system under consideration [17,18].

The problem of establishing and configuring agent-based distributed Internet and enterprise systems is stated as follows:

1. Define the set of nodes S that comprises the distributed system. Each node $S_i \in S_3$ is capable of hosting agents and provides agent communication and migration facilities.
2. Define the connectivity of nodes in S over network N.
3. Define subsets of elementary services that should be supported by each node S_{ij}, $s_i = \{es_{i1}, es_{i2}, ..., es_{ij}, ..., es_{in}\}$.
4. Deploy the service resources, i.e. software components and data, that are required to execute elementary services defined in the previous step to all nodes.

A service deployment system called the Multi–Agent Remote Maintenance Shell (MA–RMS) [19] is based on a formal model of the mobile agent network. In the prototype, agents are created using the agent platform JADE [20]. The JADE agent platform conforms to FIPA standards which define that agents and their locations are tracked and that any agent can find out the current location of any other agent in the platform [21].

The multi–agent system MA–RMS is organised as a team of agents, A_{RMS}, that share a set of nodes, S, and a network, N, with multi-agent system A which implements an Internet or enterprise system, i.e.:
(A, A_{RMS}, S, N).
One of the A_{RMS} agents is the planning agent responsible for distributing service deployment tasks to other agents in the team and co-ordinating them. A_{RMS} agents are multiservice agents capable of executing one, few or all service deployment tasks on one, several or all nodes. A single task corresponds to an elementary service, i.e. service resources supporting it at a specific node. When deploying services on a large number of nodes at remote locations using software agents, one of the major problems faced is determining the number of agents to

use, as well as distributing the required services among the agents. Deployment planning depends on a service distribution strategy. Some strategies follow:

- R1: a single agent deploys all services on all nodes;
- R2: an agent deploys a single service on one node only;
- R3: an agent deploys all services on one node only;
- R4: an agent deploys a specific service on all nodes;
- R5: tasks are assigned to the agents in order to exploit maximal parallelism: mutually independent tasks are assigned to different agents, in order to deploy services simultaneously on a node with parallel deployment support;
- R6: a hybrid solution combining R4 and R3: some agents are responsible each for a specific service on all nodes; other agents deploy all other services, each on a different node.

Inefficient distribution of services to agents can result in substantially larger deployment time. Therefore, optimisation of strategies with respect to node and network characteristics is required in order to achieve minimum configuration or reconfiguration set-up time [22]. To accomplish this goal, we propose a new strategy called RGA. This strategy uses a genetic algorithm to generate a sub-optimal distribution of services among agents. The genetic algorithm used is explained in detail in the next section.

4.2 Optimizing Service Deployment Strategies Using a Genetic Algorithm

Executing remote service software deployment on a large number of nodes can be time consuming if services are not efficiently distributed among agents. For example, the R3 distribution is only optimal when used in network topologies where all nodes are interconnected with the links of the same bandwidth. In order to optimize the deployment strategy, a heuristic approach using a genetic algorithm is proposed. The algorithm is used to optimize the number of agents and the distribution of services among them for a specific network topology with fixed network parameters.

In general, designing a genetic algorithm involves the following steps:

1. Designing a chromosome which describes the problem domain;
2. Constructing a fitness function which evaluates the effectiveness of individual chromosomes generated by the genetic algorithm;
3. Selecting which genetic operators to use;
4. Setting the parameters of genetic operators (e.g. crossover probability, mutation probability, population size, number of evolutions).

Figure 4 shows the definition of our chromosome. The chromosome is defined by $AS_{nes} = \{g_1, g_1, \ldots, g_i, \ldots, g_{nes}\}$ where g_i defines allocation of $es_{i+1} \in ES$ to $agents_k \in A_{RMS}$. The number of agents that will execute elementary services is extracted from the chromosome by analysing service distributions among agents. Only the agents that were assigned elementary services would be created. The maximum number of agents is restricted and can vary between 1 and nes.

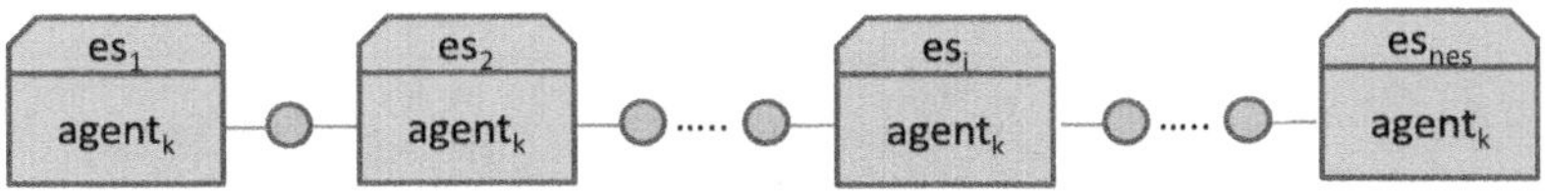

Fig. 4. Chromosome

The number of genes in the chromosome is equal to the number of elementary services that need to be executed. The locus (i.e., position) of the gene within the chromosome defines the g_i. In other words, first gene defines g_1 i.e., the agent that will execute service es_1, the second gene defines g_2 i.e., the agent that will execute service es_2, and so on up until the nes-th gene which represents g_{nes} i.e., the agent which will perform service es_{nes}. A chromosome defined as such uses direct encoding i.e., the solution is directly encoded in the chromosome. Each allele is defined as an integer.

In most cases the fitness function is a mathematical formula used to calculate the chromosome's efficiency (fitness). But due to the complexity of agent interaction, dependency between the elementary services as well as the dynamicity of the agent system a formula that could be applied in all the cases cannot be defined. To cope with this difficulty, we use the MAN Simulator [23,24,25] to calculate the efficiency of the distribution defined by the genes of the chromosome.

The MAN Simulator is capable of simulating a variety of network topologies with different network parameters. Furthermore, the simulator can be used to simulate the execution of different service distribution strategies. The simulator takes the following parameters as input: the network topology, the service distribution strategy and the set of services which have to be executed. The network topology parameter defines the network used while migrating agents to nodes. It contains information about all the nodes in the network and how they are connected. It also defines the characteristics of the network, such as network bandwidth, latency, node processing delays, etc. The service distribution strategy defines how services should be distributed among the agents. The result obtained from the simulation is the time needed for the agents to execute all the assigned services.

When the genetic algorithm needs to calculate the fitness of the chromosome, it uses the fitness function. In our case the result of the fitness function is obtained by running a simulation. The simulator accepts three parameters: the network topology, a set of services and the distribution of services among agents. The first two parameters are fixed and have to be defined before the execution of the simulator. The distribution of services among agents is obtained from the chromosome whose fitness we need to calculate. The result of the simulator is the time needed to execute all the services acording to defined parameters. The fitness returned by the fitness function is then calculated as $1/t_s$ where t_s is the time returned by the simulator. Chromosomes whose distributions produce smaller t_s values are more fitter then other.

The genetic algorithm optimizes the distribution of services in the following way:

1. The algorithm first generates the initial population. The size of the population is constant and is set to 25. The initial distribution of services in the chromosomes is selected at random;
2. The algorithm then uses genetic operators. The first is the ranking single point operator. The first step performed by this operator is to sort all chromosomes in the population according to their fitness, from lowest to highest. The next step is to calculate the selection probability for each chromosome, calculated by the following formula:

$$P_s = \frac{n_r}{\sum_{i=1}^{n} i} \tag{1}$$

   where n_r is the position of the current chromosome within the sorted list of chromosomes and n is the number of chromosomes in the population. It can be seen from formula 1 that chromosomes with higher fitness values have a larger probability of selection. The operator then selects two chromosomes from the population according to their selection probabilities. Genes between the two chromosomes are then exchanged after randomly selected locus (position). The two newly generated chromosome are then added to the population. Crossover probability is 50%; Crossover produces distributions where some agents are excluded from the distribution since they don't execute any services but it can also include new agents. This step is crucial since it is used to optimize the number of agents used as well as the distribution of services among agents.
3. The second operator is mutation operator which selects a number of genes from the population. It then selects two loci (positions) for each of them and exchanges their genes at those positions. Crossover probability is 1%; The mutation is used to prevent the algorithm to converge to a local optimum. Random mutations can create new distributions and allow the algorithm to explore new regions of the search space.
4. The last operator is best natural selector which first ranks all the chromosomes according to their fitness and assignes selection probabilities to each of them. The formula used is the same as for the crossover operator. The best chromosomes are then selected and are used in the next evolution of the genetic algorithm. The number of selected chromosomes is equal to the size of the population (25).

Steps 2–4 are repeated in every evolution. The number of evolutions is 25.

4.3 Simulation–Based Performance Assessment

In order to verify the efficiency of the proposed genetic algorithm, we performed simulations on specific network topologies. Figure 5 shows the overall network topology. The execution is started from node S0 while the services are executed

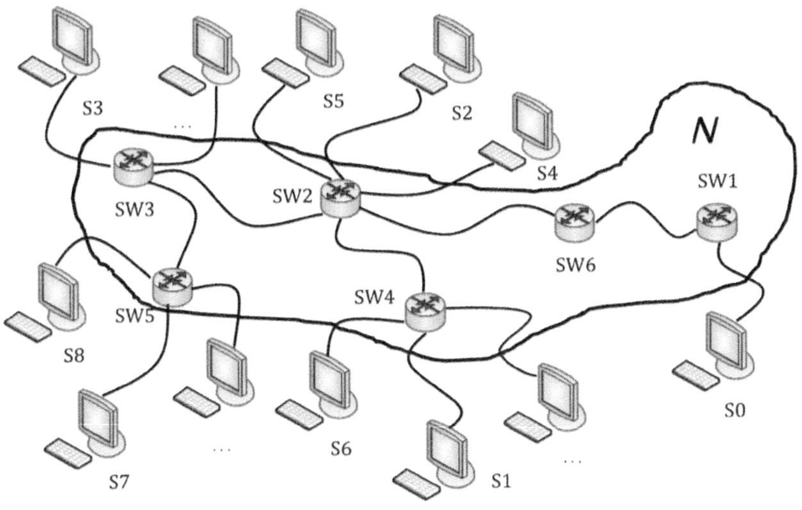

Fig. 5. The network topology used in the simulations

on nodes S1–S8. The performance assessment was performed in three simulation series. Switch network element processing time was set to Δt while service execution times were set to $125 \times \Delta t$, $326 \times \Delta t$, $92 \times \Delta t$, $131 \times \Delta t$ and $36 \times \Delta t$. All services were deployed on all nodes during the simulations.

In the first series of simulations, the bandwidth of all links was first set to C, then to $2 \times C$, and finally to $20 \times C$. In the second series of simulations, the link between switch SW6 and SW2 was kept constant at C while the remaining link bandwidths were first set to $2 \times C$, and then to $20 \times C$. In the last simulation, link bandwidths between SW3 and SW5 and between SW2 and SW4 were kept constant at C, while the remaining link bandwidths were set to $2 \times C$ and then to $20 \times C$.

The results of the simulation series are shown in figure 6 (simulation series 1), figure 7 (simulation series 2) and figure 8 (simulation series 3). Values in the graph are normalized to the best distribution in each simulation. The best distribution has a value of 100% while the values of the remaining distributions are calculated according to the following formula:

$$\frac{V_d}{V_n} \times 100\%. \qquad (2)$$

where V_n is the best distribution in the simulation and V_d is the distribution under consideration.

From these results we can see where the genetic algorithms (RGA strategy) gave better results when compared with previously defined scenarios. In the first series of simulations, we used constant bandwidth for all links in the simulated network. In this scenario, the two best distributions where the RGA and the R3. The R3 distribution strategy was better in the scenarios with $2 \times C$ and $20 \times C$

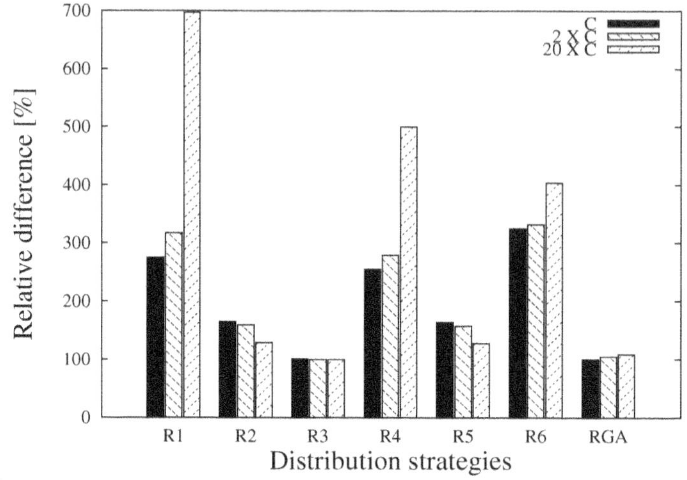

Fig. 6. Results from simulation series 1

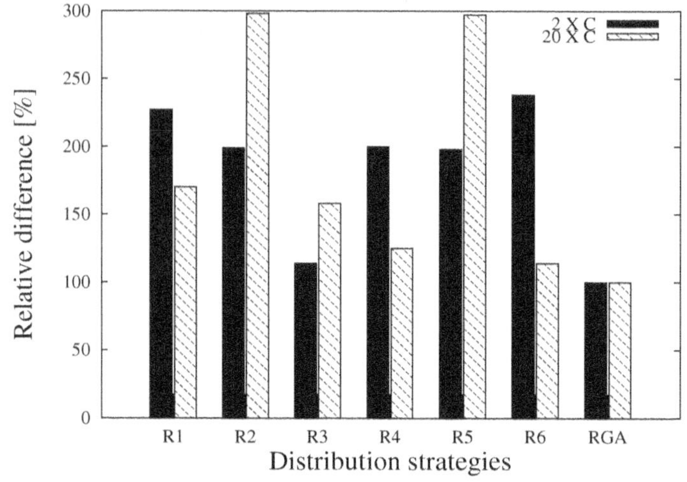

Fig. 7. Results from simulation series 2

links bandwidths by 6% and 10%, respectively. In the scenario with C links, the RGA outperformed the R3 distribution strategy by 1%. For comparison, the third (R6) and the last (R1) distribution perform on average 50% and 430% worse, respectively, for the entire scenario.

In the third simulation series, the R3 strategy obtained the best result in all simulations (2% in the with $2 \times C$ bandwidth links and 15% with $20 \times C$ links compared to RGA strategy).The third (R5) and the last (R1) distribution performed on average 66% and 357% worse, respectively, for the entire scenario.

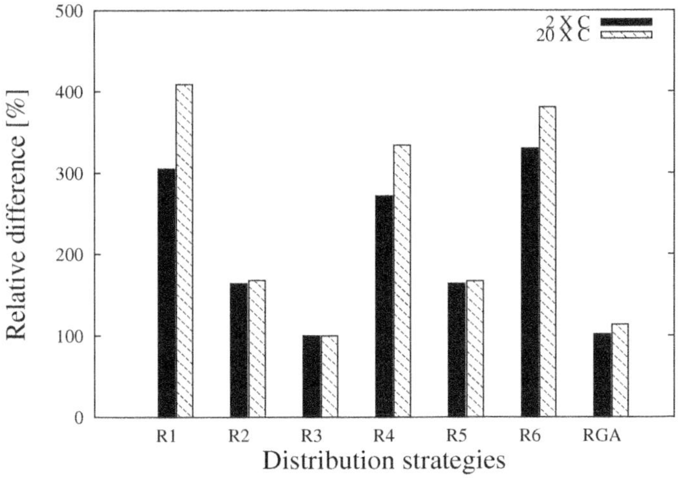

Fig. 8. Results from simulation series 3

The RGA was the most effective in the second simulation series. In all cases the RGA obtained the best results. The second best distribution strategy in the scenarios with $2 \times C$ links (R3) and the scenario with $20 \times C$ links (R6) had 14% worse results.

4.4 Performance Assessment Discussion

The results of the first simulation where as expected since it is known that the R3 distribution strategy always performs best in scenarios where the time needed to migrate all service components is significantly smaller than time needed to execute the services. However, when link speed was reduced to C, the genetic algorithm distribution (RGA) was able to generate a better distribution then the R3 strategy. The same goes for the third simulation series where the RGA did not obtain the best results, but still produced results very close to those obtained by R3. The second simulation series showed scenarios where the genetic algorithm produced the best results. The reason for this lies in the bottleneck created close to the starting node S0, making the time needed to migrate service components comparable to the time needed to execute them. In such cases, the genetic algorithm was able to produce distributions in which the first agent had more services assigned to it then the second agent. The same rule applies to all the remaining agents, i.e. the second agent had more services than the third agent. While the second agent was still migrating into the first node, the first agent was deploying its service on other nodes. Other strategies produced distributions in which agents had to wait for other agents due to long migration times.

5 Performance Analysis and Network Calculus

Performance issues related to multi-agent technology for Internet and enterprise systems include the characterisation of the multi-agent system itself, and the distributed networked environment where agents perform services and applications.

Formally, multi-agent system A operates on system nodes $S = \{S_1, S_2, \ldots, S_i, \ldots, S_{ns}\}$ interconnected by a network N allowing agent communication and migration, i.e. a mobile agent network is considered. When establishing a system, each node S_i is configured in such a way that it supports as set of elementary services, $s_i = \{es_{i1}, es_{i2}, \ldots, es_{ij}, \ldots, es_{in}\}$. An $agent_k$, with a set of assigned elementary $service_k = \{es_{k1}, es_{k2}, \ldots, es_{ki}, \ldots, es_{kn}\}$, can execute only those elementary services from the intersection $service_k \cap s_i$ at node S_i.

5.1 Modelling Processes in a Mobile Agent Network

In order to introduce performance evaluation capabilities, the mobile agent network is described as a queuing system where the agents represent information units to be served, competing to run a service on the same node (Figure 9).

User requests, as well as activation of the agents capable of handling the services, are random events. Stochastic input flow of service requests generated by service users are transformed an into input agent flow consisting of agents with assigned elementary services required to serve a request.

Generally, node S_i with processing capacity B_i receives input agent flow G_i (produced by the service users) and network agent flow X_{ij} (directed from other nodes). These flows are summarised as L_i and sent to queue Q_i organised according to the domain principle: the agent queuing discipline is First In First Out and local agent communication for exchanging data is enabled. When it is $agent_k$'s turn in queue Q_i, the processor executes an elementary service assigned to it, es_{kl}. After completion, $agent_k$ either executes its next elementary service on the same node (flow U_i) or migrates to another node and executes there. The migrating agent is placed in the migration queue, TQ_i from where it will

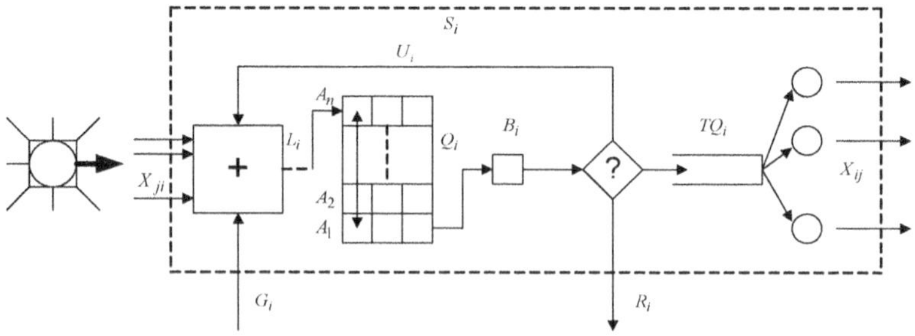

Fig. 9. Node structure

be directed towards some other node. When it is $agent_k$'s turn to migrate, it is serialised and transferred to node S_j. If $agent_k$ has completed all elementary services, it is placed in output flow R_i.

When an agent $agent_k$ is created, its elementary service list $service_k$ is set and the agent migrates to the node where the first elementary service, es_{k1}, should be executed. After elementary service execution, the agent tries to send the result to the team agent(s) that execute(s) the succeeding elementary service. Actual communication can be internal (inside of one agent), local (between agents located at the same node) or global (between agents located at different nodes). After communication, the agent takes the next elementary service from its list. If the subsequent elementary service is to be executed on a different node, the agent migrates to the node in question. This is repeated for all elementary services from $service_k$. After completing the last elementary service, es_{kn}, and delivering the result to the user, the agent is disposed of. In case of unpredicted situations, the agent tries to resolve the problem by itself. If it fails, it contacts other agents and they cooperatively solve the problem. In case of node disconnection, user requests are not served, i.e. they are lost.

In order to evaluate system performance, we consider parameters describing input flows of service requests, agents, and elementary services which can be submitted to a node from the users or originated from other nodes (Figure 10).

At the request level, the arrival process is defined by the interarrival time, t_a, a random variable that follows an exponential distribution with mean value T_a and arrival intensity $\lambda = 1/T_a$.

At the agent level, each request is represented by a set of agents. The population of agents includes all agents from the multi-agent system A, while the

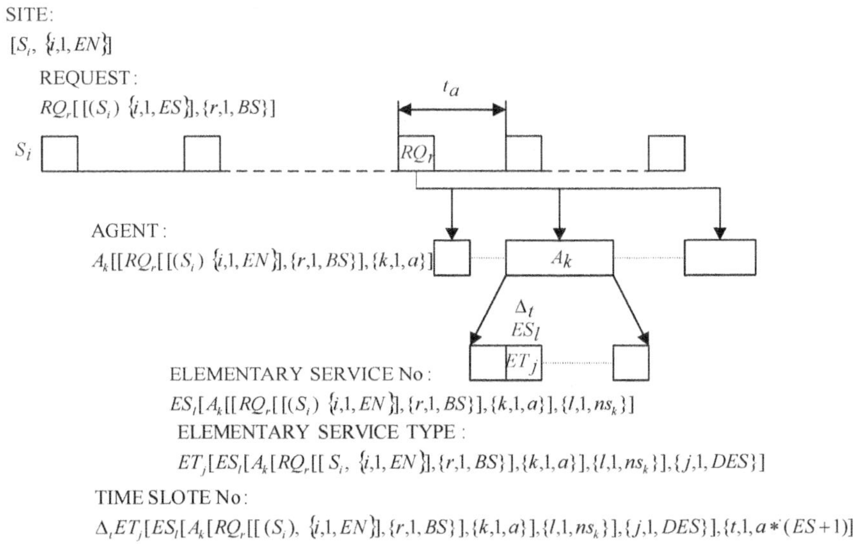

Fig. 10. Parameters describing flows of service requests, agents and elementary services

agent arrival process corresponds to the service request process. The number of agents required to serve the request is defined by a random variable a, which follows an Erlang distribution. Since simultaneous requests are submitted from different users, arrival patterns are considered to be in the form of agent bursts.

At the elementary service level, an $agent_k$ carries n_{sk} elementary services, where n_{sk} is a random variable that follows a uniform distribution. Elementary services, similar to elementary tasks used in distributed parallel processing in telecommunications [6], are treated in the following way: a) $service_k$ provided by $agent_k$ is composed of n_{sk} elementary services, es_{jk}, and b) all elementary services have a serving time of Δt.

Reasons for migrating an agent and decisions where to transfer it are excluded from the performance analysis. They are influenced by the basic properties of an agent, such as reactivity (the agent's response to changes in its environment), proactivity (the agent's initiative and goal-orientation), intelligence (the agents reasoning and learning) or autonomy (the agents ability to follow the goal without interaction from the environment). After completing the elementary service, es_{jk}, at S_i node, the node serving the next elementary service, es_{j+1k}, is selected randomly.

5.2 Analysing the Performance of a Mobile Agent Network

Network calculus is based on MinPlus algebra, with the following underlying idea: a) to regulate the input flow of bits, packets or general information units, in our case agents carrying elementary services, and b) use deterministic scheduling in order to achieve service guarantees [12,13,26].

Network calculus constrains input flow (i.e., elementary services in our problem) with an arrival curve that can be viewed as an abstraction of a regulating mechanism. In our case, it is defined by the maximum number of agents in a burst submitted for execution, the maximum number of elementary services per agent, and the burst interarrival time. The output is bounded by a scheduling service curve defining the number of completed elementary services. An example is shown in the Figure 11. Input and output flows are described by the cumulative functions defined as the number of elementary services in time interval $\{0, t\}$. A discrete time model with discrete time unit Δt is used, giving a finite set of values for input and output flows shown as increasing functions.

The three basic performance measures considered are the following: backlog, delay and elementary service rate.

The backlog performance measure, denoted as $backlog(t)$, is defined as the number of elementary services held inside the node at time t on a discrete time scale. In other words, $backlog_i(t)$ at node S_i is equal to the number of elementary tasks in queue Q_i. It is expressed in the graph by the vertical deviation between the input and output functions. The delay, denoted as $delay(t)$, is the delay experienced by an elementary service arriving at time t due to the time it takes to serve all those elementary services arriving before it, as well as itself. In other words, $delay_i(t)$ at node S_i is equal to time spent waiting in queue Q_i and its serving time. In the graph, delay is represented by the horizontal deviation

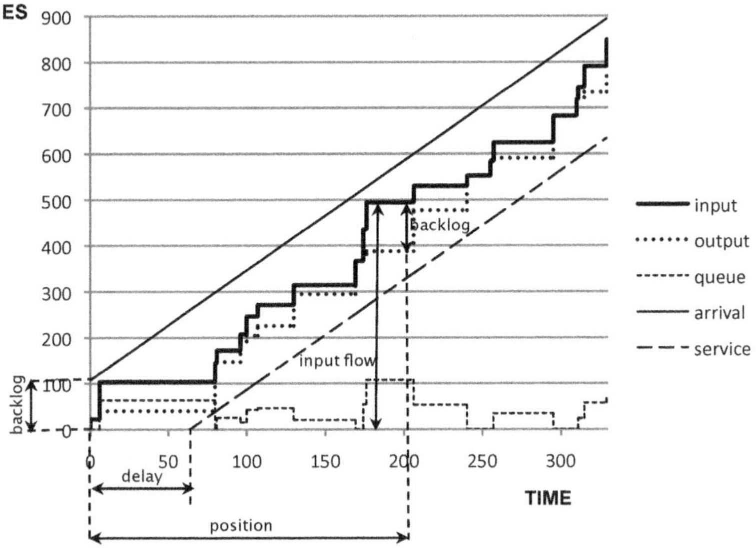

Fig. 11. An example perfornamce graph

between the input and output functions. Elementary service rate r is equivalent to the original notion of bitrate and is represented by the total number of elementary services ($EStotal$) submitted for execution up to time t, where $backlog(t) = backlogmax$.

The elementary service rate r, linear arrival and service curves are defined as follows:

$$r = \frac{EStotal}{t(backlogmax)} \tag{3}$$

$$arrival = backlogmax + r \cdot t \tag{4}$$

$$service = r \cdot (t - delaymax) \tag{5}$$

6 Case Study

Our evaluation of network calculus as a performance assessment tool for a multi-agent system and a mobile agent network is based on simulations. The considered grid network is composed of 8 nodes as shown in Figure 2.

Figure 12 shows the results of simulations for one node in the network. The following system parameters (and their defined basic values) were used:

1. the number of processing nodes, $NS = 8$;
2. processing speed, $PS = \{3, 3, 3, 3, 3, 3, 3, 3\}$;
3. the number of agent bursts, $BS = 30$;
4. agent burst arrival intensity, $\lambda = 50$;

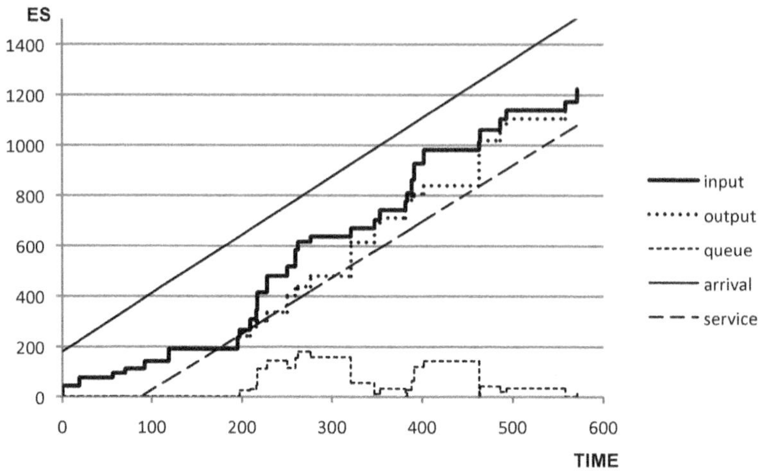

Fig. 12. Network calculus: case study result at S_1

5. mean agent burst interarrival time, $TA = 20$;
6. the maximum number of elementary services per agent, $NES = 10$;
7. transmission or communication time between two nodes, $T = 1$;

Input and output flows are constrained with arrival and service lines. The queue at the node in question is also shown, seeming to always return to 0 after some time. This implies that the system is well designed and that there were no service rejection.

Values of three basic performance measures are:

- maximum backlog, *backlogmax*, observed at $t = 276$ equals $backlog(276) = 179$;
- maximum delay observed at $t = 259$ equals $delay(259) = 88$;
- Elementary services rate $r = 2.23$.

In each of the experiments that follow, the number of agent bursts was set to $BS = 100$, and only one of the other basic parameters was changed.

In the first experiment (Figure 13), the number of processing nodes was changed from 1 to 8, keeping processing speed at 3 elementary services per Δt and agent burst interarrival time at $20\Delta t$. We can see that the *backlog*, *delay* and *bitrate* decrease as the number of nodes increases. When the number of nodes was 5 or higher, all the values decrease a bit, which means that the system is within normal parameters (i.e., can serve everything that comes into the system).

In the second experiment (Figure 14), the processing speed was examined in a network with 8 nodes and mean interarrival time set to $20\Delta t$. As seen in the figure, the backlog and delay are high for a processing speed below 3 elementary services per Δt. Thus, the minimal processing speed of all nodes should be at least 4 elementary services per Δt.

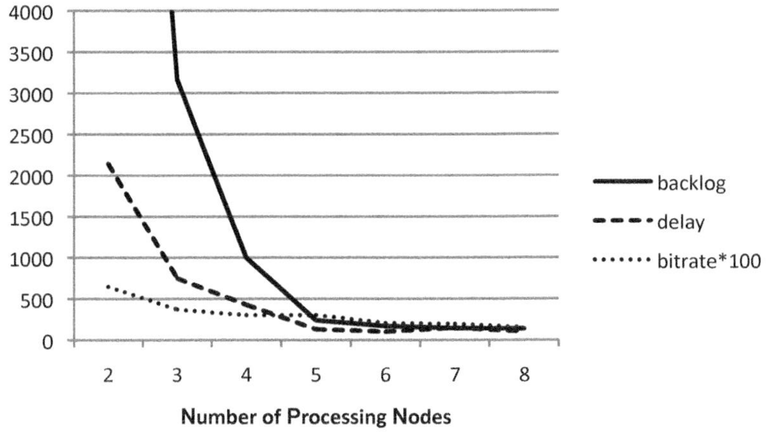

Fig. 13. Examination of Processing Nodes

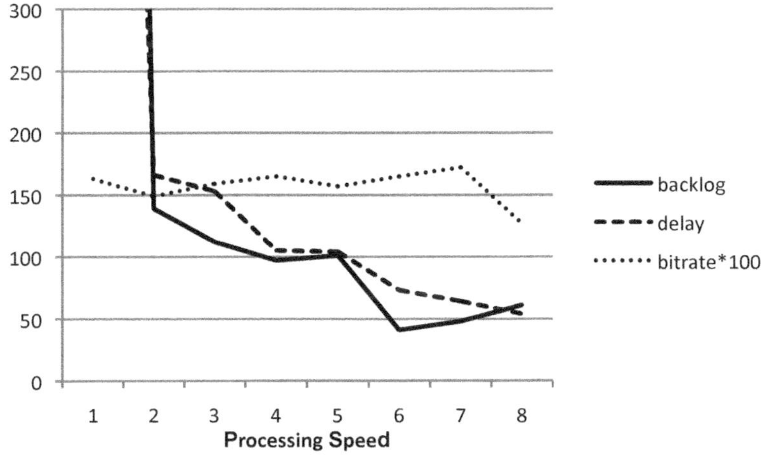

Fig. 14. Examination of Processing Speed

In the third experiment (Figure 15), the mean burst interarrival time was evaluated in a network with 8 nodes, all having processing speed of 3 elementary services per Δt. The results show that for cases when agent bursts arrive faster, the *backlog*, *delay* and *bitrate* are too high. After increasing interarrival time to over $20\Delta t$, the system worked within normal parameters.

In the last experiment (Figure 16), the maximal number of elementary services per agent was evaluated in a network with 8 nodes all having processing speeds of 3 elementary services per Δt and mean interarrival times set to $20\Delta t$. Increasing this number to over 15 generated a jump in *backlog* and *delay*. That implies that the system is within normal performance for cases when the number of elementary services per agent is below 15.

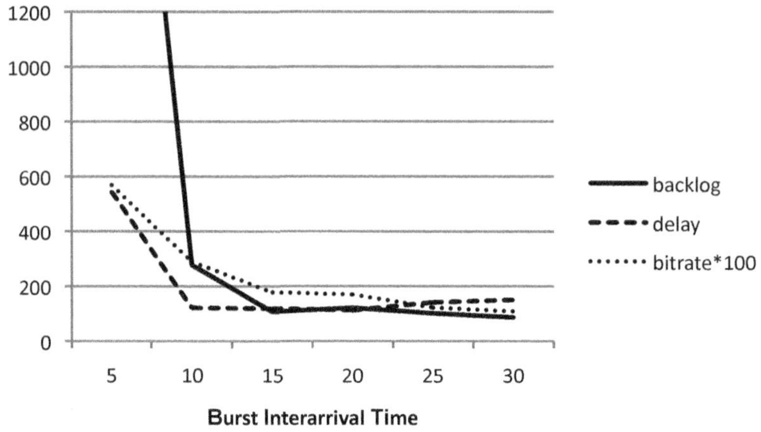

Fig. 15. Examination of Mean Burst Interarrival Time

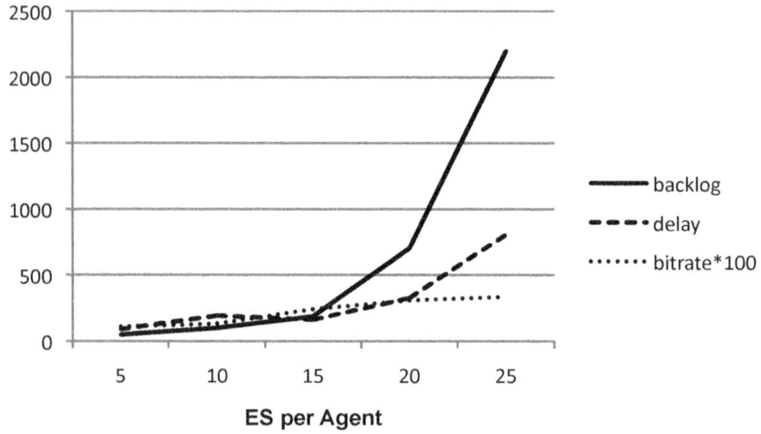

Fig. 16. The Maximum Number of Elementary Services per Agent

Response time and throughput, as basic performance measures relevant for agent-based Internet applications and enterprise systems, correspond to the delay and elementary service rate in network calculus.

7 Conclusion

Performance issues related to multi-agent systems and mobile agent network and their usage in Internet and enterprise systems are elaborated upon from two points of view: a) how to use agents efficiently for establishing, configuring and re-configuring agent-based Internet or enterprise systems, and b) how to evaluate operational performance of such systems. Key performance measures are configuration or reconfiguration set-up time with respect to the establishment of a system, and response time and throughput with respect to system operation.

Proposed is an approach based on a genetic algorithm to optimize system establishment strategies which distribute service resources to system nodes and use agents to deploy them. Network topology and link capacity, as well as node capacity, are taken into account. Simulations are used to evaluate the proposed approach. Special attention is paid to network conditions, e.g., bottlenecks due to low capacity or high traffic, and their influence on overall performance.

In order to explore operational performance, a mobile agent network is described as a queuing system and evaluated through the application of network calculus. An agent represents an information unit to be served , and nodes represent servers capable of hosting, executing and communicating with agents. Capabilities of the mobile agent network to execute user requests in Internet and enterprise systems are analysed using network calculus. User requests and activation of the agents capable of handling the services are described as random events. The stochastic input flow of service requests generated by service users is transformed into an input agent flow consisting of agents with assigned elementary services required to serve a request. Various parameters are analysed and the results are elaborated upon in order to demonstrate the applicability of network calculus for performance evaluation of multi-agent systems and mobile agent networks.

Acknowledgments

This work was carried out within research project 036-0362027-1639 "Content Delivery and Mobility of Users and Services in New Generation Networks", supported by the Ministry of Science, Education and Sports of the Republic of Croatia.

References

1. Podobnik, V., Petric, A., Jezic, G.: An agent-based solution for dynamic supply chain management. Journal of Universal Computer Science 14(7), 1080–1104 (2008)
2. Sinkovic, V., Lovrek, I.: Generic model of a mobile agent network suitable for performance evaluation. In: Howlett, R.J., Jain, L.C. (eds.) KES, pp. 675–678. IEEE, Los Alamitos (2000)
3. Kusek, M., Lovrek, I., Sinkovic, V.: Agent team coordination in the mobile agent network. In: Khosla, R., Howlett, R.J., Jain, L.C. (eds.) KES 2005. LNCS (LNAI), vol. 3681, pp. 240–245. Springer, Heidelberg (2005)
4. Huebner, F.: Performance and capacity evaluation of ip networks and systems. IT Pro 6, 38–43 (2001)
5. Sugawara, T., Fukuda, K., Hirotsu, T., Sato, S.: Multi-agent systems performance by adaptive/non-adaptive agent selection. In: IAT 2006: Proceedings of the IEEE/WIC/ACM international conference on Intelligent Agent Technology, pp. 555–559. IEEE Computer Society, Washington (2006)

6. Sinkovic, V., Lovrek, I.: A model of massively parallel call and service processing in telecommunications. J. Syst. Archit. 43(6-7), 479–490 (1997)
7. Boutilier, C.: Sequential optimality and coordination in multiagent systems. In: Proceedings of the Sixteenth International Joint Conference on Artificial Intelligence, pp. 478–485 (1999)
8. Pynadath, D.V., Tambe, M.: The communicative multiagent team decision problem: Analyzing teamwork theories and models. Journal of Artificial Intelligence Research 16, 389–423 (2002)
9. Xuan, P., Lesser, V.: Multi-Agent Policies: From Centralized Ones to Decentralized Ones. In: Proceedings of the 1st International Joint Conference on Autonomous Agents and Multiagent Systems, Part 3, pp. 1098–1105 (2002)
10. Koriem, S.M.: Development, analysis and evaluation of performance models for mobile multi-agent networks. Comput. J. 49(6), 685–709 (2006)
11. Rana, O.F.: Performance management of mobile agent systems. In: AGENTS 2000: Proceedings of the fourth international conference on Autonomous agents, pp. 148–155. ACM, New York (2000)
12. Boudec, J.Y.L., Thiran, P.: Network Calculus. LNCS, vol. 2050, p. 3. Springer, Heidelberg (2001)
13. Pandit, K., Schmitt, J., Steinmetz, R.: Network calculus meets queueing theory -a simulation based approach to bounded queues. In: Proceedings IWQOS 2004 Twelfth IEEE International Workshop on Quality of Service, Montreal, Canada, pp. 114–120. Oxford University Press, Oxford (2004)
14. Schiøler, H., Jessen, J., Dalsgaard, J., Larsen, K.: Network calculus for real time analysis of embedded systems with cyclic task dependencies. In: Proceedings of the 20th International Conference on Computers and Their Applications, New Orleans, USA, pp. 326–332 (2005)
15. Bettini, L., De, R., Loreti, N.M.: Software update via mobile agent based programming. In: Proceedings of the 2002 ACM symposium on Applied computing, pp. 32–36. ACM Press, New York (2002)
16. Jezic, G., Kusek, M., Desic, S., Labor, O., Caric, A., Huljenic, D.: Multi–agent remote maintenance shell for remote software operationsg. In: Palade, V., Howlett, R.J., Jain, L. (eds.) KES 2003. LNCS, vol. 2774, pp. 675–682. Springer, Heidelberg (2003)
17. Sherif, M.H., Ho, S.: Evolution of operation support systems in public data networks. In: Proceedings of the 5th IEEE Symposium on Computers & Communications, pp. 72–77 (2000)
18. Houssos, N., Alonistioti, A., Merakos, L., Mohyeldin, E., Dillinger, M., Fahrmair, M., Schoenmakers, M.: Advanced adaptability and profile management framework for the support of flexible mobile service provision. Special Issue on (R)Evolution towards 4G Mobile Communication Systems 10(4) (August 2003)
19. Jezic, G., Kusek, M., Desic, S., Caric, A., Huljenic, D.: Multi–agent remote maintenance shell for remote software operations. In: Palade, V., Howlett, R.J., Jain, L. (eds.) KES 2003. LNCS (LNAI), vol. 2774, pp. 675–682. Springer, Heidelberg (2003)
20. JADE: Java agent development framework, `http://jade.tilab.com/`
21. FIPA: The foundation for intelligent physical agents, `http://jade.tilab.com/`
22. Jurasovic, K., Kusek, M.: Optimizing service distributions using a genetic algorithm. In: Lovrek, I., Howlett, R.J., Jain, L.C. (eds.) KES 2008, Part I. LNCS (LNAI), vol. 5177, pp. 158–165. Springer, Heidelberg (2008)

23. Kusek, M., Jurasovic, K., Petric, A.: Simulation of mobile agent network. In: Proceedings of the 9th International Conference on Telecommunications (ConTEL 2007), pp. 49–56 (2008)
24. Kusek, M., Jezic, G., Jurasovic, K., Sinkovic, V.: Network simulation in a fragmented mobile agent network. In: Apolloni, B., Howlett, R.J., Jain, L. (eds.) KES 2007, Part I. LNCS (LNAI), vol. 4692, pp. 214–221. Springer, Heidelberg (2007)
25. Kusek, M., Jurasovic, K., Jezic, G.: Verification of the mobile agent network simulator - a tool for simulating multi-agent systems. International Journal of Software Engineering and Knowledge Engineering 18(5), 651–682 (2008)
26. Sinkovic, V., Kusek, M., Jezic, G., Lovrek, I.: Performance evaluation of a mobile agent network using network calculus. In: Lovrek, I., Howlett, R.J., Jain, L.C. (eds.) KES 2008, Part I. LNCS (LNAI), vol. 5177, pp. 174–181. Springer, Heidelberg (2008)

Resumes:

Mario Kusek received his Ph.D. degree from the University of Zagreb, Croatia, in 2005. Currently, he is an assistant professor at the University of Zagreb, Faculty of Electrical Engineering and Computing. His research interests include object oriented design and programming in distributed environments, and software agents in next generation networks and services. He is a member of the IEEE and the KES International.

Kresimir Jurasovic is Ph.D student at the University of Zagreb, Croatia, where he is working as a research associate on the project Agent-based Service Telecom Operations Management at the Department of Telecommunications of the Faculty of Electrical Engineering and Computing. His research interests are multi-agent systems with emphasis in service provisioning and next generation networks. He is a member of the IEEE and the KES International.

Gordan Jezic is an assistant professor at the Department of Telecommunications of the Faculty of Electrical Engineering and Computing, University of Zagreb, Croatia. He received Ph.D. from the University of Zagreb 2003. His research interest includes communication networks and protocols, mobile process modelling, mobile software agents and multi-agent systems. Dr. Jezic is a member of the IEEE and the KES International.

Vjekoslav Sinkovic is a professor at the University of Zagreb, Faculty of Electrical Engineering and Computing. He authored and co-authored over hundred papers published in books, journals and conferences. His research interests include information theory and information networks, telecommunication network analysis and performance evaluation, mobile agents and multi-agent systems. He is a member of IEEE and KES International.

Ignac Lovrek is with the Faculty of Electrical Engineering and Computing of the University of Zagreb, where he works as a professor at the Department of Telecommunications. He authored and co-authored over hundred papers published in books, journals and conferences. His research interests include software agent theory and applications in communication networks and information services, service modelling and processing, soft technologies for telecommunications and electronic services market. He is a member of IEEE, ACM, GI - Special Interest Group on Petri Nets and KES International.

Service Oriented Architecture and Agents: Parallels and Opportunities

Ronald L. Hartung

Chair of Computer Science, Franklin University

Abstract. This chapter explores several theses surrounding SOA and agent technical approaches. One thesis is that in spite of SOA and agent technology's different approaches, there is a common ground between the two technologies, which is explored and evaluated. Furthermore, from a design point of view, the thesis is that there are tools or design approaches applicable to both. Developers can learn from design work and tools in one technology and apply to the other. The third thesis is that the techniques developed in agent systems can be applied to achieve the promise of SOA. The ideas developed in the work with agents are powerful enablers for SOA applications. In particular, negotiation and ontology is shown as tools to provide the basis necessary to construct powerful and adaptive SOA applications.

These are presented in the chapter as well as solutions using agent techniques and agent systems to support SOA. The agents can support the SOA to become a powerful tool for handling distributed services.

Keywords: Service oriented architecture, agents, multi agent systems.

1 Introduction

Service oriented architecture, SOA, and multi-agent systems have been popular topics in recent years. SOA has been championed in the popular press and the blog sphere for enterprise architecture. Agents have grown from recent generations of artificial intelligence systems. From the work with SOA and agents, it is found that they have some interesting elements in common. This chapter explores the two concepts from several viewpoints. First of all, the design of systems in either paradigm can teach us about design in the other. Second, some of the techniques being developed in agent systems can provide capabilities to fully enable the power of SOA. Especially in some of the areas of SOA that were promising but not delivered by the current technologies.

The chapter gives a review of the principles that underlie SOA. The treatment of SOA is given considerable depth. There are a number of good references for SOA. Of these, the text by Thomas Erl [10] provides a well-organized view of SOA design. The chapter then proceeds to look at some agent system implementations. From this, analysis of possible architectural styles for agent systems is explored. Architectural styles are similar to patterns and are a useful tool for enterprise architects, and therefore these are easily extended into agent architectures. From some of the work in agent systems, extensions into the SOA architectures will be elaborated. This work remains at the level of future proposals.

A. Håkansson et al. (Eds.): Agent and Multi-agent Technology, SCI 289, pp. 25–48.
springerlink.com

2 SOA Systems

There are at least two terms that are applied to SOA systems, service oriented computing and service oriented architecture. The fundamental organizing concept is service. Many systems, including ones with humans serving as the processing elements, can be viewed as a collection of services. This is, of course, a very general concept. The preceding statements made use of the common-sense definition of a service. SOA has a very specific definition of service, which will be given presently. However, it is well demonstrated that this common-sense definition of service can be used very effectively as a modeling primitive for systems; a good example of this is the RED methodology [21].

The central promises of SOA are enhanced software architecture, improved software development time and cost, increased reuse through modularity. Pundits have pushed these promises for several years. However, recent writings and experience have cast doubt on these claims. While SOA is successfully applied in many organizations, big doubts remain when the above promises are reviewed. While proof is lacking, a plausible concern is that maintenance may suffer in the future as applications become dependent on a distributed set of services that will become uncontrolled. Two problems would most likely appear. If the maintainer of a service is unaware of the applications that depend on a service, changes may cause the applications to fail. This is less of a problem when the application is maintained in whole, but recall the goal of SOA is reuse by sharing the services and a strong commitment to loose coupling of the services. Another issue arises from the common "not invented here" syndrome common in software development. The services will proliferate beyond the design intent, because reuse fails. The proliferation increases by the discovery problems to be explored later.

Moreover, Anne Thomas Manes (http://atmanes.blogspot.com/), an analyst of the Burton Group, described SOA as dead in one blog posted on Jan 5, 2009. However, she also proclaims great prospects for SAAS (software as a service) and clouds, both built on SOA designs and concepts. Her stated reasons for the death of SOA are economic costs due to the recent recession and also a sense of disillusionment with SOA projects. While there is great debate about her position, the disillusionment with SOA projects is a real issue, even though many SOA projects are proceeding.

2.1 Service Definition

In SOA, the definition for service needs to be more precise that the common definition used at the onset of this chapter. It is also counter intuitive. Here a service is defined as a collection of capabilities. A capability is an atomic unit of work that can be performed in a service. Thus, a capability is defined more like the common definition of service. A well-designed service has a set of related capabilities that are inherent in the service. For a service to encapsulate a "grab bag" of diverse capabilities is directly analogous to similar forms of bad object design. This definition of service could fit well in object oriented-programming, in which a service is a kind of object. The capabilities are the analog of the methods of the object. What gives definition and form to services is a set of qualities. It is these qualities that set services apart as a design paradigm.

From an implementation perspective, SOA is built on web services in a distributed communication framework. Unlike many framework systems, web services are independent of underlying programming technologies. For instance, J2EE or .NET frameworks define more a more extensive programming base and/or container objects. They define support libraries, container object interfaces and configuration (e.g. Tomcat and BEA Weblogic). Web services need programming systems and containers, but leave the design of these as implementation choices, not as part of the definition of web services. Deploying web services will require containers and monitoring systems for operation of the systems. But the interface to the service is common across any of the containers or frameworks and involves only the communication protocols and message formats.

Web services generally use either SOAP or XML-RPC as a communications method for services, although others are possible, including email. In addition the web services, protocol stack includes UDDI [29] for service discovery and WSDL [30] for service description. All of these are based on HTTP as a transport protocol.

More recently, the underlying SOA communications infrastructure has been collected into an Enterprise Service Bus, ESB. The ESB provides the basic services as described by the previous protocol suite. In addition, the ESB can support more advanced interaction patterns. Some systems include the use of workflow systems to define interaction patterns and to automate complex behaviors. While this is an interesting topic, we confine the discussion here to just the basic communication facilities. In any case, use of an ESB does not change the basic properties, or qualities, of an SOA system.

2.2 Qualities of SOA

In order to achieve the goals of SOA, a set of qualities is required. Most of these qualities are not unusual or unique. Instead, these are common qualities applied in software engineering. This is reinforced by the descriptions found in Erl text [10].

Separation of concerns is a commonly invoked design principle. In order to promote modularity, separation of concerns guides the designer to isolate code into units that deal with limited functionality. However, an additional goal of SOA is a reduction of repeated or duplicated code. Services break a single application in to a set of services that are separate code units. In this case, a service can be reused in multiple applications and the functionality of the service is implemented once.

A standardized service contract is critical in several dimensions. First of all, without a solid contract using a service becomes a risky proposition because one cannot be sure what the service will do. Unlike a monolithic application where the code base is controlled and deployed only after test, services are independent units of code and subject to change independent of applications that use the service. The other dimension is discovery, a topic to be examined in detail. Discovery is the process for locating services to support and construct an application. This process can be applied at design time or at runtime, as a dynamic operation.

Service loose coupling is essential to the reduction of duplicated code [10]. Loose coupling ensures that the function performed by the service does not depend on the services that call them or that they may call. This means that services stand alone as code units with a defined and known function.

Service composibility ensures that the service can be easily assembled into full applications. In this sense, services become like functions in a functional language. Without the easy assembly, it is impossible to achieve the goal of a SOA system to implement reusable business objects. The composibility of these services is a major claimed advantage that leads to a reduction of software development costs.

Service abstraction is multi-faceted. One form of abstraction is to hide the underlying details and implementation, which is the classic approach of information hiding. This seems like a simple and easily implemented property. However, service discovery means that enough information must be exposed to aid the selection of proper services. A careful balance is required to meet proper abstraction and still enable discovery.

Service reliability is a quality that should be obvious. However, achieving a reliable implementation requires keeping reliability in mind through the whole design process.

Service autonomy is a quality strongly shared with agent-based systems. A service operates autonomously in the same way that intelligent agents and meta-agents are autonomous. The service is in control of it's own operating environment at the same time learns from the environment. It still needs application level for monitoring and controlling services. The container that supports the service's deployment and provides a service-operating environment provides monitoring.

Service statelessness is an intriguing property and one whose value can be debated. Taking the view of an enterprise architect, keeping the state in a database system and having no state in the service makes application scaling easy. One can add multiple instances of the service and even allow the instances to be distributed across multiple servers. However, it may be quite advantageous to keep state in a service, if it means reducing the need for the cost of a remote call to a database server.

Service discoverability covers the need to find and select useful services. The ultimate in discovery is an automated system that dynamically identifies and locates the required services. The opposite of automated discovery is simply locating a service that was designed into the application, that is static or manual discovery. Static discovery may still require dynamically locating the service in network space. Even when the requested/wanted service is known, the location of the service may not be known, or the service might move to another space from time to time.

2.3 SOA Metadata

While many mechanisms are required to achieve the above qualities, metadata is central and standardized in SOA implementation. This is clear when the specifications for web services are examined. There are two types of specifications, the network protocol specifications and the specifications for the service contract. All other details are tied to the container and are not part of the specification. Several metadata systems are used by SOA systems. This metadata supports service contracts, policy and discovery mechanism. The metadata, used in SOA, can also be used to define agents in a multi-agent system. This is because of the great similarity between agents and SOA services.

Service Contracts

Because services are atomic units used to construct complete applications, documentation of the service is critical to successful application construction. SOA is totally dependent on assembling services into an application. Also, in SOA, services can be changed without changing the totality of an application. The service contract is essential to both the user of the service and the developer of the service. Many organizations adopt governance policies that can restrict the dynamic behavior of SOA. This approach reduces the problems solved by metadata and become closer to having monolithic applications. However, the theme in this chapter is that such restriction defeats the promise of SOA.

A service contract has three parts: web service description language, XML schema and Web Services Policy. For current copies of these specifications see the W3C web site. The web service description language, WSDL, is defined by a W3C specification [30]. The WSDL is a XML based definition of a web service. The XML schema defines the data structures and the web services. WS policy details policy issues. Policy is a latter addition to the triad.

The WSDL provides four critical pieces of information:

- Interface information on the available capabilities
- Data types for messages
- Protocol used
- Network address for each service

The top-level element of a WSDL is the definitions element that gives the name of the service, which is the overall container for a set of capabilities. In addition, the definitions element can hold XML name space definitions. It also contains the elements of the definition and is composed of the following elements; types, service, message, port type, and binding.

The types element is used to define data types for the capabilities. However, if the needed data types are to be defined using the standard XML schema types, a types element is not needed. This is the common case, since XML schema provides a rich set of definitions to use.

The service element contains the address of the service, generally in a URL. Each message element defines a message that is either a service request or a response to a request. The service element includes the definition of the parameters for the message.

The port type element defines the combination of messages for an operation. That is, the port type defines the input data that for the request and the output data is returned as the response to the request. This is done by referencing the messages defined in the message element.

The binding element describes the transport protocol used to access the service's capabilities. It gives the details needed to access and activate the capability.

In addition to the elements listed above, documentation elements allow user readable comments to be placed into the WSDL.

Policy

Web services policy represents the capabilities and requirements for web services as policies. These policies are used to cover issues in security, addressing, logging, and other similar issues. The basic structure of policy definitions is very straightforward. The policies, themselves, are termed policy assertions and these are an ever-expanding set of specifications. A policy assertion is an atomic term, defined in a specification. For full details of the policies, the specification of the Web Services Policy 1.5 – Primer, available online at the W3C site, gives a list of referenced specifications.

The basic policy document is an XML document that lists the policies to be applied for a given web service. The document is built around a small set of tags: Policy, All, ExactlyOne and PolicyReference. Policy and All are synonyms. Policy (or All) is used to collect a group of policy assertions in a set that all apply. Alternatives are expressed by the use of ExactyOne, which specifies that one and only one of the enclosed items is to be applied. The items enclosed by the ExactlyOne tag can be either a policy (or All) or an individual policy assertion. PolicyReferences allow the inclusion of sets of policies from an external URI. This is a flexible structure for constructing combinations of the policy assertions. A structure of policy assertions, combined with the tags, forms a policy expression.

In addition, two attributes are defined for use in the policy document. One is Ignorable and the other is Optional. The Ignorable attribute is used to inform about the existence of policies that are employed by the service, but do not constrain the interaction with the service. The basic example is logging. If the web service performs logging, the user of the service may want to know if it is being logged and may choose not to use the service. However, the logging does not require the user of the service to do anything. Therefore, by placing Ignorable on the assertion, the policy need not be checked for a compatibly with the users policy.

The Optional attribute is not required. Any use of the Optional attribute can be converted into an equivalent form using only the ExacltyOne attribute. The form without the optional attribute and with policy References removed is called the normal form. A policy that uses optional and policyReferences is called compact. Normal form is useful for defining the operations over policy evaluation. The compact form is easier to read and understand, but less convenient for computation.

The policy model defines how a policy aware client can evaluate the policy expression of a service. The client will evaluate the policy expression of a compatible set of policies to use. A policy aware client is said to support a policy assertions if the client follows the behaviors represented by the assertion. The client must be able to find a set of alternatives in the service's policy that it can support.

2.4 Discovery

Discovery is, simply, the ability and mechanism to locate web services for use. Reusability hinges on the ability to locate services and capabilities that can satisfy a requirement for constructing a new application or feature. Reusability itself is the long sought grail of software engineering. Some of the recent criticisms of SOA, highlight the lack of effective discovery that has limited the effectiveness of SOA and relegated it to yet another construction technique that has failed to live up to its promised reusability. This conclusion is far from proven and may be a source of lively debate at the time of this writing.

What is required for the effective discovery are three aspects: The purpose of the service, the capabilities in it, and the limitations of the capabilities. The design of SOA is to encode the information to enable discovery in metadata that can be queried. In addition, authentication and trust must be established to make the results of queries believable.

There is an interesting tension inherent in the choices in exposing information to support discovery. Generally, one choice is to be parsimonious with the information exposed about a service, carefully isolating details about the service to promote loose coupling. But to make discovery function properly, the service contract and policy should contain as much information as possible to assist in discovery.

Discovery can be executed at design time or at runtime. Design time discovery can be static and manual. This is how many systems are implemented today. In fact, it may be the case that many systems do not have effective manual discovery. After all, developers have seldom been able to keep comments up to date in software, and therefore, updating discovery meta-data could be expected to be at least as bad in practice. The more advanced case of runtime dynamic discovery will be discussed after a description of UDDI, Universal Description, Discovery and Integration.

UDDI defines a distributed directory service to register web services and to locate the services. The data is divided into three categories: white pages, yellow pages and green pages. The White pages gives company specific data, name, description and address while Yellow Pages classifies the service offered, including the address, contact information and known identifiers. The yellow pages are structured with standard taxonomies and the Green Pages hold the technical information about a web service.

It is possible to use a public UDDI implementation or to set up a private UDDI. In the case of a public directory, access control and authentication are clearly issues to be dealt with. The white pages information is a key part of the public structure because it allows the user to associate the web services with a corporate owner. This association can act as assurance that the service is a legitimate and acceptable service to use.

UDDI provides a simple API to access the metadata. The API has eight methods:

- Find_business
- Find_binding
- Find_service
- Find_tModel
- Get_businessDetail
- Get_serviceDetail
- Get_bindingDetail
- Get_tModelDetail

The data uses XML as a format. The tModel accesses the WSDL and policy objects. The business access methods help locate the service and identify the owner of the service.

In addition to the API used to locate services, UDDI system requires a publishing interface to add services and businesses to the UDDI registry.

While automated discovery is an early promise in SOA, it still is a bit elusive. Generally, successful industrial implements have left this alone. Dynamic discovery can be side stepped with good governance practices. However, the problems in this approach are much the same as managing object-oriented design to achieve reusability. Yet this remains an extremely appealing capability. The XML based format of the meta-data invites automated techniques. A component that has been missing in enterprise SOA applications is a robust reasoning system that can mediate the discovery process. However, as will be discussed later, there are some needed elements that have not been included in the metadata of SOA.

2.5 SAAS and Clouds

Two new trends, worth mentioning, are software as a service and clouds. Both of these use SOA as a basis.

Software as a Service or SAAS is about deployment. The basic goal is to build software systems that have a web interface and require no software on the client machine. The focus of SaaS is to provide software supplied by a provider and relieve the customer's IT organization from the task of managing the software or hardware required to run it. Further, SaaS applications become components of larger collections of software, like mashups and plugins. While SAAS does not have to use SOA, SOA becomes a very useful architecture for SAAS.

Clouds are also a deployment solution. A cloud system provides the users a way to obtain outsourced scalable server systems. Scaling is real time, the cloud can adjust to load as needed and charges accordingly. The user pays for what they are using, where computing resources become a utility like electricity. The IT management is outsourced. There are a number of cloud implementations, for example Yahoo, and Google apps.

Both of these trends provide an infrastructure for SOA designs and for agent systems. This becomes attractive when agents are applied as part of commercial applications.

3 Agents

There is a solid literature on agents and the reader is referred to the works of Genesereth and Ketchpel (1994) [15], Woodridge and Jennings (1995) [31], Russell and Norvig (2003)[27] for definitions. Much of the work in this research focuses on several aspects of agents. Russell and Norvig (2003) [27] provide a nice classification of the environment of the agent. Many others, Wooldridge (2003) [31] for one, focus on the internals of an agent.

A question posed at a recent conference was: "What is all this about agents and why are they different from other software paradigms like object oriented programs?" Maybe the question was offered as a fun musing on our systems and designs, but considering this carefully is a good way to reflect on the state of software engineering. While authors listed above can site a set of significant properties and qualities of agent systems, there is a deeper answer to the question. Agents are about the quality of agency. When we define an agent, we naturally divide the problem space into the agent and the environment the agent experiences. The agent is viewed as an autonomous system reacting to and adapting to the environment it is subjected to.

While software has always had the structure of input-process-output, this structure limits the view of the software systems as one that responses to simple sets of input. The idea of agency attunes the designer to thinking of a processing entity responding to a complex environment.

Given the distributed and less than a full application that characterizes the services in SOA, the applications bear a great similarity to agents. The kinds of design and approaches from agent systems have much to offer the implementers of enterprise SOA systems. The main similarities between the two systems include distribution, interface and agency. Distribution in the sense that agent systems and SOA systems are implemented as independent units with the units separated from each other. Interface for both agents and services rely on a defined communication interface. Although the SOA systems use a standardized interface dependent on a single specification and agents use a number of interfaces determined by the actual implementation. Lastly, agency clear used and defined by agent systems, is actually implicitly use in SOA. Agency is implicit in SOA because the service responds to requests from the environment.

3.1 Agent Systems

Agents can be implemented in many ways, there are several agent implementation vehicles reviewed here. All the implementations reviewed are implemented in Java, however the real significance is the variety of approaches taken. This chapter will look at properties of each system that will be of interest as we compare SOA and agents.

JADE

JADE [12] is based on the FIPA specifications implemented in JAVA. The core specifications of FIPA cover agent communication, agent management, and agent architecture. The communication specification uses FIPA-ACL, which is an agent communication language, based on speech acts. The specification covers the communications protocol and the language itself. The communications protocol defines a seven sub-layer communication stack. The sub-layers are within the application layer of the OSI communication model. The ACL language specifies the use of an ontology to define the terms used in the communication.

Agent management defines the agent platform as the physical infrastructure that is the container for the agents. Thus the agent platform is managed by an agent management system. In order to operate, an agent must register with the agent management system. The agent platform must provide a message transport service to handle FIPA-ACL messages between the agents.

An optional component of the JADE system is the directory facilitator that supplies a yellow pages service to locate agents. This is similar to the directory structure in SOA. It provides a way to locate agents, as UDDI supports the location of services. It is a local solution; it does not provide the global discovery of the public form of UDDI implementations.

Another interesting component of JADE is the Web Services Integration Gateway. This is a component allows the agent to invoke SOA services as software components.

JADE uses LEAP as a platform for agents on a mobile platform. This is needed as JAVA JVMs on mobile platforms generally cannot support a full JADE platform. LEAP provides a JADE compatible implementation on the reduced JVM on mobile platforms.

JASON

JASON [13] is a Java-based interpreter for an extended version of AgentSpeak [5]. It contrasts with JADE in that JADE is an infrastructure while JASON is a programming language approach to a BDI implementation (Belief – Desires – Intentions) [7]. The interface to the outside environment is via a set of functions to map percepts into the internal state of the interpreter.

JASON has constructs for multi-agent systems as well. A configuration file allows the definition of a multi-agent system by specifying the agents to comprise the system and the environment they will interact with. The architecture can be specified as centralized or SACI (Simple Agent Communications Infrastructure) [22].

SACI provides a communication infrastructure using KQML (Knowledge Query and Manipulation Language) messages for communication between the agents. It implements a society model of the agents. The society is defined by a 4 tuple:

soc=<A,S,l,σ >
where:
A is the set of agent identifiers
S is the set of skills possessed by agents
l is the language shared by agents in the society
σ is a mapping from agents to skills (agents can possess multiple skills)

Under SACI, the society can change over time. The society can change in population by agent's entrance or exit. Agents in the society can change their skills. SACI has a facilitator that maintains the society. Facilitators handle communication between the societies. For an agent to communicate with an agent in another society, the agent desiring to communicate contacts the facilitator of the other society with a request to enter the society. It is also possible to have multiple societies in an environment. After the entrance, communication can proceed.

SACI implements a class, MBox, to provide the communications interface. Communication is handled by the methods in MBox, isolating the agent itself from the communications details.

In order to start agents, SACI has launcher classes, one for agents and one for societies. This launch society class will operate under a web browser, as a Java application or by an independent, user written, agent launcher.

More recently, JASON has been employed as a construction method for agents of JADE.

DECAF

Distributed Environment-Centered Agent Framework, DECAF [16], is an agent system that provides services for a course grained agent implementations, communication, planning, scheduling, execution, monitoring, and coordination. To these services, DECAF also plans to add learning and self-diagnosis. DECAF takes the approach of building an operating system like structure to support the operation of the agents.

The agents, which are BDI based, are built with a GUI plan-editor. The agents are implemented with a plan file constructed in the plan editor.

The operating systems portion of DECAF provides Middle Agents that support the operation of the problem solving agents.

- Matchmaker – a yellow pages service for finding services
- Broker – a white pages for collections of services
- Proxy – enabling web page based Java applet to communicate with agents
- Agent Management Agent – for monitoring and control
- Agent Name Server – a DNS for agents

The system provides concurrent processing in a multi-threaded model that allows the extension from single processors to multi-processor platforms. The DECAF operating system is designed to extend to any problem-solving domain, by crafting a domain independent system. The internal system has several modules that process events from queues. An agent initialization module reads the plan files and creates agents. A dispatcher handles both incoming message and pending actions. It creates objectives feeding the objective queue to the planner module. The planner creates the tasks sent to the scheduler, via the task queue. The scheduler feeds the executor module. Execution of the agents can feed action results back to the scheduler or pending actions back to the dispatcher. While this is a bit more complex than a typical process scheduling system in an operating system (OS), it is recognizable as a system that provides OS like operation of processes. One of the complexities of this structure is the explicit role of planning to determine what tasks (processes) can be executed.

DECAF was given some scaling tests and the results are published. The result looks relatively respectable, but the tests do not push above 20 agents.

MASON

This MASON system [25] is designed for swarm applications, especially in a social context. The heart of the system is a discrete event scheduler. It is intended to scale up to 100,000 agents. The event scheduling is intended to implement time such that the agents operate producing the same results independent of platform and system loading. In addition the system supports check pointing to allow simulation runs to be restarted and to continue past runs. In addition to the agents, the system supports the generic concept of a space, called fields, containing objects. Fields can be arbitrary constructs. These can be as simple as 2D or 3D spaces (discrete or continuous), directed graphs, or something more complex. There can be more than one field and an object can be in multiple fields. MASON has a graphical visualization layer. Using graphical visualization does limit the number of agents.

The biggest limitation of MASON is the lack of distributed processing. The use of an discrete event scheduler makes the distribution across multiple processor much more difficult.

4 SOA for Agents

This section will discuss what SOA can offer to the builders of agent systems. The similarity between SOA and multi-agent architectures indicates that SOA could be a very good implementation vehicle for multi-agent systems. To make a solid analysis of this thesis, we need to analyze the needs and varieties of multi-agents systems. The proceeding survey of several implementation approaches and frameworks for multi-agents systems builds part of the basis for this argument.

Agent Styles and Agent System Properties

In order to evaluate agent systems, styles and properties are introduced as a representation of the agent system at an abstract level. These provide a classification of agent systems. Some of these represent existing agent systems, but others are possible agent systems. Te possible agent systems are those that have not been seen (at least by the author), but appear possible and useful.

Architecture styles have been used to describe software systems at a very high level of abstraction [3]. Styles are a more abstract representation than patterns. Software patterns, defined by the so called Gang Of Four [14], describe specific solutions to specific programming problems. Architectural styles describe the pattern of interactions between components in systems. For example, pipeline where a series of components pass the data stream though successive stages to solve the problem. When describing styles, the operational properties of the agents become part of the style, but the internal operation or implementation of the agent is not specified. This is a departure from most of the agent literature, where how to construct the internal logic of the agent is the common topic. The implementation approach used to construct an agent is not described in this work for two reasons. Firstly, it is well covered elsewhere in the literature. Secondly, when considering frameworks, the interaction styles supported or enabled is the more interesting dimension.

While many of the styles found in [3] are applicable to multi agent systems, there are other styles that are possible and useful in multi agent systems. Styles need not be used alone – they can be combined. That is, a system may use multiple styles by combining them in several ways. For example, styles may operate in hierarchal arrangements were one style works at one level of a system decomposition and another style is used on a different level of the decomposition. Another approach uses different styles in different portions of the system. When styles are mixed, the resulting system is referred to as having heterogeneous styles.

In addition to styles, agent systems can exhibit properties. These properties modify the application of the style. The distinction between styles and properties is subtle. Styles generally define the interaction between the agents; properties represent modes of operation or interaction within the styles. Another dimension is the communications models used by the agents. For purposes of the current work, this aspect is not included, however the reader is directed to [9] for a good discussion.

Styles

A commonly found style of agent systems is the Swam. Swam system solve the problem at hand as an emergent behavior of the collection of agents. Ant systems are a common example of a swarm system. A common attribute of swarms is a large collection of identical agents operating in an environment. Communication is often accomplished indirectly by modifications in the environment that is observed by other agents.

Event driven agent systems operate by events or messages passing between the agents. The important aspect here is that agents wait on events. This is different from agent systems were agents are actively working on tasks and communicate as part of the processing, but do not wait on messages in order to proceed. [20]

Event simulation models implement agents using a classic event scheduler. They can be constructed on either discrete or continuous schedulers. In this style, time is a controlled property of the system; the agents share a common concept of time.

Meta-agent systems use a hierarchy of agents where the meta-agents are operating at a conceptual level above the other agents. [18][19]

Heterogeneous agents, not to be confused with heterogeneous styles, are systems where agents of multiple types are employed. This may seem like a property, but it really implies a style of problem solving where different agents take on different tasks. Note that this is not the same as having agents with different behaviors like those found in social simulation agent system. Heterogeneous agent systems have agents that perform different tasks, where the social systems have agents that approach the same tasks with different goals or approaches.

Central supervision has a special agent or component that has oversight and control over the agents in the system. The control will include agent creation and destruction It may also include task assignment for the agents.

Social models use agents that mimic the social interactions of a human, or animal, community. Often these systems are built to study or observe the social interactions, but social behavior may also be used as a computational model for the system to employ, whatever the intended purpose of the system.

Coalition formation is a style of operation where the agents are directed to join forces with other agents to perform problem solving. [4]

Non-communicating agent systems have agents that interact with the environment, but not with each other. Blackboard or database centric systems are good examples of this in the styles from [3].

Properties

Open vs. Closed systems. Most agent systems observed in the literature are closed systems, meaning that agents are part of the system and all the agents are defined and built by the system designers. From a future perspective, open systems will operate in an environment where other agents can be found and recruited to perform work for the system. Likewise, SOA has promoted the view of dynamic service discovery entailing the ability to find and use services anywhere in the web and utilize the offered service. The issue of discovery is still a problem in SOA systems.

Trusted agents imply that agents can trust the interactions and messages of other agents. Note that social systems often do not use this property.

Law abiding in similar to trust, but slightly different. Agents that must follow laws can have predicable behavior that is consistent with the laws of the environment. However, this does not preclude agents that may deceive one another to achieve their goals.

Fault tolerance achieved by replication of the agents is a modification of the basic agent paradigm. These systems are constructed to have multiple agents that duplicate the functions so that higher reliability is possible.

Mobile systems have agents that can move in spatial location or move within the computing fabric of the system.

Systems with complete information, or completely observable systems, allow the agents to have complete information of the environment.

Static vs. dynamic systems of agents describe the creation and destruction of agents. All systems tend to do some creation and destruction. This property is applied to the nature of problem solution. If a system starts up a collection of agents at the beginning of problem solving, it is static under this definition. Dynamic systems create agents as part of the problem solving, see [17].

When we apply the styles to the 4 frameworks, we get a view as in table 1.

Table 1. Styles and Properties in the four Frameworks

	JADE	JASON	DECAF	MASON
Event driven	Can be programmed	Can be programmed	Yes	Yes
Event simulation	Not directly	Not directly	Yes	Yes, strongly
Meta-agent	Not directly	Can support this	Not directly	Not directly
Heterogeneous	Yes	Yes	Yes	Yes
Central supervision	Yes	Weak	Yes	
Social models	Yes, by programming extensions	Yes	Yes, by programming extensions	Yes

Table 1. *(continued)*

Coalition formation	Can be programmed	By society formation	Can be programmed	No
Non-communicating	No designed for this, could be built	No designed for this, could be built	No designed for this, could be built	Yes, environment is a built in concept
Open vs. closed	Mostly closed	Supports open	Mostly closed	closed
Fault tolerant	No	No	No	No
Complete Information	No	No	No	By use of the environment
Dynamic	Can support this	Yes	Yes	No

While the frameworks are rather similar, there are some differences that become significant for some applications. Several of the styles can be obtained by programming extensions or using libraries of functions in the agents., This does require extra work. MASON is most diverse and that is to be expected when dealing with swarm applications. The issue that is central in this work is what if we build our agent systems in an SOA framework.

4.1 Building Agents in SOA

SOA supports construction of services in a distributed framework. For many of the styles of agent systems, as described above, it is easy to map an implementation approach in SOA. One excellent feature of SOA implementations is the ability to spawn multiple parallel invocations of a service's capability. It also supports reliability by redundancy. The one change in design required to implement agents in SOA is to account for persistence. This is done by maintaining the state in a persistent structure or in a database.

Many of the agent toolkits provide a monitoring capability; likewise SOA systems have logging and monitoring abilities as well. These are necessary in any production environment. A difference will be found is the control system, especially the FIPA specification as implemented in JADE. The JADE control system is better for experimental systems. In an experimental system the agents needs to be stopped and started frequently. Also, debugging is greatly simplified by tight controls. A production systems is more likely to run by deploying agents that persist and respond to user driven activities or to operate in a continuous activity. The levels of control structures found in SOA are fully adequate for production use.

The control system specified in the FIPA specs is less useful for an open system, since agents will enter into interactions without being under the control of a centralized monitoring system. This is a really exciting possibility with SOA allowing dynamic systems where agents discover each other and use planning and negotiation systems to problem solve. The frameworks are not as strong in this regard as they are designed to build the kind of experimental systems studied in the research domain. However, JADE actually has an SOA adapter to join to SOA systems. However, designing the agent system directly in SOA reduces the need for an adapter and will improve performance. The adapter only allows agents to use SOA services, however SOA services cannot use the agents, a further limit to adapter approach.

While a swarm could be built in an SOA system, swarm systems may be better accommodated in swarm specific toolkits. Especially when the toolkit is set up to provide distribution over multiple CPU's or on a grid. Swarm systems often depend on a large number of identical agents. Supporting such large numbers will depend on the scalability of the SOA infrastructure. This is clearly an issue that has to be analyzed in terms of the infrastructure capabilities. However, scaling to a large number of simulations services is often required in commercial applications of SOA. Another feature common to swarms is a shared environment that the swarm agents interact through. In an SOA system, construction the common environment can be approached in several ways. One would be to use a database as the environment. The second would be to build an SOA service that is the environment. The standard databases, like MySQL or Oracle, are designed for scalability. The SOA service scaling will depend on the scalability of the SOA container, hence using a service as the environment may be a limiting approach with respect to performance. The database approach is generally the stronger alternative.

Agents systems built on either events or providing a standard message processing method in the service can be built in SOA. Depending on the implementation language, many libraries exist to make this a trivial exercise. However, the more useful approach may be to use the SOA service invocation as the interface to pass events or messages.

Meta-agents or systems with multiple agent types are not a difficulty for SOA based agent systems. Merely adding a variety of agents of different types into the SOA substrate is easy. Social models and coalition systems prove no real challenge to SOA implementation.

More problematic are the simulation-based systems. SOA is not structured to support the event scheduler need for these implementations. Of course, not all the frameworks provide this ability either. It is clearly a specialized requirement.

Central supervision based systems can be build into SOA. However, this requires development and is not automatically available. Here an existing agent framework may be a better approach.

When attention is turned to the properties, we can see some of the opportunity with SOA based agent systems. First of all open systems can be nicely envisioned and supported on an SOA substrate. Closed systems are not an issue here at all. The big challenge here is to support dynamic service discovery to enable fully open systems.

Fault tolerant systems can be supported nicely on SOA, commercial application require the implementations of SOA to use replication for reliability. From an agent perspective Fedoruk *et. al.* [11] discuss fault tolerance by replication.

Mobile systems are not immediately plausible. Most SOA implantations do not provide mobile services. Similarly, the use of systems that dynamically create agents may not be good candidates for SOA. This depends on the nature of the dynamic behavior. If the dynamic agent creation is purely for scaling, then SOA will function well. If the agents are truly different, then SOA is probably the wrong implementation, until effective service discovery can be implemented.

It is an easy conclusion to see that agent systems can easily be constructed with SOA systems. This allows several advantages.

Agents written in an SOA framework can become components in application systems. This is a viable path for agents to move beyond research and become successful application components. Further, this allows agent applications to become industrial applications and SOA infrastructures provide a supportable and commercially viable substrate for agents. SOA construction will allow commercial applications to easily make use of the agents.

SOA frameworks are well developed and are supported by either open source or commercial development efforts. This allows a much larger support base for the frameworks and may deliver higher quality.

Third, the use of SaaS and clouds can offer very flexible frameworks that support scaling to large collections of agents. Although for a research effort, a cloud grid arrangement may be of real use as a large scale shared research network.

5 Agents Systems Support for SOA Systems

Agent technologies are all ready exploring some of the ideas and tools that would be very helpful to SOA systems.

5.1 Ontology

While ontology is not a solely an agent system, the work with agents combined with ontology has some ideas to offer to SOA. The metadata specifications for SOA are ripe for ontology assistance. Indeed, this is just a true for non-automated and static design as it is for dynamic. The automated discovery aspects of ontology will be discussed separately. For this section, we start with how ontology can be applied to the design of SOA systems.

One major advantage to be gained with an ontology is stronger definition and consistent usage in describing services. This is clearly an advantage across organizations as it is very easy to have variations in the definition of terms. However, introspection about the ability to have code commented with meaningful comments shows that this is a more general and wide spread issue. As poorly as comments are kept meaningful in source code, having to maintain metadata that is decoupled from the code itself is generally worse.

There are three places were ontology can be used. One is the construction of WSDL, one is policy and finally, the directories in UDDI. Of these, the WSDL is most complex, so it is the first considered. The WSDL has several parts including the documentation elements that are designed as human readable elements. The critical elements of the WSDL are:

- Service which is the name of a set of capabilities,
- Message which is the name for a capability of the service
- Types used for the data types in the message

The issue is that a simple name alone is not a reliable definition of the operation. One solution to this is in the optional documentation elements attached to the elements of the WSDL definitions. A better approach is to build an ontology for the names of the service, message, and the data element names.

A second extension would be to apply formal expression to the documentation elements in the WSDL. Using terms from the ontology with first order logic to form additional descriptions of the constraints on the service. This looks like an interesting future work and could be build on the work in formal requirement specifications.

Automated generation of WSDL is supported in some of the toolsets for SOA. To use the ontology extensions as described here, these tools should be extended to support the ontology. This will require two kinds of extension. One is to add the new terms into the ontology, which will require some human assist. And second to use existing ontology to error-check the generated WSDL for consistency.

Policy, like WSDL, can be well supported by use of an ontology. This may prove to be less critical, since policies are defined by standards. But, a real task to look at is using reasoning techniques to examine the stated policies of a service against the requirements for the application. This is especially true is policies are not independent, but have relationships (for example polices that are subsets of other policies).

5.2 Discovery

This is arguably the largest disappointment in SOA. Most current systems still use manual assembly processed. In the author's opinion, agents offer a real chance to fix this issue. There are a number of approaches to discovery reported in the agent literature [1] [2] [8].

Discovery needs to be automated for several good reasons. In the case of current systems that are being built manually, automated discovery offers the chance to do a more exhaustive search for SOA component services. Even within a single enterprise, finding all the possible services can become a difficult problem as services proliferate. When services may be used from multiple enterprises, even those within a single corporation are increasingly difficult to find and evaluate the possible services.

However, the use of manual discovery is hampered by a number of realities. First of all the information in the WSDL is incomplete. The name of the service is supposed to connote the definition of the operation. Sadly, designers do a generally poor job of naming. This is evident when you try to reuse any code. The name alone is sometimes enough, but it can leave the definition of the operation unclear. Software libraries need documentation in a natural language to clarify the operation. Second, the parameter types should help define the inputs and outputs. Here there is a double problem. The naming deficiency applied to the function name applies to the parameters as well. Worse, unit of measure information is generally not given either. Is distance in meters, feet, inches centimeters? Recall the ill fated Mars mission that failed due to units of measure in metric in one system and English units in another, units are clearly needed to full apply discovery.

Of course, with a manual system there remains yet another problem. A purely human failing, the "not invented here" effect. When we rely on human processes, it is often easier to build what you need, not spend time looking for ready to use components. Indeed, it can be argued, validly in many cases, that just building what you need is faster than searching and verifying some other piece of code.

There are two aspects to automated discovery. The first is locating a usable service. The second is dealing with policy. As will be shown, they are different kinds of problems, with policy being a simpler problem.

WSDL defines the service definition found in registries thus it defines the possible data to use for discovery. The services are defined by operation names and messages. There is a provision for documentation elements to provide human readable comments. This is the basic data on which a system can work. The operation name is the first clue about the service. The name is generally a string of nouns and can be separated by simple means. The operation also has input and output message parts. These are given names and types.

Policy is in a very different situation from the area of service discovery. Policy is given a more precise format. Policies are defined by specifications managed by the W3C consortium. The policy definitions in SOA are given as Boolean formulas with the defined policy features. This makes it easy to work with the policies. The main addition would be an ontology to help define partial ordering over policies to handle relations between the policies. This would include cases where one policy can subsume another, or where a pair of policies is contradictory.

Policy negotiation has been explored in several recent papers.

An approach found in [28] is Semantically Enhanced Unified Service Discovery. The approach includes the design of an query system and language used to locate services. The query engine relies on ontologies to analyze the service offering. Their architecture requires a plug-in for registries that support searching from the query engine. These handle syntactic, semantic and quality of service aspects (e.g. policy) of the discovery process. The query covers the application domain, service capability, the input and output parameters and finally the quality of service required. The query language supports semantic operations controlling the match. These include exact matching, the obvious choice. But the also define abstraction which is a matching that allows the element to subsumed by the parameter in the match. And, extension, which requires the parameter's value must subsume the value in the match.

A further extension to this approach is the consideration of service types other than web services. This extension is described as still evolving in the referenced paper. Grid services and P2P are given as possible examples. One could equally well consider SAAS and Clouds as other possible service types.

There is some interesting work around negotiation and web services. Brzostowski Et al [6] describe a negotiation mechanism for service level agreements. This approach has great merit when the task is to construct a complex system of web services to achieve an objective. This approach is applied to quality of service. The total quality of service is formed from the quality of service supported by each of the components. The negotiation looks at the trade-offs between different possible compositions of services in order to identify a set that will meet the required quality of service. The work is performed in a two-layer framework of agents.

Another interesting approach is found in [24]. This algorithm was originally developed for logistic supply chains. An application constructed by a sequence of SOA services and capabilities and it can be viewed as a supply chain. Each service provides a output for the chain and consumes some input in order to produce the output. There are a set of costs defined in the model which are used to optimize the selection of services, including the cost to use a service, the cost in delay to move objects between the service nodes.

There is a cost associated with each service in the application. Delay is one of the costs for a service. This may be unknown at initial discovery but could measured and stored whenever the service is invoked as part of any application. Another possible cost can be the monetary cost for services that are pay for use. However, even when the service does not a charge for use, the reality is that no service is really free. There is a cost to any organization that runs a service. This cost is the CPU and I/O costs consumed by the execution of the service, but also the energy costs of running the systems. Energy is becoming an critical cost for some data centers. A more difficult cost to assess is the opportunity cost, that is the cost of lost opportunity to use that service in some other application, because it is used in the current one.

The other costs are transport costs to move data between the nodes of the application. Of course the transport costs in a supply chain are higher than in a SOA application. Still, the cost to move data around the net is definitely measurable.

The algorithm, in the paper, is a Monte Carlo that optimizes the choice of the algorithm.

There is also work on adaptive negotiation systems using genetic algorithms [23]. This paper gives a very nice formalism to view service and policy negotiation. While the previous mentioned works view the service from a cost view, this paper views the negotiation as offers and preferences. An offer is a set of attribute values or intervals of values. This easily fits the policy definitions for web services. The actual service provided can require a more complex valued attribute. The simplest value is to give the name of the method as the attribute value and then apply ontology to help evaluate its fit to the required operations. Similarly, the attributes for parameters can be simple. Although as argued earlier in this work, the units of measure should be included in web service definitions. The other extreme would be to use the WSDL as the value of the attribute. The paper then defines a preference. A simple preference is to have acceptable ranges of attribute values. However a stronger system is to apply utility functions to the attributes. The utilities values are normalized to 1. This provides a both an easy way to evaluate the offer and even more it provides a way to determine counter offers or concessions in the negotiation process.

A genetic algorithm process uses the utility function as a fitness measure. It also includes time pressure and the counter offers in the negotiation process as part of the fitness function. This approach generates solutions that are close to the ideal case. The ideal case is defined as finding the joint payoff that depends on the utility functions of both the offerer and the accepter.

An aspect that could also be applied to discovery is to use the documentation tags that are allowed in the WSDL. There are two ways that can be envisioned for this approach. First is to apply natural language techniques to the documentation to aid the discovery process by providing additional detail beyond the simple name based approach in the formal WSDL definitions. A second approach is to develop a more formal description approach to add to the WSDL using the documentation tags.

A possible approach for the formal specification can be found in [26]. In their approach an abstract machine is used to define the service. This includes constraints, pre and post conditions, invariants and other formal semantic constructs. And in their approach type checking is applied. This approach is appealing and might be equally well applied to agent systems that use dynamic composition. However, there is a real weakness as well. Programmers are generally not skilled in formal semantics. Even though some of the formal semantics can be checked against the actual implementation, it still appears to be an obstacle to use in industrial practice.

6 Designing New Systems Using SOA and Agent-Systems

The strength of agent systems is the view of the agent as an active entity that reacts and responds to the environment. In SOA, the view is a service that implements a piece of business logic, such that it has composibiliy supporting application development by combinations of the services. At least a segment of the development community has concluded that SOA has not achieved the desired level of re-use. The question is what can be done to change this? And of necessity, the dependent question is there a good reason to try to fix SOA, rather than eek the next software development model?

The required SOA implementation components will not go away since they are very lightweight. This is one of the great advantages of SOA. Java J2EE or the .NET frameworks may continue or be extensively modified. They may go out of support and use. But the required support for an SOA system is easily implemented in dozens of possible languages and systems, PHP for one example. Since services are normally small-contained units of software, porting them to a new infrastructure is generally a simple process. In terms of survivability of services, this is a great advantage. As long as a business entity is willing to maintain a service, other systems can continue to use it and new applications can still find it. Considering other systems, this could allow SOA a very long life span. Given this advantage, improvements to SOA would be well worth the effort.

Thus, if we look at services as agents, what would be done differently? First of all, an agent is critically dependent on its environment. The agents have precepts and responses. Since services are normally stateless with any state stored in a database, the database becomes part of the environment. It becomes part of the design of the service to characterize the database aspects that the agent will read or write. The other part of the environment is the set of messages that the service will respond to. This second part is the set of precepts and the outbound messages are responses of the agent. Of course reads from the database are also precepts for the agent. And writes are responses of the agent. There is a difference between the database and the messages. The database is accessed by active observation by the agent.

Likewise a great deal of work has been expended on design methodologies for SOA. This can be profitably used in agent design. To date most agent systems have been constructed with a reality small number of agent types. Some may use a large number of actual agents of just a few different types. As agent systems are used for full applications development in a commercial environment, then a large number of agent types may be required. If so, then increased software engineering would be a wise idea. Using the SOA metadata will become an increasing useful method to document and to locate agents.

Meta data to define contracts and communication interfaces used in SOA systems can become a very useful and powerful description of the agents in a system. This will form the basis of a sound engineering approach to agent reuse and system construction.

When multi agent systems transition into business applications and large teams are used to construct the agents, them the design approaches used in SOA will become a usable basis for agent engineering and design.

Likewise, testing methodologies like test driven design or the use of testing systems like jUnit can be leveraged to improve development effectiveness. Using a rigorous testing practice can be a critical to producing long-lived systems. It can markedly increase the reliability and modifiability of systems.

7 Conclusion and Future Work

The goals for this chapter are to consider both SOA and agents for synergy. To that end, several conclusions are reached. It is certainly the case that SOA and agent systems have some common synergies. Also, SOA can provide a strong framework for agent systems development and deployment.

The future work needs to include building some good examples of agents systems in an SOA framework.

A second interesting concept explored briefly is the use of styles and properties to describe agent systems. This is a common practice among software architects and can be well extended into agent systems. The work here is just a start and can be extended and improved. It was undertaken here as a way to think about SOA implementation of agents.

References

1. Agarwal, S.: Goal Specification Language for Automated Discovery and Composition of Web Services. In: Proceedings of the first international joint conference on autonomous agents and multiagent systems, AAMAS 2002 (2002)
2. Anton, A.I.: Goal-based requirements analysis. In: ICRE 1996: Proceedings of the 2nd International Conference on Requirements Engineering, ICRE 1996 (1996)
3. Bass, L., Clements, P., Kazman, R.: Software Architecture in Practice, 2nd edn. SEI Series in Software Engineering. Addison-Wesley Professional, Reading (2003)
4. Beer, M.D., Whatley, J.: A Multi-Agent Architecture to Support Synchronous Collaborative Learning in an International Environment. In: Proceedings of the International Joint Conference on Autonomous Agents and Multiagent Systems AAMAS 2002, p. 505 (2002)
5. Bordini, R.H., Hübner, J.F., Wooldridge, M.: Programming Multi-Agent Systems in AgentSpeak Using Jason. John Wiley & Sons, Ltd., Chichester (2007)
6. Brzostowski, J., Baruwal, M., Kowalcyzk, R.: Three Decision-making Mechanisms to faciliate Negotiation of Service Level Agreements for Web Service Compositions in Negotiation and Market Engineering. In: Jennings, N., Kersten, G., Ockenfels, A., Weinhardt, C. (eds.) Internationales Begegnungs- und Forschungszentrum Informatik (IBFI), Schloss Dagstuhl, Germany (2007)

7. Cholvy, L., Belin, E., Garion, C.: Desires, norms and constraints. In: The Third International Joint Conference on Autonomous Agents and Multi Agent Systems, pp. 722–729 (2003)
8. Chu, W., Qain, D.: Semantics Based Enterprise Modeling for Automated Service Discovery and Service Composition. In: The 2nd IEEE Asia-Pacific Service Computing Conference, December 2007, pp. 439–445 (2007)
9. Di Stefano, A., Santoro, C.: Modeling Multi-Agent Communication Contexts. In: Proceedings of the First International Joint Conference on Autonomous Agents and Multiagent Systems AAMAS 2002, p. 174 (2002)
10. Erl, T.: SOA Principles of Service Design. Prentice-Hall, Englewood Cliffs (2008)
11. Fedoruk, A., Deters, R.: Improving Fault-Tolerance by Replicating Agents. In: Proceedings of the First International Joint Conference on Autonomous Agents and Multiagent Systems AAMAS 2002, p. 737 (2002)
12. Bellifemine, F., Claire, G., Greenwood, D.: Developing multi-agent systems with JADE. Wiley, Chichester (2007)
13. JASON (2009), `http://jason.sourceforge.net`
14. Gamma, E., Helm, R., Johnson, R., Vlissides, J.: Design Patterns: Elements of Reusable Object-Oriented Software. Addison-Wesley Professional, Reading (1994)
15. Genesereth, M.R., Ketchpel, S.P.: Software Agents Communications of the ACM 37(7), 48–53 (1994)
16. Graham, J.R., Decker, K.S., Mersic, M.: DECAF – A Flexible Multi Agent System Architecture. Autonomous Agents and Multi-Agent Systems 7(1-2), 7–27 (2003)
17. Guessoum, Z., Ziane, M.: Monitoring and Organizational-Level Adaptation of Multi-Agent Systems. In: Proceedings of the first International Joint Conference on Autonomous Agents and Multiagent Systems, AAMAS 2004, pp. 514–521 (2004)
18. Håkansson, A., Hartung, R.L.: Calculating optimal decision using Meta-level agents for Multi-Agents in Networks. In: Apolloni, B., Howlett, R.J., Jain, L. (eds.) KES 2007, Part I. LNCS (LNAI), vol. 4692, pp. 180–188. Springer, Heidelberg (2007)
19. Håkansson, A., Hartung, R.: Using Meta-Agents for Multi-Agents in Networks. In: Arabnia, H., et al. (eds.) Proceedings of The 2007 International Conference on Artificial Intelligence (ICAI 2007), The 2007 World Congress in Computer Science, Computer Engineering, & Applied Computing, vol. II, pp. 561–567. CSREA Press, U.S.A. (2007)
20. Håkansson, A., Hartung, R.: An approach to event-driven algorithm for intelligent agents in multi-agent systems. In: Nguyen, N.T., Jo, G.-S., Howlett, R.J., Jain, L.C. (eds.) KES-AMSTA 2008. LNCS (LNAI), vol. 4953, pp. 411–420. Springer, Heidelberg (2008)
21. Hartung, R., Ramanthan, J., Bolinger, J.: Ontology for Enterprise Modeling. In: Lovrek, I., Howlett, R.J., Jain, L.C. (eds.) KES 2008, Part III. LNCS (LNAI), vol. 5179, pp. 799–807. Springer, Heidelberg (2008)
22. Hübner, J.F., Sichman, J.S.: SACI - Simple Agent Communication Infrastructure (2009), `http://www.lti.pcs.usp.br/saci/`
23. Lau, R.Y.K.: Adaptive Negotiation Agents for E-business. In: ICEC 2005 (2005)
24. Levchenkov, A., Gorobetz, M.: The Algorithm of Negotiation of Multi Agents for Planning in Geographically Distributed Logistic Supply Chains. In: Proceedings of the 19'th European Conference on Modeling and Simulation, ECMS (2005)
25. Luke, S., Cioffi-Revilla, C., Panait, L., Sullivan, K.: MASON: A New Multi-Agent Simulation Toolkit (2009), `http://cs.gmu.edu/~eclab/projects/mason/`
26. Milanovic, N., Malek, M.: Adative Search and Learning-based Approaches for Automatic Web Service Composition in Web Services Reasearch and Practices. LLang-Jie Zhang Cybertech Publishing (2008)

27. Russell, S.J., Norvig, P.: Artifical Intelligence: A Modern Approach, 2nd edn. Prentice Hall, Englewood Cliffs (2003)
28. Tsalgatidou, A., Pantazoglou, G.A.: Semantically Enhanced Unified Service Discovery (2005), http://www.w3c.org/2005/FSWS/Submissions/10/paper.html
29. Universal Description, Discovery and Integration UDDI (2009), http://uddi.xml.org/
30. Web Services Description Language (WSDL) 1.1, http://www.w3.org/TR/wsdl
31. Woodridge, M.J., Jennings, N.R.: Intelligent Agents Theory and Practice Knowledge. Engineering Review 10(2), 115–152 (1995)

Service Fusion in Mobile Contexts

Jun Miao Shen[1], Michael J. O'Grady[2], and Gregory M.P. O'Hare[2]

[1] School of Computer Science & Informatics, University College Dublin (UCD), Belfield, Dublin 4, Ireland
[2] CLARITY: Centre for Sensor Web Technologies, University College Dublin (UCD), Belfield, Dublin 4, Ireland
sjm2218@hotmail.com, {Michael.J.OGrady,Gregory.OHare}@ucd.ie

Abstract. Service provision for mobile users poses several challenges for those managing enterprise information systems. Some of these are ubiquitous, for example, privacy and security. Others give rise to integration and interoperability issues. The agent paradigm, though offering potential in fixed networked scenarios, has not been seriously considered as a mechanism for incorporating mobile clients, primarily due to their computational intractability on mobile devices. However, recent developments in agent and mobile computing technologies – both from a hardware and software perspective, have rendered the mobile phone a viable platform for agents. One implication of this is that agents now offer a viable platform for service delivery to mobile users. This chapter reflects on some of the pertinent issues necessary for this, including Agent Oriented Software Engineering (AOSE). The discussion is illuminated with a systematic description of EasyLife – a framework that harnesses the agent paradigm to construct and deliver services to mobile users.

Keywords: Ambient Intelligence, Mobile Computing, Agent-oriented Information Systems, Location-aware computing.

1 Introduction

As the boundaries between mobile telecommunications and computing continue to blur, the demand for access to services and data from mobile devices increases. This development is not surprising; the process has been ongoing since the production of the first laptop, or at least one augmented with a wireless modem. However, it can give rise to significant problems to service providers of various hues. As an example of this consider the difficulties experienced and the energy expended in making the internet accessible to mobile users. The difficulties are exacerbated by the nature of the archetypical mobile device. When compared with the traditional workstation or desktop PC, a mobile device is not as computationally powerful, possibly several orders of magnitude smaller. Nevertheless, many people prefer to use their mobile phone for service access. Though the experience may be leave something to be desired, the convenience of anytime, anywhere access compensates in many instances.

Enterprises are not immune to developments in mobile computing, and the option of being able to access company data remotely is one that executives and sales

A. Håkansson et al. (Eds.): Agent and Multi-agent Technology, SCI 289, pp. 49–67.
springerlink.com

professionals amongst others find particularly useful. Of more interest to enterprises however is revenue generation, and providing services to the mobile computing community. Such services may be specialised in nature, or may demand cooperation with other enterprises or third-part service providers. Either way, it frequently demands a radical review of business and IT management practice.

2 The Quest for the Enterprise Mobility

Enabling Enterprise mobility and mobile service provision demands an innate understanding of the mobile computing domain. A firm grasp of its limitations must also be obtained if these are to be circumvented.

2.1 Mobile Computing Paradigms

Two approaches exist for supporting mobile users. In the first instance, an infrastructure may be embedded in a physical environment such that computational resources can be accessed as needed - ubiquitous computing adopts this approach. In the second instance, the user carries the necessary computational apparatus about their very person. Wearable computing would be an example of such an approach. However, for many people, conventional mobile telephony would be the de facto interpretation of mobile computing.

A more recent interpretation of mobile computing is Ambient Intelligence (AmI) [1]. AmI incorporates ubiquitous computing and seeks to provide seamless and intuitive interaction to services. Though an environment may be saturated with embedded computational artefacts, enabling intuitive interaction with such artefacts is essential if the services provided by the environment are to be availed of. Intelligent User Interfaces (IUIs) [2] are proposed as a means of achieving such interaction. How the intelligence required for such interfaces may be realised in practice for embedded devices, or indeed lightweight mobile devices such as mobile phones is an open question. One computationally tractable solution is that of embedded agents [3]. Though such agents may drive the logic for the IUI, there is no reason why they cannot be harnessed for other purposes, and form the mobile component of enterprise-based Multi-Agent Systems (MASs). When viewed in this light, it can be seen that such agents offer an apt solution in Enterprise computing, enabling the construction of innovative services in Business (B2B), Business to Consumer (B2C) and Business to Employee (B2E) scenarios.

2.2 The Mobile Phone – Services on Demand

The mobile phone represents the default mobile computing platform of choice for many people. From an enterprise perspective, a case could be made for the laptop; however, in comparison with the fixed workstation, the wireless connection with its poor latency and data rates is the key differentiator. In contrast, the situation is more complex with mobile phones. The limited screen real estate, typically ¼ VGA, as well as the unwieldy methods for interaction, usually a combination of 5-way navigation pad, soft keyboard and a compact alpha-numeric keypad, are characteristic of mobile devices, and are not designed for long interaction sessions. These characteristics, or from a software engineering perspective – constraints, significantly influence the design of services and may be the sole cause of re-engineering the user interface component. Another constraint is the paucity of raw computational resources available on such devices. Again, this has many implications, one being that the ability to manipulate and visualise data on the device is seriously compromised. Thus service composition and adaptation, for example, personalisation, usually occurs on fixed network servers.

2.3 The Question of Heterogeneity

Mobile devices are manifested in a multitude of capabilities and form factors. This heterogeneity may be perceived as a great strength as the needs of many market segments are addressed; the commercial success of mobile telephony is testimony to this. From a service provider's viewpoint, and indeed that of software engineer, heterogeneity introduces major complexity into the design and implementation process. Two approaches can be considered. In the first case, a service can be designed for the lowest common platform specification. In this case, the customer base is widened, but the potential for service differentiation and augmentation is limited. A second approach is to develop for a platform of a certain configuration. For example, the host device must possess an interface of ¼ VGA, and support the rendering of MP3. This is invariably a higher spec platform and will only be available to a subsection of the customer base. However, it does allow for the provision of high quality, high revenue generating services. For B2B and B2E solutions, this strategy is attractive. In the case of B2C, the case is less clear cut, with the application domain and profile of the end-user being key determinants.

From a software engineering perspective, harnessing internet technologies offers a number of advantages. Internet Protocol (IP) is supported by practically all network operators, irrespective of their deployed networking technology. Many phones have micro browsers that enable pages constructed in HTML, or some variant, be viewed. In the case of multimedia, SMIL [4] viewers have been ported to mobile devices. In essence, if a service can be implemented using the rich array of internet technologies available, even those that may have to simplified for mobile devices, the development process is simplified. However, this is not always feasible or desirable.

Service composition may demand the sourcing of data or use of other services from disparate sources. This is another source of heterogeneity in the mobile service provision process. Such services may use a series of highly specialised protocols and data formats that may be incompatible with those adopted by the service provider, or more problematically, may have previously adopted a non-standardised approach. Either way, the problem of service integration for composite service provision remains. Potential solutions to the problem exist, but may require the expenditure of significant effort. For example, Simple Object Access Protocol (SOAP) [5], an implementation on XML, enables the exchange of structured information. Another approach that has been documented involves the use of agents as service wrappers. In this case, legacy services would be encapsulated by an intelligent agent, and this agent would provide the service to other agents that constituted the service being provided.

2.4 Why Agents?

Agents represent a paradigm for modelling and implementing software. Their characteristics are reasonably well defined but how these are harnessed is at the discretion of the software engineer. The degree to which an agent may be considered intelligent is open to interpretation, and implementation. Those in the AI community would envisage agents as encompassing a sophisticated reasoning engine. Others might view agents as a series of collaborating entities or even software objects. Irrespective of one's viewpoint, agents have been harnessed in a multitude of domains including those pertinent to this discussion, namely enterprise information systems and mobile computing.

Many researchers have demonstrated the applicability of the agent paradigm in various aspects of enterprises information systems. Kishore et al [6] investigate the agent paradigm as a means of realising integrated business information systems. Nahm & Ishikawa [7] adopt a hybrid multi-agent approach for inter-enterprise function, resource integration and collaboration in networked manufacturing systems. Agents have been harnessed for workflow management as they are seen as providing a robust and open solution to deliver dynamic and automatic workflow adaptations – a necessity in modern business environments [8][9]. In a similar vein, BDI agents have been adopted for realizing agile business processes [10]. In some cases, internet technologies have been harnessed, For example Li [11], describes AgentStra, an internet agent based system for strategic decision making. The potential of agents for Customer Relationship (CRM) management has been investigated [12], as has their applicability in WWW services [13].

While there are documented mobile computing systems that interacted with agent-enabled services, those that incorporate agents on the mobile devices are rare, not surprisingly. The mobile tourist domain has been the focus of two pioneering efforts. CRUMPET [14] harnesses agents to monitor the QoS of the wireless connection. In addition, it utilizes legacy services via agent wrappers. Gulliver's Genie [15] is concerned with the timely delivery of multimedia services to tourists, and uses agents to capture and interpret the tourist's contextual state. Ad-me [16] captures end-user biometric feedback to identify opportunities to deliver advertisements to users. EasiShop [17] encompasses a suite of agents, some mobile, to bargain with those traders that the shopper is in the vicinity of.

2.5 Agents: Integrating Enterprises with Mobile Subscribers

Deploying agents on mobile devices offers one potential avenue for enterprises to provide services to mobile users, be they customers or employees. From a software modelling perspective, the MAS may encompass companies' fixed networked server nodes and well as mobile nodes. This offers a standardised method for modelling the various components and the interactions between them. While agents offers some advantages, such as managing the heterogeneity discussed previously, it is the enablement of a single platform for service deployment that is of most importance, and offers most potential. The key factor in this has been the development of new agent frameworks that can operate on mobile devices. The availability of a common software platform across many devices, namely J2ME, is also a key contributor.

In the next section, the software engineering cycle is considered from an agent perspective, concluding with a brief discussion on agent frameworks for mobile devices.

3 Agent Oriented Software Engineering

Software engineering of agent systems follows the same steps as traditional software engineering. One common concern about agents would naturally concern their incorporation into methodologies for software development. To meet this, a number of researchers have considered various stages of the software engineering cycle and proposed methodologies and frameworks for each from an agent perspective. As an illustration of this, some proposed approaches in the various stages are now considered.

3.1 AOSE: Requirements Engineering

TROPOS [18] covers five phases of the software development cycle including Early Requirements, Late Requirements, Architectural Design, Detailed Design and Implementation. It is particularly concerned with addressing a perceived lack of support for mentalistic or knowledge level notions in other proposed methodologies. Recall that the traditional goal of a software designer is to achieve a sufficient level of abstraction from which classes, methods and other attributes can be easily identified. However, in doing this, one may lose sight of any mentalistic notions that were inherent in the original system design and these may need to be "resurrected" during implementation. Thus TROPOS sets itself the goal of supporting knowledge level specifications throughout the development cycle. AUML [19] diagrams are adapted for the design phase. TROPOS has been illustrated using a sample e-commerce application, which was developed using JACK [26].

3.2 AOSE: Design Methodologies

Prometheus

Prometheus [20] was motivated by a desire to develop a process that could be taught to industry practitioners and students who might not have a background in intelligent agents. In particular, it is concerned with the development of agents of a BDI-like genre. While conceding that this makes the methodology less attractive to those working with other agent architectures, the authors were keen to address a perceived deficiency in existing methodologies where agents were treated as simple interacting software processes, or as a black box. Prometheus supports three distinct though non-sequential development phases:

1. System Specification Phase
   One of the first issues addressed is the identification and specification of how the agent system both perceives and interacts with its environment. In parallel, the high

level objectives, termed functionalities of the system, must be identified. Other methodologies refer to these as roles. Finally, standard UML use case scenarios must be constructed.

2. Architectural Design Phase
   The primary objective of this phase is the identification of the agents that will comprise the system. This is achieved by considering the functionalities identified during the specification phase and deciding how these might be optimally assigned to agents. After this, each agent is assigned a number of events to listen to and to react to. In addition, shared data objects are identified at this stage. Finally, a system overview diagram that lists the agents, events and shared data objects is constructed. This is arguably the most important artefact in the entire design process. Using this diagram as a basis, interaction diagrams and interaction protocols are constructed. Prometheus uses AUML notation for specifying interaction protocols.

3. Detailed Design Phase
   During this phase the internal structure of each agent and how it will achieve its tasks in the system are specified. It is at this stage that Prometheus becomes specific to BDI-like agents and capabilities, internal events, plans and data structures for each agent are defined. These will provide the necessary details for implementation. Note that the detailed design will be heavily influenced by the choice of deployment platform.

SODA

SODA (Societies in Open and Distributed Agent spaces) [21] is a methodology for the development and design of internet-based Multi-Agent Systems. Most of the prevailing methodologies concentrate on intra-agent issues. In contrast, SODA concentrates on the inter-agent issues or the social aspects of Multi-Agent Systems. In addition, the agent's environment is of particular importance. SODA is based on the premise that agents cannot be considered in isolation from their proposed environment and advocates consideration of environmental issues from the very start. Therefore SODA is particularly concerned with the engineering of agent societies and their environment and does not concern itself with the actual agent specifics for which another methodology should be used. SODA supports the two standard phases of analysis and design.

1. Analysis Phase

After undertaking normal analysis tasks, for example, studying the application domain, identifying technology resources and constraints and so on, three models are constructed:

1. Role Model: the goals are modeled in terms of the tasks to be performed and these are then associated with roles and groups.
2. Resource Model: the environment is modeled in terms of the services available.
3. Interaction Model: All interactions are modeled in terms of interaction protocols and any interaction rules governing interaction within the system are identified.

2. Design Phase
During design, the abstract models produced during analysis are transformed into design abstractions that can be mapped on a one-to-one basis with actual components in the deployed system. Three models are produced:

- Agent Model: Individual and social roles are mapped onto agent classes.
- Society Model: Groups are mapped onto societies of agents. These are then designed and organized around a suitable co-ordination abstraction e.g. an Agent Communications Language (ACL).
- Environmental Model: Resources are mapped onto infrastructure classes and associated with topological abstractions e.g. domain names.

Agent Object Relationship (AOR) Modelling
Agent Object Relationship (AOR) [22] was proposed for the design of organisational information systems. Its raison d'être is simple: to remedy a deficit in traditional modelling approaches for handling both the dynamic and deontic aspects of organisational information systems. While AOR was inspired in part by Shoham's agent-oriented programming proposal [23], the methodology itself is heavily influenced by two popular database modelling methodologies, namely the Entity Relationship (ER) meta-model and the Relational Database (RDB) model. One of the objectives for AOR is to facilitate the transformation of an AOR model into an appropriate database schema in as simple a manner as possible. Clearly this implies the use of formal semantics when defining the elements of the AOR model.

In AOR, an entity can either be an agent, an object, an event, an action, a claim or a commitment. AOR is quite specific about the difference between agent and objects. In AOR, an object is a passive entity, a definition that contrasts with much of the language used in the literature. Only agents can perceive events, perform actions, communicate or make commitments. AOR assumes agents communicate via an Agent Communication Language (ACL) and uses a mentalistic model of knowledge representation. It also considers the concepts of perception, for example, of the environment, and memory, for example, of past events and actions, quite important.

3.3 AOSE: Implementation Frameworks

Agent Factory
Agent Factory [24] provides a cohesive framework for the development and deployment of agent-based applications. In particular, it provides extensive support for agents of the BDI genre. The Agent Factory runtime environment includes a Message Transport System which manages the delivery of messages between agents on the same platform and between agents on different platforms. A Migration Manager oversees the migration of agents between different platforms. A dedicated agent, the White Pages Agent, supervises the agent platform and manages the creation and deletion of agents. Occupying a layer above the agent runtime environment is the agent development environment. This contains a number of tools for enabling the fabrication of agents as well as providing features for monitoring and dynamically configuring various aspects of an agent's behaviour.

Jade

JADE (Java Agent Development Framework) [25] is a FIPA compliant agent platform that facilitates agent development by providing a set of agent-related services. Conceptually, it may be regarded as a kind of agent "middleware" solution. The platform itself can be distributed across several hosts, the only requirement being that a JVM is available on each. JADE does not subscribe to any particular agent theory or architecture and leaves the implementation details to the developer. However, a general agent model has been developed that can be easily customised to an architecture of the designer's choosing.

Jack

The JACK Intelligent Agent™ framework [26] was developed by Agent Oriented Software Pty. in Australia with the aim of bringing the intelligent agent concept into mainstream commercial software engineering as well as Java environments. JACK is designed primarily for use as a component in a larger software environment as this is seen as the most appropriate way of incorporating it with legacy code. Therefore a JACK agent is regarded as just another object by non-agent software. Conversely, the agent is equipped with a number of mechanisms for easily interfacing with other components.

3.4 AOSE: Frameworks for Embedded Agents

LEAP

LEAP (Lightweight and Extensible Agent Platform) [27], originally an EU's Information Society Technologies (IST) project, is a FIPA compliant platform that can be used to deploy agents on a heterogeneous network of fixed and mobile devices. LEAP has evolved from the JADE platform and can interact with legacy JADE agents. The LEAP platform itself is modular and consists of the following modules:

- Kernel Module: This module is mandatory as it manages the platform and enables the realization of the agents.
- Communications Module: A mandatory module for handling the heterogeneous communications protocols.
- Administration module: An optional module that contains a GUI for managing and visualising agents.

AFME

AFME (Agent Factory Micro Edition) [28] was specifically designed for deploying intentional agents on mobile devices. It is based on Agent Factory and can interoperate with it. Designed for mobile phones, it requires the availability of the Constrained Limited Device Configuration (CLDC) Java platform augmented with the Mobile Information Device Profile (MIDP). AFME agents follow a sense-deliberate-act cycle. A perceptor feature is used to monitor the state of the agent's environment. A belief resolution function enables the identification of what it is that the agent can achieve within the current cycle. Appropriate actuators are then harnessed to realize the necessary change within the environment.

4 EasyLife

EasyLife [29] provides a framework that enables service providers to deliver customised combinations of location-aware services to their subscribers. Many examples of location-aware services have been documented, for example, AccesSights[30], HIPS [31], CoMPASS [32] and ACCESS [33]. However, EasyLife, like ACCESS, adopts an agent-based approach, and is concerned with service fusion through a suite of heterogeneous agents.

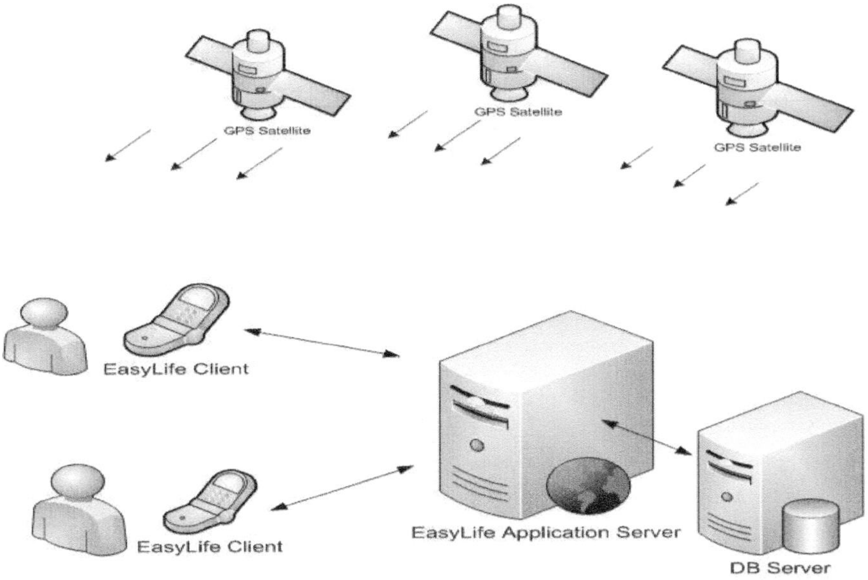

Fig. 1. Overview of EasyLife

4.1 Overview of EasyLife

EasyLife is a multi-tier, location-aware, service-oriented system (Figure 1). It uses agents for realizing different services. At present, it consists of three services, a weather service (location-aware weather using the REST API and a mashup for the Yahoo weather services), a restaurant service (a location-aware recommending system) and a shopping service (a location-aware advertising system). With these services, users can instantly get the weather forecast for their location. Users can also determine where the nearest restaurant is. And enterprises can promote their services or products by delivering their advertisements to those people within a certain range. EasyLife harnesses GPS, Bluetooth, J2ME and AFME for its mobile client. It harnesses Wi-Fi as the wireless carrier for communicating with the service host. On the server side, it harnesses servlets, Jade, Hibernate and MySql for service delivery. EasyLife is modeled on a suite of heterogeneous agents that collaborate to deliver the required service.

4.1.1 Characteristics of EasyLife

Three characteristics define EasyLife:

1. EasyLife is truly a Software-as-a-Service. Customers may easily subscribe and use the services in an on-demand fashion. Service providers can focus on service development and service access.
2. EasyLife is more than just an ad-hoc service delivery application, providing a framework for building other customized services. It is a generic composite application that enables easy integration of services – either those provided by a third-party or ones based on core enterprise functions.
3. EasyLife is extensible and scalable. Its agent-based architecture is essential for realizing these features.

4.1.2 EasyLife Services

EasyLife implements three simple but useful services, all using location as the primary context, and leveraging Web 2.0 service by the mobile client (Figure 2).

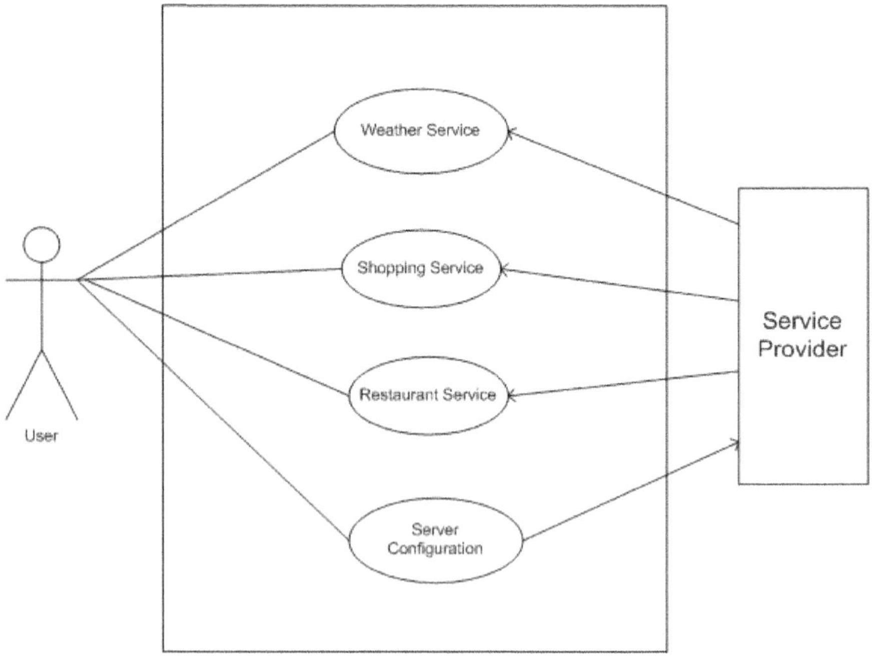

Fig. 2. Use Case scenario for EasyLife

Some scenarios for using EasyLife are as follows:

- **Scenario 1:** A subscriber is planning on playing some outdoor sports, and would like to know the weather forecast. By selecting the weather service, the subscriber could instantly obtain short and longer term forecasts.
- **Scenario 2:** While shopping, the subscriber would like the location of the nearest restaurant. By selecting the restaurant service, the corresponding agent-based service finds the nearest restaurant registered with EasyLife and informs the subscriber of its location. Should this be unsatisfactory, other nearby restaurants will be recommended.
- **Scenario 3:** While shopping on a busy street, those shops registered with EasyLife may send advertisements and details of special offers to the subscriber, but only if in the vicinity of the shop.

4.1.3 Accessing EasyLife Services

Figure 3 illustrates the workflow for EasyLife services. After a service has been selected, the gateway agent is informed via a message encoded with the location and required service. This agent forwards the message to the income controller agent, which parses the message for location and desired service, and forwards the request to the appropriate service agent. The service agent processes the request, and assembles a message for the subscriber. This in turn is forwarded to the outcome controller agent and then to the gateway agent for dispatch to the mobile client. The message will contain some combination of text, multimedia, or a URL, and will be rendered on the mobile device for viewing.

4.2 Architecture of EasyLife

From an architectural perspective, EasyLife consists of three key components:

- EasyLife client;
- AgentProxy_app;
- AgentServices.

The benefits for dividing the system into these three different parts are that the resultant system is loosely coupled and each component is well abstracted and easily extended. Each component is now briefly described.

4.2.1 EasyLife Client

An illustration of the architecture of the EasyLife client may be seen in Figure 4. It is composed of five key elements:

- **Agent:** controls the behavior and workflow of the EasyLife client. It also listens for and handles all generated events.

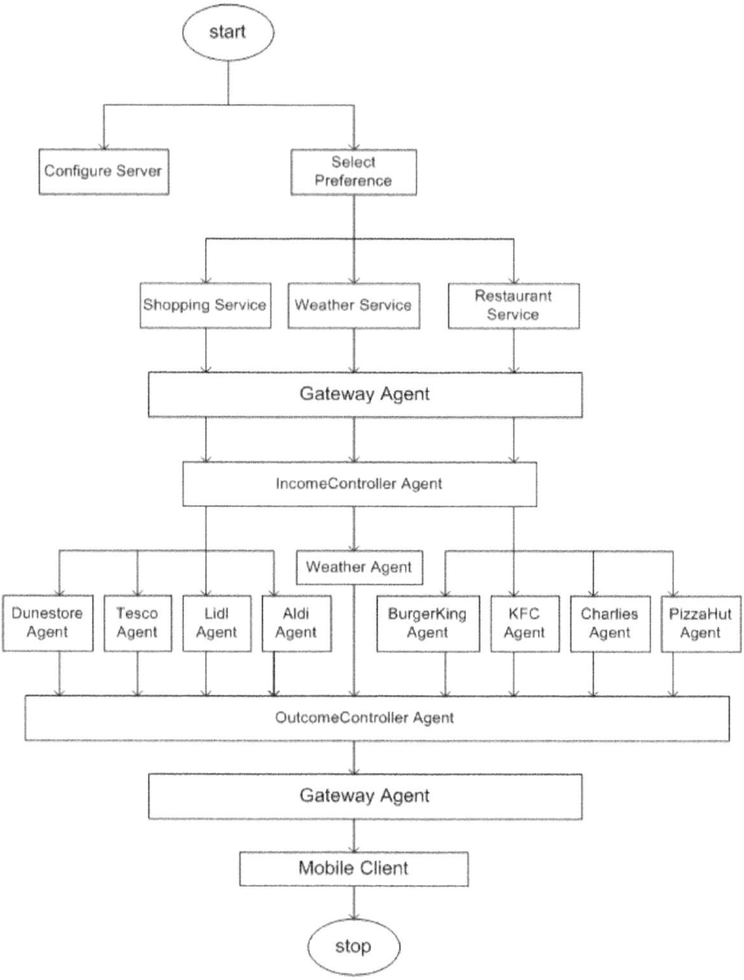

Fig. 3. Workflow for an EasyLife service

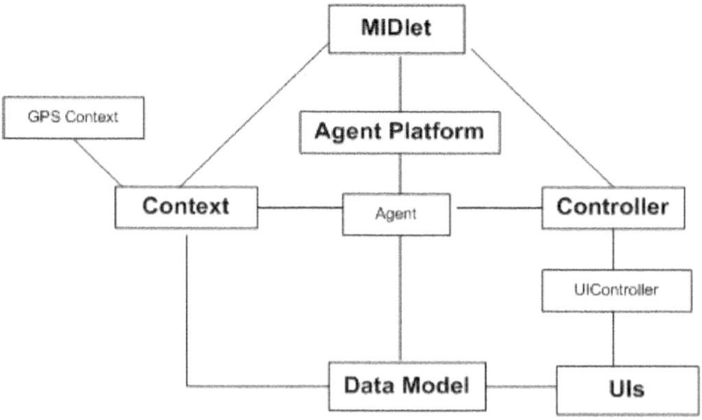

Fig. 4. EasyLife – Client Architecture

- **Context:** is responsible for obtaining the context from the external physical environment. In particular, it gets the location information from the external Bluetooth GPS receiver. It can be extended to handle other sensors if need be.
- **Controller:** provides an access point for certain functions – one example being the UIController. It provides an API for calling different UIs. For extensibility purpose, other controllers could be developed.
- **UIs:** provides the Graphic User Interface (GUI) for the EasyLife client.
- **Data Model:** represents an abstraction of the data from different domains, such as the GPS data model, the weather data model and so on.

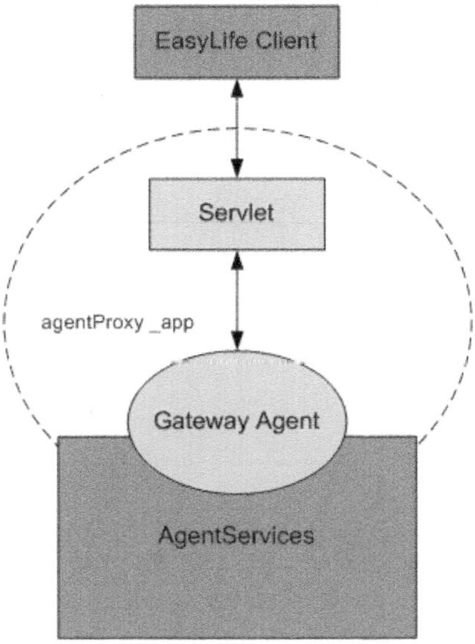

Fig. 5. AgentProxy_app architecture

4.2.2 AgentProxy_app

The AgentProxy_app is best viewed as a virtual router. It has a Gateway Agent which parses service requests, and forwards them to the correspondent agent service. It also relays messages to the mobile client. From an implementation perspective, agentProxy_app is implemented as a servlet (Figure 5). It communicates with the EasyLife client via WiFi, and standard TCP/IP with the Gateway Agents which acts as the interface to the Agent Services.

4.2.3 AgentServices

AgentServices consists of many different agents, each representing an enterprise and providing different services. There are controller agents which could be seen as service brokers interacting with other systems. AgentServices consists of two parts, controller agents and service agents (Figure 6).

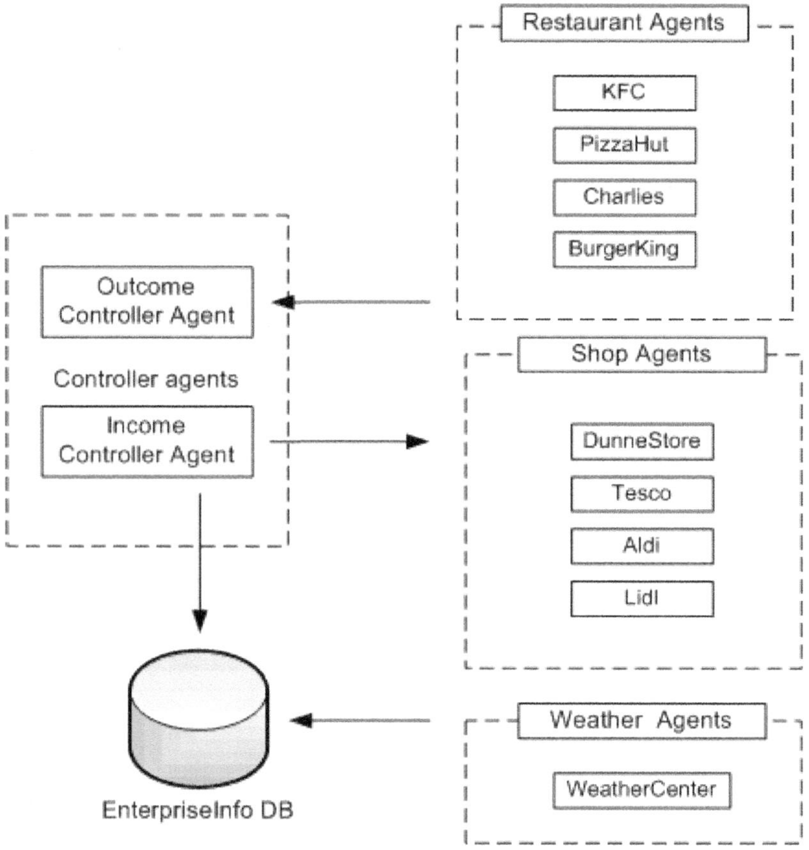

Fig. 6. AgentServices architecture

Controller Agents
Income Controller Agents are the entry points for the agent service. They are responsible for parsing the request and dispatching it to the correspondent agent according to the location and the type of the service. They also inform the Outcome Controller Agent of the address of the Gateway Agent, so that the Outcome Controller Agent knows where to send the service data.

Outcome Controller Agents obtains the address of the Gateway Agent from the Income Controller Agent. All the service agents send their data to the outcome agent and this dispatches the service data to the correspondent Gateway Agent.

Service Agents
In general, all the service agents perform a similar role including the following functions:

- Event Listening;
- Service Data Constructing;
- Sending Messages.

In the case of the Restaurant service, four agents were constructed - each representing different restaurants, namely the KFC Agent, the PizzaHut Agent, the BurgerKing Agent and the Charlies Agent. The restaurant service demonstrates the location-based recommending service.

Four agents for the Shop service were created, representing different shops. The Shop service demonstrates a location-based advertisement service.

One agent for the weather service was created. This leverages the Yahoo weather service by calling two REST APIs - one for getting the code for the location, the other for getting weather forecast information from the Yahoo website.

4.3 Implementation

EasyLife harnesses a number of technologies but intelligent agents from the key constituent components. Two different agent development environments are used – JADE for the resource intensive server components and AFME for the lightweight mobile devices. The EasyLife client is implemented using CLDC 1.1 and MIDP 2.0, as most mobile phones currently available in the marketplace supports these standards. Figure 7 illustrates the class hierarchy for EasyLife client. The servlet engine is deployed using Tomcat. The server agents are all implemented in JADE; databases are implemented in mySQL and Hibernate for object persistence. A summary of the technologies used is as follows:

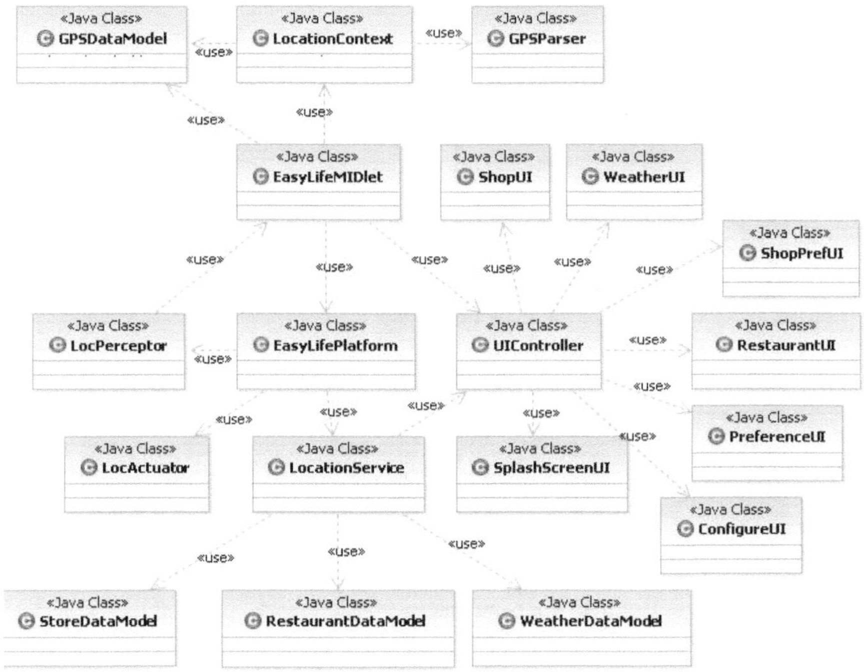

Fig. 7. EasyLife Client Class Diagram

EasyLIfe client
- JSR-139: Connected Limited Device Configuration 1.1
- JSR-118: Mobile Information Device Profile 2.0
- JSR-82: Java APIs for Bluetooth technology 1.0
- JSR-135: Mobile Media API 1.0
- AFME: Agent Factory Mobile Edition

AgentProxy_app
- Servlet 2.5: Tomcat runtime environment
- JADE 3.5: Java Agent Development Framework

AgentServices
- JADE 3.5: Java Agent Development Framework
- Hibernate 3.2: O/R mapping
- Mysql-connector-java-5.0.7: JDBC connector

4.4 The EasyLife User Interface

EasyLife has been deployed on a Nokia N91 platform. Figure 8a illustrates the EasyLife welcome screen. The subscriber must then select the required service from the three standard EasyLife services (Figure 8b).

(a) (b)

Fig. 8. EasyLife Welcome screen (a) and service options (b)

Figure 9a depicts what an archetypical weather service report looks like. The restaurant locator service identifies the nearest restaurant (Figure 9b).

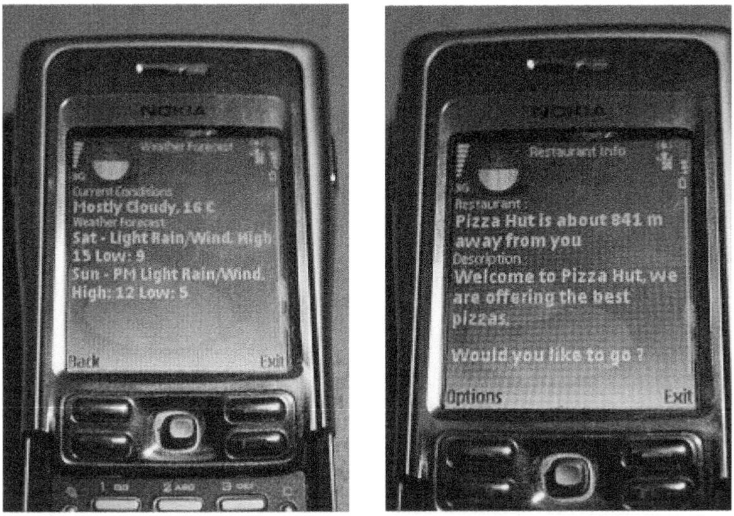

(a) (b)

Fig. 9. Example of the EasyLife weather service (a) and the restaurant locator (b)

5 Conclusion

EasyLife demonstrates the viability of the agent paradigm for integrating heterogonous information sources for service provision to mobile users. The robustness of the agent platform is verified through the use of two separate agent toolkits being harnessed to deliver EasyLife services. Developments in Agent-Oriented Software Engineering (AOSE) and agent frameworks are resulting in an increasing range of options for those considering harnessing the characteristics of agents for their applications and services. The more recent development of agent platforms for mobile devices represents a new frontier for Multi-agent Systems. From an enterprise perspective, it validates MASs as a candidate solution of extending the reach of legacy services to the mobile computing community, for combining existing services in new and innovate ways or indeed, delivering new original services to mobile users.

Acknowledgments. This material is based upon works supported by the Science Foundation Ireland (SFI) under Grant No. 07/CE/I1147.

References

1. Vasilakos, A., Pedrycz, W. (eds.): Ambient Intelligence, Wireless Networking, Ubiquitous Computing. Artec House (2006)
2. Hook, K.: Steps to take before intelligent user interfaces become real. Interacting with Computers 12(4), 409–426 (2000)

3. Doctor, F., Hagras, H., Callaghan, V.: A fuzzy embedded agent-based approach for realizing ambient intelligence in intelligent inhabited environments. IEEE Transactions on Systems, Man and Cybernetics Part A 35(1), 55–65 (2005)
4. Ayars, J., Bulterman, D., Cohen, A., Day, K., Hodge, E., Hoschka, P., Hyche, E., Jourdan, M., Kim, M., Kubota, K., Lanphier, R., Layaida, N., Michel, T., Newman, D., van Ossenbruggen, J., Rutledge, L., Saccocio, B., Schmitz, P., ten Kate, W., Michel, T.: Synchronized multimedia integration language (SMIL 2.0), W3C recommendation, 2nd edn. (2005)
5. Louridas, P.: SOAP and Web Services. IEEE Software 23(6), 62–67 (2006)
6. Kishore, R., Zhang, H., Ramesh, R.: Enterprise integration using the agent paradigm: foundations of multi-agent-based integrative business information systems. Decision Support Systems 42(1), 48–78 (2006)
7. Nahm, Y.-E., Ishikawa, H.: A hybrid multi-agent system architecture for enterprise integration using computer networks. Robotics and Computer-Integrated Manufacturing 21(3), 217–234 (2005)
8. Muller, R., Greiner, U., Rahm, E.: AW: A workflow system supporting rule-based workflow adaptation. Data & Knowledge Engineering 51(2), 223–256 (2004)
9. Ehrler, L., Fleurke, M., Purvis, M.A., Savarimuthu, B.T.R.: Agent-Based Workflow Management Systems (WfMSs): JBees - A Distributed and Adaptive WFMS with Monitoring and Controlling Capabilities. Information Systems and E-Business Management 4(1), 5–23 (2006)
10. Burmeister, B., Arnold, M., Copaciu, F., Rimassa, G.: BDI-Agents for Agile Goal-Oriented Business Processes. In: Seventh International Conference on Autonomous Agents and Multiagent Systems (AAMAS), Estoril, Portugal, pp. 37–44 (2008)
11. Li, S.: AgentStra: an Internet-based multi-agent intelligent system for strategic decision-making. Expert Systems with Applications 33(3), 565–571 (2007)
12. Pani, A.K., Venugopal, P.: Implementing e-CRM using intelligent agents on the internet. In: International Conference on Service Systems and Service Management, pp. 1–6 (2008)
13. Dickinson, I., Wooldridge, M.: Agents are not (just) Web Services: Considering BDI Agents and Web Services. In: Proc. of the 2005 Workshop on Service-Oriented Computing and Agent-Based Engineering (SOCABE 2005), Utrecht, The Netherlands (2005)
14. Poslad, S., Laamanen, H., Malaka, R., Nick, A., Zipf, A.: Crumpet: Creation of user-friendly mobile services personalized for tourism. In: Proceeding of the Second IEE International Conference on 3G Mobile Communication Technologies, London, UK (2001)
15. O'Grady, M.J., O'Hare, G.M.P., Sas, C.: Mobile Agents for Mobile Tourists: A User Evaluation of Gulliver's Genie. Interacting with Computers 17(4), 343–366 (2005)
16. Hristova, N., O'Hare, G.: Ad-me: Wireless Advertising Adapted to the User Location, Device and Emotions. In: Proceedings of the 37th Hawaii International Conference on System Sciences, Hawaii (2004)
17. Keegan, S., O'Hare, G.M.P., O'Grady, M.J.: EasiShop: Ambient Intelligence Assists Everyday Shopping. Journal of Information Sciences 178(3), 588–611
18. Bresciani, P., Perini, A., Giorgini, P., Giunchiglia, F., Mylopoulos, J.: Modeling early requirements in Tropos: a transformation based approach. In: Wooldridge, M., Ciancarini, P., Weiss, G. (eds.) AOSE 2001. LNCS, vol. 2222, pp. 151–168. Springer, Heidelberg (2000)
19. Odell, J., Van Dyke Parunak, H., Bauer, B.: Extending UML for Agents. In: Proceedings of the Agent-Oriented Information System Workshop at the 17th National Conference on Artificial Intelligence, Austin, USA, pp. 3–17 (2000)

20. Padgham, L., Winikoff, M.: Prometheus: A Methodology for Developing Intelligent Agents. In: Proceedings of Autonomous Agents and Multi-Agent Systems (AAMAS 2002), Bologna, Italy (2002)
21. Omicini, A.: SODA: Societies and Infrastructures in the Analysis and Design of Agent-Based Systems. In: Ciancarini, P., Wooldridge, M.J. (eds.) AOSE 2000. LNCS, vol. 1957, pp. 185–193. Springer, Heidelberg (2001)
22. Wagner, G.: The Agent-Object-Relationship Meta-Model: Towards a Unified View of State and Behavior. Information Systems 28(5), 475–504 (2003)
23. Shoham, Y.: Agent-oriented Programming. Artificial Intelligence 60(1), 51–92 (1993)
24. Collier, R., O'Hare, G., Lowen, T., Rooney, C.: Beyond prototyping in the factory of the agents. In: Mařík, V., Müller, J.P., Pěchouček, M. (eds.) CEEMAS 2003. LNCS (LNAI), vol. 2691, pp. 383–393. Springer, Heidelberg (2003)
25. Bellifemine, F., Caire, G., Greenwood, D.: Developing Multi-Agent Systems with JADE. John Wiley & Sons, New Jersey (2007)
26. Busetta, P., Ronnquist, R., Hodgson, A., Lucas, A.: JACK Intelligent Agents - Components for Intelligent Agents in Java. Technical Report, Agent Oriented Software Pty. Ltd, Melbourne, Australia (1998)
27. Bergenti, F., Poggi, A.: LEAP: A FIPA Platform for Handheld and Mobile Devices. In: Meyer, J.-J.C., Tambe, M. (eds.) ATAL 2001. LNCS (LNAI), vol. 2333, pp. 303–313. Springer, Heidelberg (2001)
28. Muldoon, C., O'Hare, G.M.P., Collier, R., O'Grady, M.J.: Agent Factory Micro Edition: A Framework for Ambient Applications. In: Alexandrov, V.N., van Albada, G.D., Sloot, P.M.A., Dongarra, J. (eds.) ICCS 2006. LNCS, vol. 3993, pp. 727–734. Springer, Heidelberg (2006)
29. Shen, J.M., O'Grady, M.J., O'Hare, G.M.P.: EasyLife: A Location-aware Service Oriented Mobile Information System. In: Lovrek, I., Howlett, R.J., Jain, L.C. (eds.) KES 2008, Part I. LNCS (LNAI), vol. 5177, pp. 229–236. Springer, Heidelberg (2008)
30. Klante, P., Krösche, J., Boll, S.: AccessSights - a multimodal location-aware mobile tourist information system. In: Miesenberger, K., Klaus, J., Zagler, W.L., Burger, D. (eds.) ICCHP 2004. LNCS, vol. 3118, pp. 287–294. Springer, Heidelberg (2004)
31. O'Grady, M.J., O'Rafferty, R.P., O'Hare, G.M.P.: A tourist-centric mechanism for interacting with the environment. In: Proceedings of the First International Workshop on Managing Interactions in Smart Environments, pp. 56–67. Springer, Dublin (1999)
32. Weakliam, J., Lynch, D., Doyle, J., Bertolotto, M., Wilson, D.: Delivering Personalized Context-Aware Spatial Information to Mobile Devices. In: Li, K.-J., Vangenot, C. (eds.) W2GIS 2005. LNCS, vol. 3833, pp. 194–205. Springer, Heidelberg (2005)
33. Strahan, R., O'Hare, G.M.P., Phelan, D., Muldoon, C., Collier, R.: ACCESS: An Agent based Architecture for the Rapid Prototyping of Location Aware Services. In: Proceedings of the 5th International Conference on Computational Science (ICCS 2005), Emory University Atlanta, USA (2005)

Environmental Impact Assessment by Multi-Agent Systems

Marina V. Sokolova[1,2] and Antonio Fernández-Caballero[1]

[1] Universidad de Castilla-La Mancha, Departamento de Sistemas Informáticos
Campus Universitario s/n, 02071-Albacete, Spain
marina.v.sokolova@gmail.com, caballer@dsi.uclm.es
[2] Kursk State Technical University, Kursk, ul.50 Let Oktyabrya, 305040, Russia

Abstract. Environmental impact assessment (EIA) is one of the main indicators for human health evaluation as well as for further issues in occupational medicine, public health and planning. Certain difficulties to evaluate EIA appear due to, on the one hand, the nature of contaminants and health characteristics, and, on the other hand the difficulties found in data processing, such as heterogeneity of the information sources, lack of data quality and need to use numerous data mining methods. The application of multi-agent systems (MAS) to EIA helps to operate with distributed decentralized information and to generate general decisions and solutions. In order to facilitate the workflow of MAS planning and implementation in decision support systems (DSS) applied to EIA, a generalized multi-agent system technology (MAST) is proposed. This workflow is general enough to enable the related MAS to be used in any application area requiring decision making. We demonstrate that by only changing the domain ontology the MAS is easily oriented towards another problem area. The functional organization and the roles executed in the MAS conform to the logical sequence of the data transformation flow, which includes information retrieval and fusion, data pre-processing, data mining and modelling, and simulation and decision making.

1 Introduction

Recent advances and continuous changes in computer science, cybernetics and data management have penetrated into industries and businesses, and nowadays have become essential for many applications. Environmentalists are accustomed to using information technologies for environmental management, engineering and simulation. There are some explanations to this fact, being one of them the complexity of the environmental issues, which requires interdisciplinary approaches and their fusion, usage of distributed methods and techniques for their study. As shown by recent research reports, environmental problem solving benefits from the application of such methods and frameworks, and intelligent agency is a striking example of such a paradigm.

Sustainable development is a broad concept that spans over and across the social, economic, environmental, medical, demographic and political spheres. That is why

A. Håkansson et al. (Eds.): Agent and Multi-agent Technology, SCI 289, pp. 69–89.
springerlink.com

the concept of "sustainable development" is an abstract and highly complex one. It includes components (in agreement with the renewed EU Strategy of Sustainable Development)[1] that deal with climate change, sustainable transport, production and consumption, natural resources management, social issues (demography, migration, etc), and public health. The link between sustainable development and public health is obvious and even does not need to be emphasized.

The concept of "health" represents a complex system that includes physical, social, mental, spiritual and biological well-being, spanning across all the spheres of our lives. As it was noted in the European Scientific Committee's issues[2], the processing of data for HIA "has turned out to be very laborious and resource-demanding", as "there is a limited access to data on environmental exposures and health status in specific geographical areas".

Complex systems inherit all the necessary characteristics to be modelled with the multi-agent systems (MAS) approach. Their characteristics are modularity, decentralization, changeability, ill structure and weak predictability. Multi-agent approach seems to be the only technique which can help to reduce the complexity of the system by creating modular components, which solve private subtasks that constitute together the whole goal. Every agent uses the most effective technique for solving the subtask and does not apply the general approach, acceptable for the system as a whole but not optimal for a concrete subtask. If the subtasks are interconnected, agents have to communicate with each other and the system can be represented as a community of semi-autonomous and autonomous distributed intelligent entities, solving the private tasks and constituting a whole complex system.

In the case of EIA, all the advantages of intelligent agents become crucial. EIA is an indicator that enables to evaluate the negative impact upon human health caused by environmental pollution. Environmental pollution, as one of the factors with dominant and obvious influence, causes direct and latent damage. This harm has to be evaluated and simulated in order to create a set of preventive health-preserving solutions. That is why linking all the named components in one system and its study leads to the analysis of potential and current health problems, the retrieval of the new ones, and to working out thoroughly the situation development, strategies and activities aimed at the management and control. Large amounts of raw data information describe the "environment-human health" system, but not all the information is useful though. It transforms from the initial "raw" state into the "information" state, which suggests organized data sets, models and dependencies. And finally into the "new information" state, which has the form of recommendations, risk assessment values and forecasts.

After studying some previous related works (section 2), the proposal of a multi-agent system (MAS) for an environmental impact assessment support system (section 3) and the description of the agent teams are provided. Section 4 shows a significant example of the implementation of the obtained model using JACK Intelligent Agents [1].

---

[1] Areas and topics related to Sustainable Development.

[2] `http://ec.europa.eu/employment_social/health_safety/acsh_en.htm`

2 Related Works

In [2] it is reported about the application of the software agents' paradigm to an environmental monitoring informational system created by a three-layered MAS, which aims to provide measurements of meteorological information and air pollution data, to analyze them and to generate alarm signals. The authors present a multi-agent system for environmental monitoring that is created by means of an intelligent platform Agent Academy. The system has a three-levelled organizational structure, where data preprocessing, its manipulation and distribution are carried out. The necessary steps of data transformation are served by four types of agents: Diagnosis agents, Alarm agents, Database agents and Distribution agents. Diagnosis agents capture information from sensors and validate it. Alarm agents check input signals with permitted levels and produce an alarm signal, if necessary. Database agents update data storages with the monitoring data, and, finally, distribution agents transfer emergency alerts to users.

In another article [3], the authors report about the application of the agent paradigm to the evaluation of socially-oriented advertising campaigns, aimed at affecting consumers' behaviour. The authors create social communication models to simulate public response to mass-media influence, and introduce a social grid populated with autonomous consumer agents. This grid is also applied to evaluate possible outcomes of the public campaign on water demand control. Situation assessment has also been realized by an agent-based system dedicated to the creation of a general approach to situation assessment, which can be applied to different problem areas [4]. The authors accept the JDL model [5] as a basis for situation awareness; the new approach to situation assessment learning is described and the structure of the MAS is presented as well. In another paper [6], the authors present the framework of a decision support system for water management in the Mediterranean islands coupling a multi-agents system with a geographic information system. The platform developed makes it possible for users to better understand the current operation of the system, to apprehend the evolution of the situation, and to simulate different scenarios according to the selected water policies and the climatic changes hypothesis. Recently, the development and experimental evaluation of an Internet-enabled multi-agent prototype called AgentStra [7] for developing marketing, competitive and associated e-commerce strategies has been introduced.

On the other hand, the specialists working in environmental sciences and public health store huge volumes of relevant monitoring factual information about pollutants and human health that are thought to be analyzed. Continuous processing and maintenance of the information requires essential efforts from the practitioners and professionals not only while handling and storing data, but also when interpreting them. Actually, it seems very hard to handle all the data without using data mining (DM) methods [8]. They can autonomously dig out all the valuable knowledge that is embedded in a database without human supervision and provide a full life-cycle support of data analysis, using such techniques as clustering, classification, logical and association rule-based reasoning, and other methods, which are highly demanded for a comprehensive environmental data analysis. For instance, DM techniques for knowledge discovering and early diagnosis were used for early intervention in developmentally-delayed children [9]. In [10] it is reported about the MAS named "Instrumented City Data Base Analyst", which is aimed at revealing correlations between human health and environmental stress factors (traffic activity, meteorological data, noise monitoring information and health statistics) by using a wide range of DM methods, including regression analysis, neural networks, ANOVA and others. The architecture of the system counts with a number of modules, placed within four levels. The multi-agent structure includes specific modelling agents that create models for the environmental stress factors, and then are harmonized by the model coordination agent. The Data Abstractor – an agent that gets information from sensors fuses and pre-processes it. Interaction with humans is provided by the Reception agent.

In [11] the authors suggest the use of intelligent agents within an agent-based intelligent decision support system (IDSS) in the area of clinical management and research. The IDSS aims to improve the decision quality and to be used in urgent cases. The system uses Web services to display and supervise the decision making process, what requires the other layers. That is, the solution manager service, which analyzes information from the Information Integrator layer and builds solutions.

A group of researchers presents a number of works dedicated to decision making support in wastewater treatment plant (WWTP) [12]. In order to provide an intelligent modelling and control for WWTP, the authors use eight kinds of autonomous agents that support the decision-making process. The agents that interact with WWTR and with users are: the Monitoring Agent (MA) and the Actuation Agent (AA). The first one captures some data from WWTP through the system of sensors, supplies retrospective information, displays it and supplies a basic alarm system. The AA acts over some WWTP parameters, suggests orders, processes alarms and reacts to solve them. And, lastly, there are User Agents (UA), which are responsible for the organization of the human-user interface (HCI), which accept orders and transform them into goals for the multi-agent system, receive and display information and recommendations, and supply justifications of certain actions. The MA, AA and UA interact with the "nucleus" of the multi-agent system through the communication system. Then the data is analyzed and transformed sequentially by the Modeling agent (MO), the Predictive Agent (PA), the Numerical Control Agent (NC) and next it is validated and used by the Expert Reasoning Agent (EA) and the Experience Reasoning Agent (EX). The MO accepts a selection of simulation and control models, enables and disables the simulation and automatic control of the WWTP, and, as it incorporates a number of modelling methods, it applies them to simulate specific processes of WWTP and suggests control actions. The PA dedicates exclusively to task prediction, interpreting the results of simulations and selecting the best models. The NC has to determine the best control algorithm among the ones available from MO. The EA contains all the expert knowledge and serves as a rule-based data base that gives its acceptance of new knowledge. The EX learns new experiences provided by an expert or another source of knowledge whenever a new problem is solved.

Another approach [13] continues and develops the idea and realization of the DSS noted above and offers environmental decision support system related to domain ontology. The authors aim to check if incorporation of ontology and DSS based on rule-based reasoning (RBR) and case-based reasoning (CBR) can improve the decisions it generates. The article describes the OntoWEDSS architecture, which is derived from the model of the DSS for WWTP, but having an ontology embedded and a reasoning module that serves to facilitate knowledge sharing and reuse. The reasoning module is modified, and it contains RBR, CBR and ontology-based reasoning.

Since all these works have demonstrated novel and promising theoretical and practical outcomes, it seems to be valuable to create an agent-based decision support system (MAS) for knowledge discovery and assessment environmental tension upon the population by detail analysis of endogenous and exogenous diseases cases. Thus, the main practical aim of this paper is to create a situation assessment agent-based system, monitoring the environment pollution and following the correspondent changes in human health, generating a set of alternatives for successful and sustainable situation management.

3 The General Approach for Multi-Agent System Creation

3.1 Information Change

Large amounts of raw data information describe the "environment-human health" system, but not all the information is of use, though. For the situation modelling we orient to factual and context information, present it in data sets and use intelligent agents to extract it. So, the information transforms from the initial "raw" state into the "information" state, which suggests organized data sets, models and dependencies, and finally, into the "new information" which has the form of recommendations, risk assessment values and forecasts. The way the information changes is given in Figure 1.

The hidden information is discovered by agents. But for new information construction not only intelligent agents, but also the knowledge of a decision maker or expert are involved. The agent-based decision support system that we are creating provides these information changes [14-16].

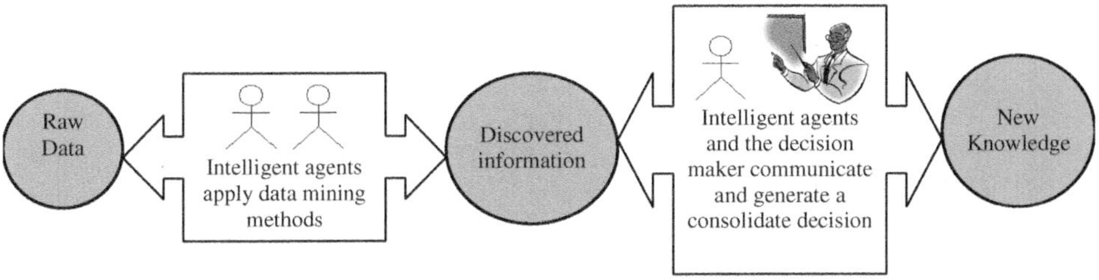

Fig. 1. The information transformation, which changes from a weakly organized and heterogeneous view into knowledge

The process of information change, shown on Figure1, corresponds to the MAS lifecycle flow, which in the case of MAS counts the following steps:

1. **Domain Analysis** - is related to the analysis of the project idea, problem definition, extraction of aims, creation of goal trees, sequencing of tasks and subtasks to be solved. This stage also implies the domain ontology creation, which covers the problem area, the set of relations between the concepts and the rules to incorporate new knowledge. The experience of domain area experts is required on this stage.
2. **Software Elements Analysis** - this stage also deals with private ontologies creation; but now ontologies are created for the system and its elements. The sets of goals and tasks are related to the sets of system functions (roles), required resources (commonly in form of informational files), interactions, and so on.
3. **Specification** - is the written description of the previous stages, which results in system meta-ontology creation.
4. **Software Architecture** - implies the abstract representation of the system to meet the requirements. The software architecture includes interfaces for human-computer communication.
5. **Implementation** - is the iterative process of program creation.
6. **Testing** - program testing under normal and/or critical conditions.
7. **Deployment and Maintenance** - program application and support until the software is put into use. Sometimes training classes on the software product are made.
8. **End of Maintenance** - is the final stage of the software life cycle.

The workflow of tasks, which has to be solved for information integration (see Figure 1) contains four sequential states of data transformation: (1) initial heterogeneous data sources, (2) storages of extracted data, (3) mapped (fused) meta-data, (4) shared global ontology of the problem area (domain ontology) and three flows/processes that provide and organize the transformations: (a) data retrieval and extraction, (b) data mapping (fusion), and, (c) filling in the ontology of the problem area (domain ontology).

3.2 Multi-Agent System Organization and Architecture

We have implemented an agent-oriented software system dedicated to environmental impact assessment. The system receives retrospective statistical information in form of direct indicator values - water pollution, solar radiation - and in form of indirect indicator values - types and number of vehicles used, energy used annually and energy conserved, types and quantity of used fuel, etc. The indirect indicators are used in accordance with the ISO 14031 "Environmental Performance Evaluation" standard, in order to estimate air and soil pollution [17]. The population exposure is registered as number of morbidity cases, with respect to the International Statistical Classification of Diseases and Related Health Problems, 10th review (ICD-10).

In order to provide the system design, we use the Prometheus Development Tool (PDT), which opens a wide range of possibilities for MAS planning and implementation [18]. These are the system architecture, the system entities, their internals and communications within the system and with outer entities. The most important advantages of PDT are an easy and understandable visual interface and the possibility of generating code for JACK™ Intelligent Agents, which is used for MAS implementation, verification and maintenance. The process of multi-system creation, in accordance with the Prometheus methodology, consists of three phases, which are:

System Specification. The Prometheus methodology focuses particularly on specification of goals, and on scenario description. As part of the interface specification, Prometheus also addresses interaction with any external data stores or information repositories. The aspects developed in the System Specification phase are: (1) specification of system goals with associated descriptors, (2) development of a set of scenarios that have adequate coverage of the goals, (3) definition of a set of functionalities that are linked to one or more goals and which provide a limited piece of system behaviour, and, (4) description of the interface between the agent system and the environment in which it is situated. Goals are central to the functioning of the intelligent software agents that are going to implement the system. Functionality is the term used for a chunk of behaviour that includes a grouping of related goals, as well as precepts, actions and data relevant to this behaviour. Scenarios are complementary to goals in that they show the sequences of steps that take place within the system. Possible scenario steps are achieving a goal, performing an action, receiving a percept, or referring to another use case scenario.

Architectural Design. The three aspects that are developed during the Architectural Design phase are: deciding on the agent types used in the application, describing the interactions between agents using interaction diagrams and interaction protocols, and, describing the system structure through the system overview diagram.

Detailed Design. In the Detailed Design, for each individual agent it is decided which capabilities are needed for the agent to fulfil its responsibilities as outlined in the functionalities it contains. The process specifications to indicate more of the internal processing of the individual agents are developed. And when getting into greater detail, the capability descriptions to specify the individual plans, beliefs and events needed within the capabilities are developed. Then the views that show processing of particular tasks within individual agents are developed. It is during this final phase of detailed design when the methodology becomes specific to agents that use event-triggered plans in order to achieve their tasks.

The further analysis of the system has resulted in obtaining and describing the system roles and protocols. There, the proposed system is logically and functionally divided into three layers; the first is dedicated to meta-data creation (information fusion), the second is aimed at knowledge discovery (data mining), and the third layer provides real-time generation of alternative scenarios for decision making.

The goals drawn in Figure 1 repeat the main points of a traditional decision making process, which includes the following steps: (1) problem definition, (2) information gathering, (3) alternative actions identification, (4) alternatives evaluation, (5) best alternative selection, and, (6) alternative implementation. The first and the second stages are performed during the initial step, when the expert information and the initial retrospective data are gathered. The stages 3, 4 and 5 are solved by means of the MAS, and the 6th stage is supposed to be realized by the decision maker.

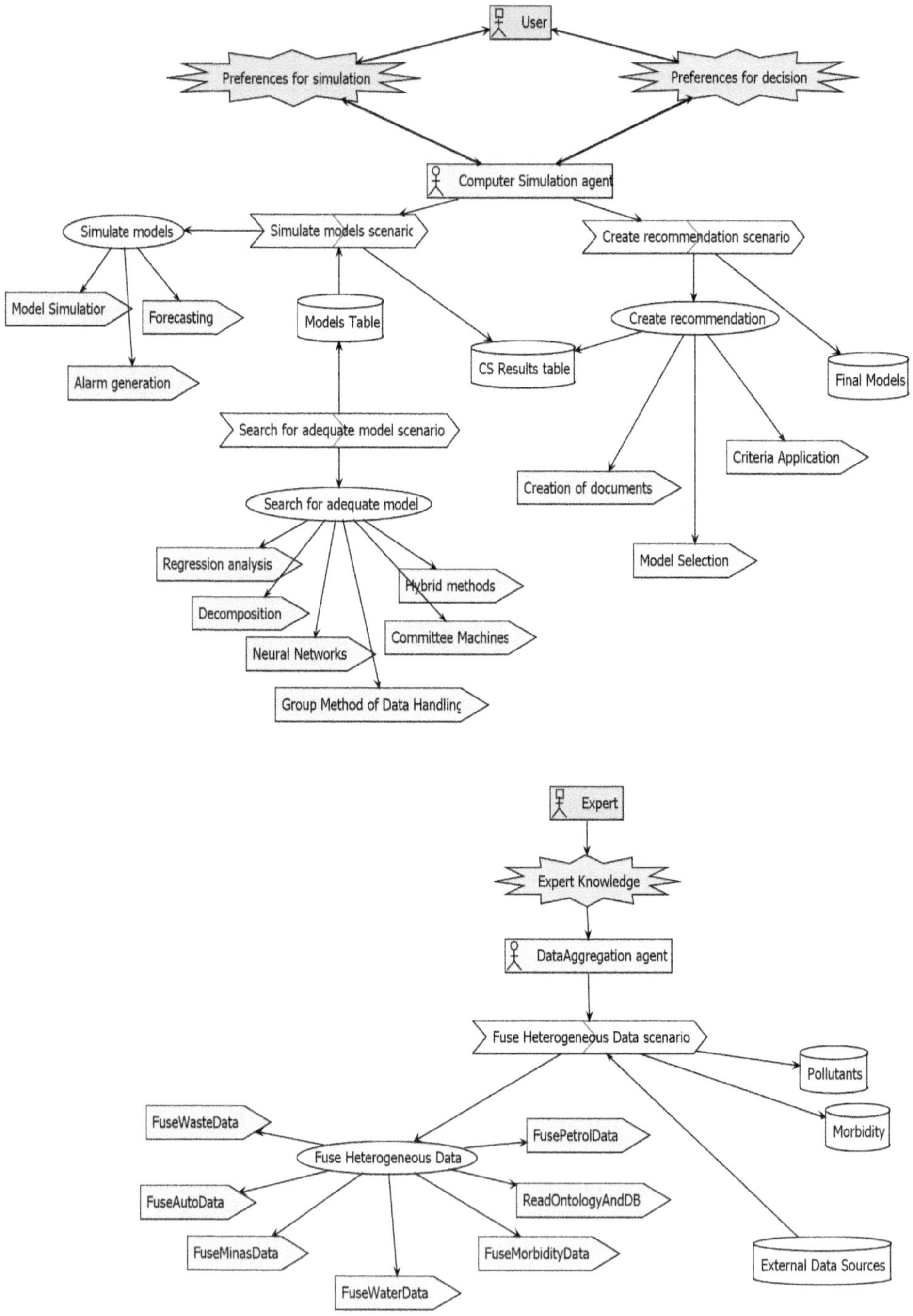

Fig. 2. The Prometheus Diagram of MAS interaction with actors

Being implemented by means of the Prometheus Design Tool, the Analysis Overview Diagram of the MAS enables to see the high-level view composed of external actors, key scenarios and actions (see Figure 2).

The proposed MAS presupposes communication with two actors. One actor is named as "Expert" and it embodies the external entity which possesses the information about the problem area - in more detail, it includes the knowledge of the domain of interest represented as an ontology - and delivers it through protocol ReturnEI to the MAS.

The data source, named "The CS Results", stores the results of the simulation and forms a knowledge base (KB). Through the *Simulate Models* scenario user interacts with the KB, and gets recommendations if they have been previously simulated and stored before, or creates and simulates the new ones. As a result of the interaction within the *FuseHeterogeneousData* scenario, the raw information is being read, and it is shown as "Heterogeneous Data Sources" data storage. There are also "Pollutants" and "Morbidity" data sources created.

The second actor, named "Decision Maker", is involved in an interactive process of decision making and choosing the optimal alternative. This actor communicates with agents by message passing through protocol ReturnSUI, stating the model, simulation values, prediction periods, levels of variable change, etc. It accepts the best alternative in accordance with its beliefs and the MAS. The flow of works, which are essential for decision making, include three scenarios: the Simulate models scenario, the Create recommendation scenario and the Search for the adequate model scenario; and three goals, which related to every scenario and have similar names. Each goal has a number of activities, and within each scenario informational resources in form of data sources are used, modified or created.

In addition to the accepted MAS architecture and in order to gain time of the recommendation generation process and optimize interactions between agents, we use local agent teams, which coordinate and supervise task execution and resource usage. Agent teams allow synchronizing the work of the system, planning execution in a concurrent mode and strengthening the internal management by local decision making. We use four agent teams within the system: two within the first level, one on the second and third levels, and each "main" agent plays several roles (as described in Table 1).

3.3 Data within the System

During the system work cycle, agents manipulate diverse income and outcome information flows: data transmission protocols, messages, income and outcome data, etc. These information sources differ by its "life time": can be permanent and temporary, by the assessment levels – some can be used modified or deleted by agents, and the decisions about others have to be taken by a system user.

Table 1. The roles played in the MAS

Logical level	Main agent	Subordinate agent	Role
Data Fusion	Data Aggregation agent	Domain Ontology agent Traffic Pollution Fusion agent Water Data Fusion agent Petroleum Data Fusion agent Mining Data Fusion agent Morbidity Data Fusion agent Waste Data Fusion agent	Data Fusion
	Data Pre-processing agent	Normalization agent Correlation agent Data Smoothing agent Gaps and Artefacts Check agent	Data Clearing
Data Mining	Function Approximation agent	Regression agent ANN agent GMDH agent Committee Machine agent Decomposition agent Evaluation agent	Impact Assessment Decomposition Function Approximation
Decision Making	Computer Simulation agent	Forecasting agent View agent Alarm agent	Computer Simulation Decision Making Data Distribution

Describing data sources, which are created and used by the agents, on Figure 3, it is shown the Prometheus data coupling diagram. Here on Figure 3, the Heterogeneous Data Storage is raw data storage and the CSResultsTable is a knowledge base. The data sources between them form and show the change which is required to transform the raw heterogeneous data to the view of knowledge.

The "Morbidity" and "Pollution" data storages contain fused and homogenized information. The "dataX", "dataY", "ranges" and "Correlation Table" are a set of data, ready for machine learning. These files are generated as an outcome of the first logical level (by DAA and DPA). Then, FAA calls for machine learning and data mining procedures, and the subordinate agents create information files: "decomposition results" (by Decomposition agent), "IA Results", "Models table" (by ANN agent, GMDH agent, Regression Agent and Evaluation agent) and "Final Model List" (by Committee Machine agent). These information files are then read by CSA and accepted as a background for recommendation. Recommendations are written into the knowledge base "CS Results List".

Fig. 3. Data Coupling diagram in Prometheus

4 Description of the Agents

4.1 The Data Aggregation Agent

DAA has a number of subordinate agents under its control; these are the Domain Ontology agent (DOA) and the *fusion agents*: the Water Data Fusion agent (WFA), the Petroleum Data Fusion agent (PFA), the Mining Data Fusion agent (MFA), the Traffic Pollution Fusion agent (TFA), the Waste Data Fusion agent (WDFA) and the Morbidity Data Fusion agent (MFA). Remember that this information fusion is the result of a meta-data creation (see Figure 4).

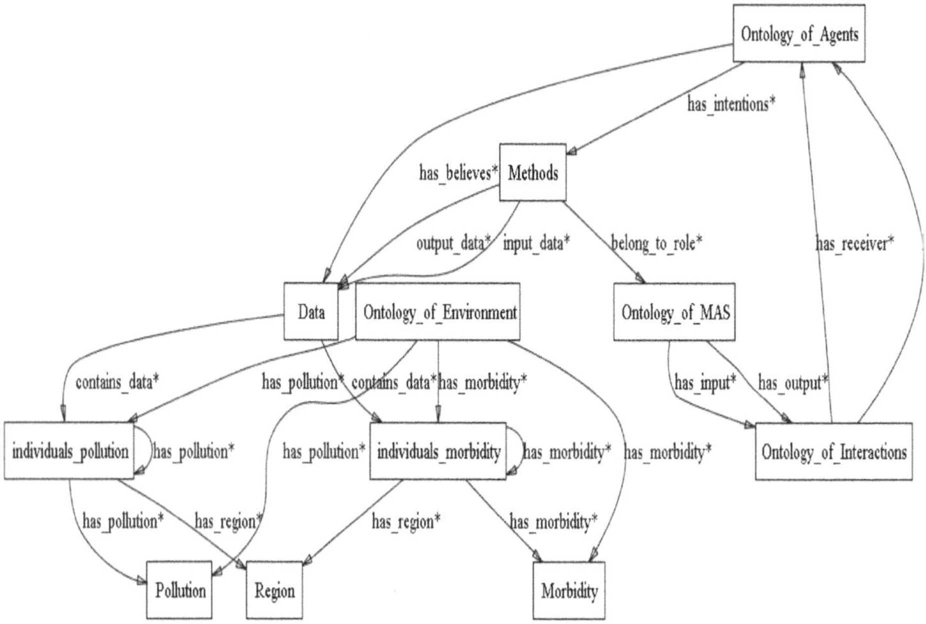

Fig. 4. Meta-ontology as a result of private ontologies mapping

First, the DAA sends the message *ReadOntology* to the DOA, which reads the OWL-file, which contains information about the ontology of domain, and make it available to the DAA. The DOA terminates its execution, sending the message *OntologyIsBeingRead* to the DAA.

Next, the DAA sends the message *Start Fusion* to the fusion agents, which initiate to execute. When it starts to execute each fusion agent searches for the files that may contain information about the concept of its interest. Each fusion agent works with one or few concepts of the domain ontology: WFA searches for the information about water contaminants and their properties, PFA about the use of petroleum and related concepts, MDF retrieves data about the contamination related to mining industry activity, the WDFA retrieves data about wastes and its components, the TFA retrieves data about transport vehicles activity, and the MFA looks for data about morbidity and their properties. When it finds the information file, the agent retrieves the information about the concept and its values, and changes their properties (in order to get rid of heterogeneity and to homogenize information) and sends it to the DAA, which pools retrieved information together.

Finally, DAA fills the domain ontology with data, and put data into a standard format. After that, the data files are ready to be pre-processed, and the DAA through the protocol ReturnDF says to the DPA that data is fused and pre-processing can be started.

4.2 The Data Pre-processing Agent

DPA provides data pre-processing and has a number of subordinate agents which specialize in different data clearing techniques: Normalization agent (NA), Correlation agent (CA), Data Smoothing agent (DSA), Gaps and Artefacts Check agent (GAA). They perform all data pre-processing procedures, including outliers and anomalies detection, dealing with missing values, smoothing, normalization, and so on.

Figure 5 gives a look at the first logical level, within which act the Data Aggregation agent and the Data Pre-processing agent.

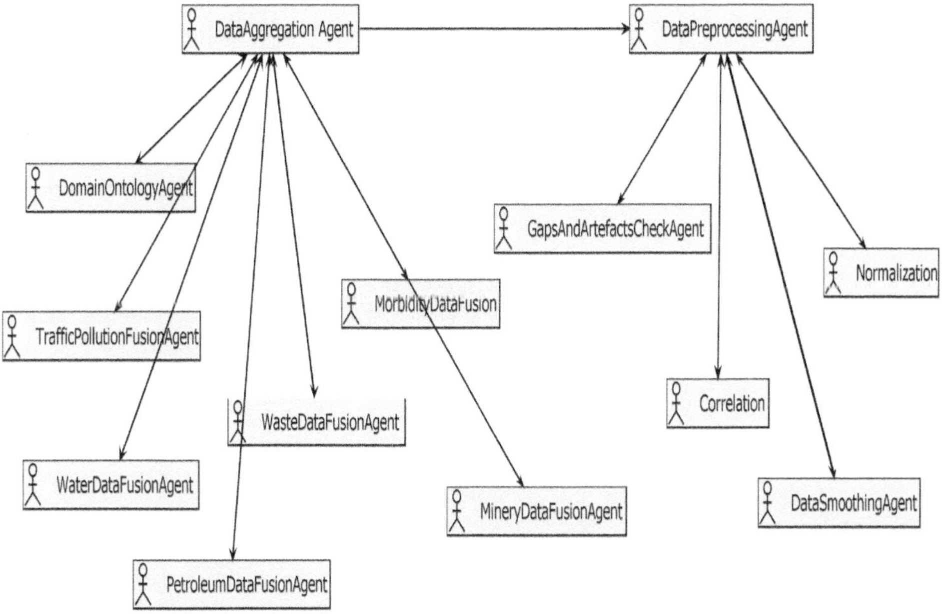

Fig. 5. Interaction between the Data Aggregation agent and the Data Pre-processing agent

DPA starts to execute as soon as it receives a triggering message from DAA. The main function of the DPA is to coordinate the subordinate agents and decide when they execute and in which order. Starting its execution, DPA sends the *StartDataConsistenceCheck* message, which triggers the GAA. Then the DAA eliminates artefacts, searches for the double values and fills the gaps. Having finished execution, GAA sends to DPA a message.

Then, DPA through the message *StartSmoothing* calls for DSA, which can execute exponential and weighted-average smoothing and terminates sending the *SmoothingIsFinished* message to DPA. Then, NA and CA are called for in their turn. The outputs of the DPA work are: data, ready for further processing and modelling, and additional data sources with correlation and normalization results. Figure 6 shows a detailed diagram of the DAA and DPA.

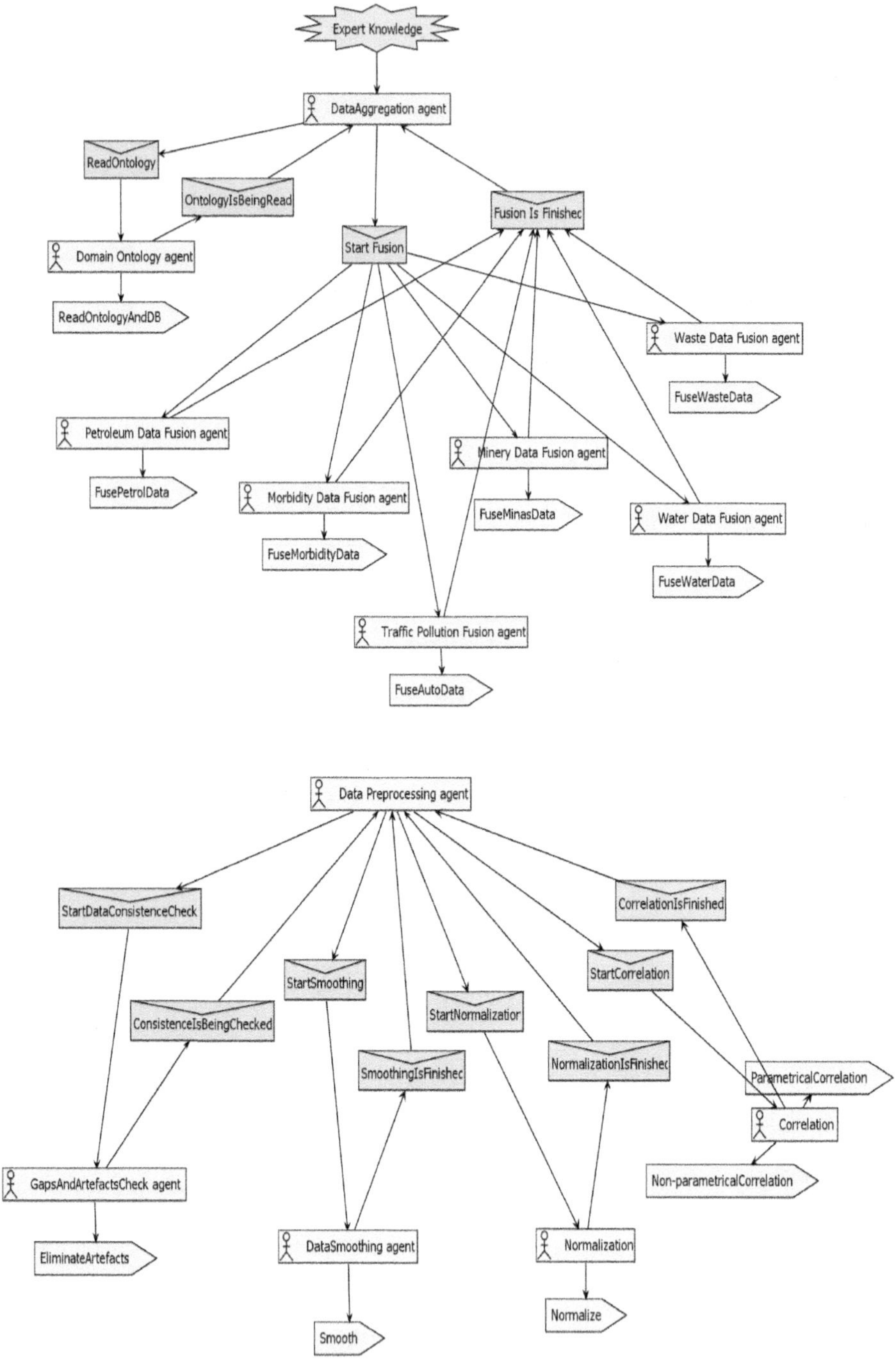

Fig. 6. Detailed diagram of the DAA and DPA

4.3 The Function Approximation Agent

FAA has a hierarchical team of subordinate agents (see Figure 7), which serve to support the roles "Impact Assessment", "Decomposition" and "Function Approximation". FAA has under its control a number of *data mining agents*: the Regression agent (RA), the ANN agent (AA), and the GMDH agent (GMDHA), which work in a concurrent mode, reading input information and creating models. Then, if any agent from this group finishes modelling, it calls for the Evaluation agent (EA), which evaluates received models and returns the list of the accepted ones, while the others are banned and deleted. The FAA pools the output of the agents work, creates the list with the accepted models and then, once RA, AA and GMDHA have finished their execution, calls for the Committee Machine agent (CMA), which creates the final models in form of committees for each of the dependent variables, and saves them.

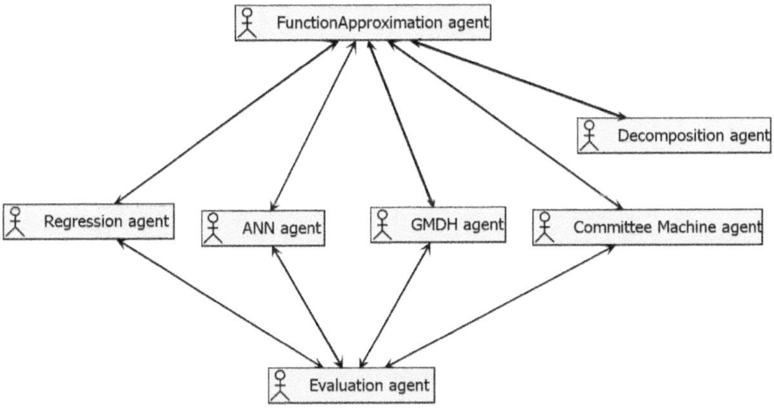

Fig. 7. Function Approximation agent and its team

The FAA working cycle is the following one: FAA sends the *StartDecomposition* message and waits until DA finishes its execution. Then, having received the *StartDataMining* message, the data mining agents start execution in a concurrence mode. Each one of them has plans with a particular tool, and in the case of AA, it has neuralNetwork and evaluateImpactAssessment plans, where the first plan is oriented to artificial neural network (ANN) creation and training, and the second plan aims to evaluate the environmental impact by means of ANN with determined structure and characteristics. EA is called by each of the data mining agents to evaluate created models, and to check the adequacy of the model to the experimental data. EA is triggered by the *StartEvaluation* message from a data mining agent, and, wherever it is not busy, starts to execute. Having terminated the execution, it is ready to receive tasks and handle them. CM is the last to be called by FAA, as CM creates final hybrid models for every dependent variable. Each hybrid model is based on the previously created and evaluated models from the data mining agents, and uses the data sources created by them: Models Table, and IAResults. the complete information is provided in Figure 8.

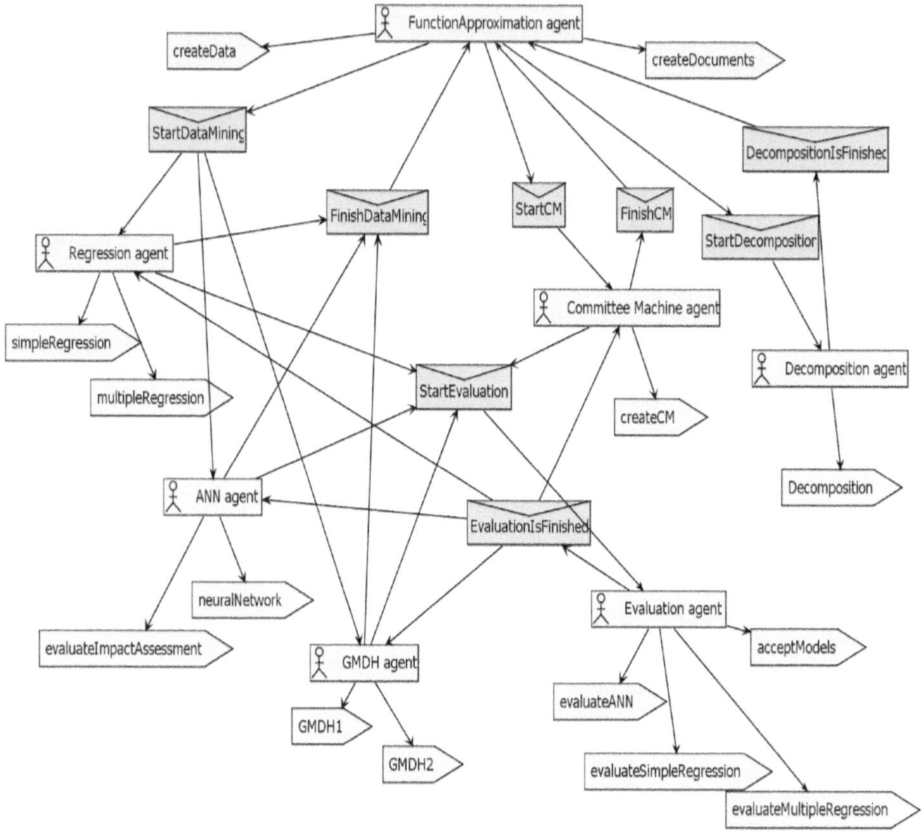

Fig. 8. Function Approximation agent and its team

4.4 The Computer Simulation Agent

The CSA (see Figure 9) interacts with the user and performs a set of task within Computer Simulation, Decision Making and Data Distribution roles. It has the agent team, which includes Forecasting agent (FA), Alarm agent (AmA) and ViewAgent (VA).

The CSA execution cycle (see Figure 10) starts by asking for user preference. To be more precise, it asks for the information about the disease and pollutants of interest, the period of the forecast, and the ranges of their values change. Once the information from the user is received, CSA sends a message *SimulateAlternative* to FA, which reasons and executes one of the plans, which are Forecasting, ModelSimulation, and CriterionApplication. When the alternative is created, CSA sends the *StartAlarmCheck* message to AmA. The AmA compares the simulation and forecast data from the FA with the permitted and alarm levels for the correspondent indicators. If they exceed the levels, AmA generates alarm alerts.

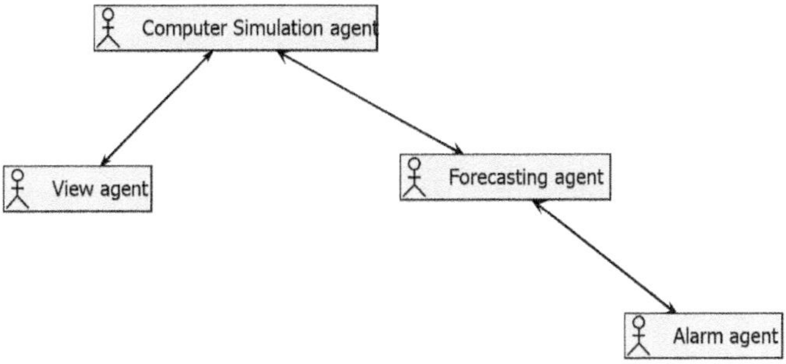

Fig. 9. Computer Simulation agent and its team

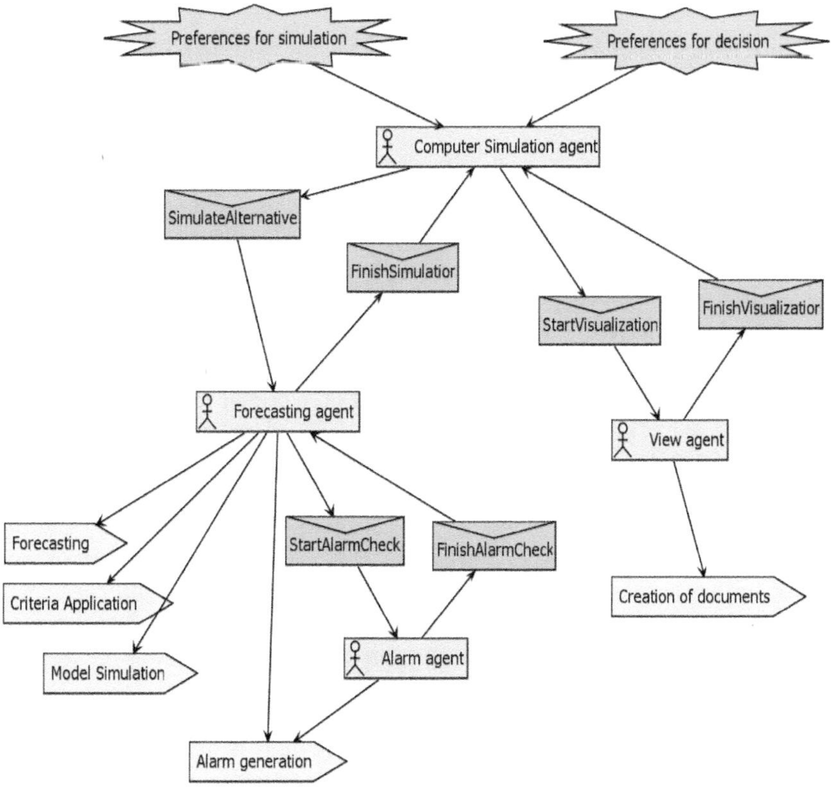

Fig. 10. Computer Simulation agent and its team

5 Results and Conclusions

The MAS has an open agent-based architecture that allows us an easy incorporation of additional modules and tools, enlarging the number of functions of the system. The system belongs to the organizational type, where every agent obtains a class of tools and knows how and when to use them. Nowadays, such types of systems have a planning agent that plans the orders of the agents' executions. In our case, the main module of the Jack program carries out these functions. The ViewAgent displays the outputs of the system functionality and realizes interaction with the system user. As the system is autonomous and all the calculations are executed by it, the user has only access to the result outputs and the simulation window.

For example, the Data Aggregation agent is built up with a constructor:

"DataAggregationAgent DAA1 = new DataAggregationAgent ("DAA")",

and then some of its methods are called, for example, DAA1.fuseData(). The DataPreprocessingAgent is constructed as

DataPreprocessingAgent DCA = new DataPreprocessingAgent ("DPA", "x.dat", "y.dat")"

where "*x.dat*" and "*y.dat*" are agents believes of "global" type. This means that they are open and can be used by the other agents within the system. Finally, the ViewAgent, which displays the outputs of the system functionality and performs interaction with the system user, is called.

He/she can review the results of impact assessment, modelling and forecasting and try to simulate tendencies by changing the values of the pollutants.

To evaluate the impact of environmental parameters upon human health in Castilla-La Mancha, in general, and in the city of Albacete in particular, we have collected retrospective data since the year 1989, using open information resources offered by the Spanish Institute of Statistics and by the Institute of Statistics of Castilla-La Mancha. As indicators of human health and the influencing factors of environment, which can have a negative effect on the noted above indicators of human health, the factors described in Table 2 were taken.

The MAS has recovered data from plain files, which contained the information about the factors of interest and pollutants, and fused in agreement with the ontology of the problem area. Some changes of data properties (scalability, etc.) and their preprocessing have been necessary. After these procedures, the number of pollutants valid for further processing has decreased from 65 to 52. This significant change was caused by many blanks related to several time series, as some factors have started to be registered recently. After considering this as an important drawback, it was not possible to include them into the analysis.

The human health indicators, being more homogeneous, have been fused and cleared successfully. The impact assessment has shown the dependencies between water characteristics and neoplasm, complications of pregnancy, childbirth and congenital malformations, deformations and chromosomal abnormalities.

Table 2. Diseases and pollutants studied in this research

Type of Disease / Pollutant	Disease class
Endogenous diseases:	Certain conditions originating in the perinatal period; Congenital malformations, deformations and chromosomal abnormalities.
Exogenous diseases:	Certain infectious and parasitic diseases; Neoplasm; Diseases of the blood and blood- forming organs and certain disorders involving the immune mechanism; Endocrine, nutritional and metabolic diseases; Mental and behavioural disorders; Diseases of the nervous system; Diseases of the eye and adnexa; Diseases of the ear and mastoid process; Diseases of the circulatory system; Diseases of the respiratory system; Diseases of the digestive system; Diseases of the skin and subcutaneous tissue; Diseases of the musculoskeletal system and connective tissue; Diseases of the genitourinary system; Pregnancy, childbirth and the puerperium; Symptoms, signs and abnormal clinical and laboratory findings, not elsewhere classified; External causes of morbidity and mortality.
Transport:	Number of Lorries, Buses, Autos, Tractors, Motorcycles, Others.

Part of Table 3 shows that within the most important factors apart from water pollutants, there are indicators of petroleum usage, mines output products and some types of wastes.

Table 3. Part of the Table with the outputs of impact assessment

Disease Class	Pollutant with influence on the disease
Neoplasm	Nitrites in water; Miner products; DBO5; Dangerous chemical wastes; Fuel-oil; Petroleum liquid gases; Water: solids in suspension; Asphalts; Non-dangerous chemical wastes;
Diseases of the blood and blood- forming organs, the immune mechanism	DBO5; Miner products; Fuel-oil; Nitrites in water; Dangerous wastes of paper industry; Water: solids in suspension; Dangerous metallic wastes
Pregnancy, childbirth and the puerperium	Kerosene; Petroleum; Petroleum autos; Petroleum liquid gases; Gasohol; Fuel-oil; Asphalts; Water: DQO; DBO5; Solids in suspension; Nitrites.
Certain conditions originating in the prenatal period	Non-dangerous wastes: general wastes; mineral, constriction, textile, organic, metal. Dangerous oil wastes.
Congenital malformations, deformations and chromosomal abnormalities	Gasohol; Fuel-oil; DQO in water; Producing asphalts; Petroleum; Petroleum autos; Kerosene; Petroleum liquid gases; DBO5 in water; Solids in suspension and Nitrites.

The MAS has a wide range of methods and tools for modelling, including regression, neural networks, GMDH, and hybrid models. The function approximation agent selects the best models, which are: simple regression – 4381 models; multiple regression – 24 models; neural networks – 1329 models; GMDH – 2435 models. The selected models are included into the committee machines. We have forecasted diseases and pollutants values for the period of four years, with a six month step, and visualized their tendencies, which, in common, and in agreement with the created models, are going to overcome the critical levels. Control under the "significant" factors, which cause impact upon health indicators, could lead to the decrease of some types of diseases.

As a result, the MAS provides all the necessary steps for standard decision making procedure by using intelligent agents. The levels of the system architecture, logically and functionally connected, have been presented. Real-time interaction with the user provides a range of possibilities of choosing one course of action from among several alternatives, which are generated by the system through guided data mining and computer simulation. The system is aimed at regular usage for adequate and effective management by responsible municipal and state government authorities.

We use as well as traditional data mining techniques, other hybrid and specific methods with respect to data nature (incomplete data, short data sets, etc.). The combination of different tools enables us to gain in quality and precision of the reached models, and, hence, in the recommendations based on these models. Received dependencies of interconnections and associations between the factors and dependent variables helps to correct recommendations and avoid errors.

As the work appeared to be very time consuming during the modelling, we are looking forward to both revising and improving the system and deepening our research. We are also considering the possibility of making more experiments varying the overall data structure, and trying to apply the system to other but similar application fields.

Acknowledgements

This work was partially supported by the Spanish Ministerio de Ciencia e Innovación under project TIN2007-67586-C02-02, and by the Spanish Junta de Comunidades de Castilla-La Mancha under projects PII2I09-0069-0994 and PEII09-0054-9581.

References

1. Winikoff, M.: JACK^{TM} Intelligent Agents: An industrial strength platform. Multi-Agent Programming: Languages, Platforms and Applications. Multiagent Systems, Artificial Societies, and Simulated Organizations 15, 175–193 (2005)
2. Athanasiadis, I.N., Mentes, A.K., Mitkas, P.A., Mylopoulos, Y.A.: A hybrid agent-based model for estimating residential water demand. Simulation 81(3), 175–187 (2005)
3. Athanasiadis, I.N., Mitkas, P.A.: Social Influence and Water Conservation: An Agent-Based Approach. Computing in Science and Engineering 7(1), 65–70 (2005)

4. Gorodetsky, V., Karsaeyv, O., Samoilov, V.: Multi-agent and data mining technologies for situation assessment in security-related applications. Advances in Soft Computing, 411–422 (2005)
5. Ly, T.C., Greenhill, S., Venkatesh, S., Pearce, A.: Multiple hypotheses situation assessment. In: Proceedings of the Sixth International Conference of Information Fusion, vol. 2, pp. 972–978 (2003)
6. Urbani, D., Delhom, M.: Water management policy selection using a decision support system based on a multi-agent system. In: Bandini, S., Manzoni, S. (eds.) AI*IA 2005. LNCS (LNAI), vol. 3673, pp. 466–469. Springer, Heidelberg (2005)
7. Li, S.: AgentStra: An Internet-based multi-agent intelligent system for strategic decision-making. Expert Systems with Applications 33(3), 565–571 (2007)
8. Sokolova, M.V., Fernández-Caballero, A.: Data mining driven decision making. In: International Conference on Agents and Artificial Intelligence, pp. 220–225 (2009)
9. Chang, C.L.: A study of applying data mining to early intervention for developmentally-delayed children. Expert Systems with Applications 33(2), 407–412 (2006)
10. Chen, H., Bell, M.: Instrumented city database analysts using multiagents. Transportation Research, Part C 10, 419–432 (2002)
11. Foster, D., McGregor, C., El-Masri, S.: A survey of agent-based intelligent decision support systems to support clinical management and research. In: First International Workshop on Multi-Agent Systems for Medicine, Computational Biology, and Bioinformatics (2006)
12. Riaño, D., Sánchez-Marré, M., R.-Roda, I.: Autonomous agents architecture to supervise and control a wastewater treatment plant. In: Monostori, L., Váncza, J., Ali, M. (eds.) IEA/AIE 2001. LNCS (LNAI), vol. 2070, pp. 804–811. Springer, Heidelberg (2001)
13. Ceccaroni, L., Cortés, U., Sànchez-Marrè, M.: OntoWEDSS: Augmenting environmental decision-support systems with ontologies. Environmental Modelling & Software 19(9), 785–797 (2004)
14. Sokolova, M.V., Fernández-Caballero, A.: Modeling and implementing an agent-based environmental health impact decision support system. Expert Systems with Applications 36(2), 2603–2614 (2009)
15. Sokolova, M.V., Fernández-Caballero, A.: Agent-based decision making through intelligent knowledge discovery. In: Lovrek, I., Howlett, R.J., Jain, L.C. (eds.) KES 2008, Part III. LNCS (LNAI), vol. 5179, pp. 709–715. Springer, Heidelberg (2008)
16. Sokolova, M.V., Fernández-Caballero, A.: An agent-based decision support system for ecological-medical situation analysis. In: Mira, J., Álvarez, J.R. (eds.) IWINAC 2007. LNCS, vol. 4528, pp. 511–520. Springer, Heidelberg (2007)
17. ISO/IEC 12207 home page, `http://www.iso.org/iso/`
18. Sokolova, M.V., Fernández-Caballero, A.: Facilitating MAS complete life cycle through the Protégé-Prometheus approach. In: Nguyen, N.T., Jo, G.-S., Howlett, R.J., Jain, L.C. (eds.) KES-AMSTA 2008. LNCS (LNAI), vol. 4953, pp. 73–82. Springer, Heidelberg (2008)

Multi-Agent Forex Trading System

Rui Pedro Barbosa and Orlando Belo

Department of Informatics, University of Minho, Portugal
{rui.barbosa,obelo}@di.uminho.pt

Abstract. Automated trading is a novel field of study in which computer programs are put in charge of deciding when and how to trade financial instruments. Intelligent agents, with their ability to act autonomously and to adapt and interact with the environment, seem like an obvious choice for the development of automated trading systems. The aim of this article is to analyze how well intelligent agents suit this task. We implemented a set of autonomous currency trading agents, using an architecture that consists of an ensemble of classification and regression models, a case-based reasoning system and an expert system. A total of six trading agents were implemented, each being responsible for trading one of the following currency pair in the Forex market: EUR/USD, EUR/JPY, EUR/CHF, USD/JPY, USD/CHF and CHF/JPY. The agents simulated trades over a period of 23 months, having all achieved a reasonable profit trading independently. However, their strategies resulted in relatively high drawdows. In order to decrease the risk inherent to these high drawdowns, the same simulation was performed while making the agents share the monetary resources. As expected, this strategy of investment diversification originated better results. Still, when the trading costs were taken into consideration, the overall trading performance was less than impressive. That was due to the fact that each agent performed too many trades, and the cost associated with the trading commissions became prohibitively high. We were able to lessen the impact of the trading costs in the total profit by integrating the agents in a multi-agent system, in which the agents communicated with each other before opening new trades. This allowed them to calculate the intended exposure to the market, which in turn enabled them to avoid redundant trades. Under simulation and using low leverage, this multi-agent system obtained a 55.7% profit in 23 months of trading, with a 9.0% maximum drawdown.

1 Introduction

In order to obtain an edge over their competitors, financial institutions and other market participants have been devoting and increasing amount of time and resources to the development of automated trading systems. For obvious reasons, most of these systems are proprietary, and their implementation is shrouded in secrecy. It is, therefore, quite difficult to find publicly available information regarding the architecture and performance of this type of systems. We will try to partially fill this void by describing the step-by-step implementation of an autonomous Forex trading system. This system will consist of a set of intelligent agents, each trading a different currency pair. In order to make it enticing from a practical point of view, our main objective will be to make it as profitable as

A. Håkansson et al. (Eds.): Agent and Multi-agent Technology, SCI 289, pp. 91–118.
springerlink.com

possible. As several studies have shown in the past, data mining models can be of use in this endeavor. Financial time series prediction mechanisms have been implemented using data mining models such as RBF recurrent networks [1] and other types of neural networks [2][3][4], association rules [5], nearest neighbor models [6], support vector machines [7] and more complex hybrid approaches [8][9]. These studies have shown that data mining models can be used to develop potentially profitable trading strategies. However, while the models' predictions help answer the question of when to buy or sell a financial instrument, they do not provide any insight regarding how much to invest in each trade, when to close open trades and when to stop trading. In other words, they do not deal with money management and risk management, which are extremely important parts of a trading strategy, and a requirement for the development of an autonomous trading system. For this reason, our trading agents will be using data mining models for price prediction, but will resort to other artificial intelligence methodologies to be able to perform money management. In a bid to minimize the overall number of trades and the capital requirements, we will integrate these agents in a multi-agent system. As previously mentioned, we want to focus on the practical utility of the system, thus we will be using the profit and the maximum drawdown to evaluate its performance. The maximum drawdown measures the maximum accumulated loss experienced while trading. It is, therefore, an important measure of how risky a trading strategy was in the past. We will demonstrate how the system's performance, as measured by these two performance gauges, can be optimized using two different strategies. To minimize the maximum drawdown, we will resort to investment diversification. To increase the profits, we will use leveraged investments, i.e., investments done on margin with borrowed funds.

2 Trading Agents' Architecture

The agents will need to be able to make several decisions in order to trade autonomously. These include deciding when to buy or short sell a financial instrument, how much to invest in each trade and when to close open trades. In a previously published article [10], we described an agent architecture that can be used to implement trading agents with the ability to make this sort of decisions. This architecture is represented in Figure 1. It defines software agents that are inserted in a trading environment from which they obtain periodic price changes and results of previous trades, and that can act on this environment by placing new trades. The architecture consists of three modules:

- *Intuition Module* – this module is responsible for predicting if the price of the financial instrument will increase or decrease. This prediction is accomplished by an ensemble of classification and regression models that search for hidden patterns in the instrument's price data.
- *A Posteriori Knowledge Module* – this module uses information from previously executed trades to suggest how much to invest in the next trade. This suggestion is done by a case-based reasoning system. Each case in this system corresponds to a trade executed by the agent and its final result (return in percentage).
- *A Priori Knowledge Module* – this module is responsible for making the final trading decision, using the prediction from the *Intuition Module* and the suggestion from the *A Posteriori Knowledge Module*. This decision is made by a rule-based expert system, which contains several rules regarding when to invest and when to close trades.

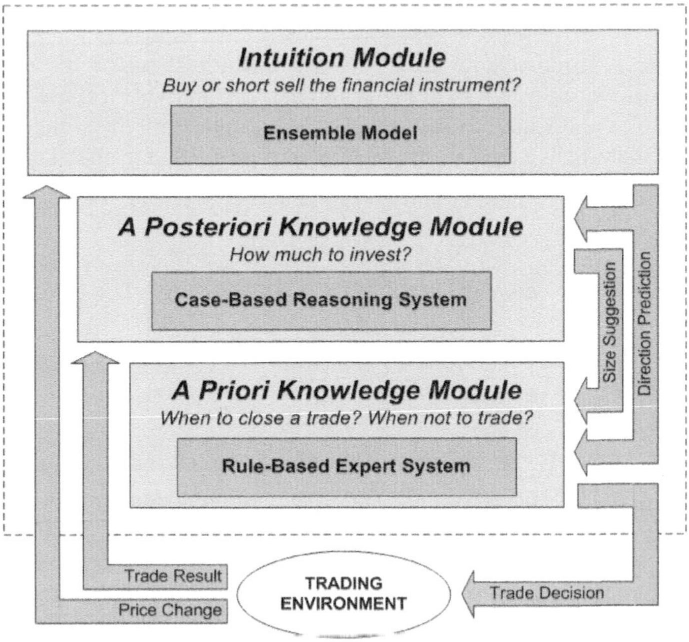

Fig. 1. Agent architecture for implementing intelligent trading agents

2.1 Intuition Module

Before placing a trade, the trading agent needs to decide if it should buy or short sell a financial instrument. In order to accomplish this, it needs to possess some sort of intuitive ability to predict what will happen to the price of the instrument in the future. We implemented this "intuition" as an elaborate pattern recognition mechanism. This mechanism is used for price prediction, and consists of an ensemble of classification and regression models with the ability to find hidden patterns in financial data. The ensemble works as a weighted voting system, in which the weight of each vote is based on the profitability of each model.

The data mining models in the ensemble do not try to predict the exact price of the financial instrument in the future, they simply try to predict what will happen to the price. Each model's prediction thus corresponds to one of two classes: "the price will increase" or "the price will decrease". While this is straightforward for classification models, for regression models the price predictions need to be converted into one of the classes: if the predicted price is higher than the current price then the predicted class is "the price will increase", otherwise if it is lower than the current price the predicted class is "the price will decrease".

To ensure that the agents will keep learning over time, the models in the ensemble need to be retrained with new data as it becomes available. However, we only want the retrained models to become a part of the ensemble if they can perform at least as well as they did before retraining. To accomplish this, before each prediction the *Intuition Module* splits all the available instances into two datasets: the test set, with the most recent N instances, and the training set, with all the instances left. Using these two sets of data, the following sequence of steps is applied to each model in the ensemble:

1. The model is retrained using the training set and tested using the test set.

2. A trade is simulated for each instance in the test set (if the model predicts "the price will increase" a long trade is simulated, otherwise if it predicts "the price will decrease" a short trade is simulated). The results from the simulated trades are used to calculate the overall profit factor, long profit factor and short profit factor of the retrained model:

$$Overall\ PF = \frac{\sum returns\ of\ profitable\ trades}{\sum |\ returns\ of\ unprofitable\ trades\ |} - 1 \quad (1)$$

$$Long\ PF = \frac{\sum returns\ of\ profitable\ long\ trades}{\sum |\ returns\ of\ unprofitable\ long\ trades\ |} - 1 \quad (2)$$

$$Short\ PF = \frac{\sum returns\ of\ profitable\ short\ trades}{\sum |\ returns\ of\ unprofitable\ short\ trades\ |} - 1 \quad (3)$$

3. If the overall profit factor of the retrained model is higher or equal to the overall profit factor of the model before retraining, then the retrained model replaces it in the ensemble. Otherwise, the retrained model is discarded.

4. The model in the ensemble makes its prediction: if it predicts "the price will increase", the weight of its vote is its long profit factor; if it predicts "the price will decrease", the weight of its vote is its short profit factor. If the weight is a negative number then it is replaced with zero, which effectively means the model's prediction is ignored.

After all the models have made their predictions, the ensemble prediction is calculated by adding the weights of the votes of all the models that predicted "the price will increase", and then subtracting the weights of the votes of all the models that predicted "the price will decrease". If the ensemble prediction is greater than zero then the *Intuition Module's* final class prediction is "the price will increase", otherwise if it is lower than zero the final prediction is "the price will decrease".

There are several reasons why we decided to perform the predictions using an ensemble in conjunction with the previously described algorithm:

- Some models are more profitable under certain market conditions than others. We are able to take advantage of this factor by continuously updating the weight of the vote of each model in the ensemble. This enables the prediction mechanism to adapt to changes in the market conditions: as a model becomes more profitable, its vote becomes more important. On the other hand, if a model becomes unprofitable, its vote continuously loses weight, up to a point where its predictions are completely ignored.
- Some models are better at predicting when the price will go up, while others are better at predicting when the price will go down. By using different weights for a model's vote when it predicts a price increase or decrease, the prediction mechanism is able to combine the qualities of the best models at predicting long trades with the qualities of the best models at predicting short trades.

- A new instance becomes available whenever a trade is closed. Our algorithm ensures that this new data will eventually become a part of the training set that is used to retrain the models. Retraining the models before each prediction is key to the trading agents' autonomy, because it enables them to keep learning even as they trade.

The *Intuition Module* was implemented with profit optimization in mind. Equation (1) is used to decide if a model should be replaced by a more profitable retrained version of itself, while Equations (2) and (3) define each model's vote weight based on the profitability they displayed in the recent past. In a previously published article [11], we demonstrated that these same equations could be defined with accuracy optimization in mind. We verified that, in a domain where profit is the main objective, profit optimization will result in better practical results.

2.2 A Posteriori Knowledge Module

Deciding when to buy or short sell a financial instrument is a very important part of successful trading. But there is another equally important decision: how much to invest in a trade. One way to make this decision is to try to predict how profitable the trade will be, before it is placed. Using this prediction, the agent can optimize the profit by investing more money if it expects the trade will be particularly profitable, or investing less if it predicts the trade will be riskier.

In order to calculate how profitable a trade is expected to be, we will be looking at the individual predictions of the models in the ensemble. Intuitively, we might expect that the probability of a trade being successful will be higher if all the models make the same prediction (all predict "the price will increase" or all predict "the price will decrease"), compared to a trade in which the models' predictions are mixed (some predict "the price will increase" and some predict "the price will decrease"). Empirical evidence demonstrates that these expectations are well founded. Certain combinations of individual predictions really are more profitable than others. Our agents' money management strategy is based on this empirical observation.

We implemented the *A Posteriori Knowledge Module* using a case-based reasoning system, in which a case represents a trade previously executed be the trading agent. The following information is contained in each case in the system's database: the predicted class, the trade result (return in percentage) and the predictions of all the models in the ensemble. Before a trade is open, these cases are used to try to predict how profitable it will be, by performing the following sequence of steps:

1. The *Intuition Module* makes the ensemble prediction and sends the individual predictions of the models in the ensemble to the *A Posteriori Knowledge Module*. This module's case-based reasoning system retrieves from its database all the cases with the same ensemble class prediction and the same set of individual predictions.
2. If the number of retrieved cases is not higher or equal to a predefined minimum number of cases, the case-based reasoning system removes the last prediction in the set of individual predictions and retrieves the cases again. This process is repeated until enough cases are retrieved.
3. The case-based reasoning system calculates the overall profit factor of the retrieved cases using Equation (1). This measures how profitable the trade is expected to be.

4. If the overall profit factor is greater or equal to a predefined value the agent doubles the investment; if it is lower than another predefined value the agent does not trade; otherwise, the normal investment amount is used.

A new case is inserted in the case-based reasoning system's database whenever a trade is closed, thus becoming available for the prediction of the expected profit of future trades. This enables the *A Posteriori Knowledge Module* to improve its accuracy as more trades are executed.

2.3 A Priori Knowledge Module

No matter how "smart" the agents are, there is still some trading knowledge they will not be able to pick up from their empirical trading experience. For this reason, the module responsible for making the final trading decisions consists of a rule-based expert system, in which important rules can be inserted by trading experts. Some of these rules can be quite simple. For example, we will probably want to define rules to prevent the agents from trading in low liquidity days, such as those around Christmas or New Year's Day, to avoid paying higher spreads. Other more important rules are those where the settings for take profit orders and stop loss orders are defined. These are necessary so that the agents know when to close an open trade. A take profit order is used to close a trade when it reaches a certain profit percentage, to guarantee that profit. A stop loss order is used to close a trade when it reaches a certain loss percentage, to prevent the loss from widening.

Before each trade, the rule-based expert system receives the prediction from the *Intuition Module* and the suggested investment amount from the *A Posteriori Knowledge Module*. It then uses the rules defined by experienced traders to make the final decision regarding the trade direction, investment amount and exit conditions. The agents can be made completely autonomous by making them use a broker's proprietary API to send the investment decisions directly into the market.

3 Software Implementation

In order to facilitate the development of trading agents, we implemented the previously described architecture as a software "shell". This software considerably reduces the time needed to create an agent that can trade a financial instrument. We named it *iQuant*, short for "intelligent quantitative analyst". A screenshot is shown in Figure 2.

To build a trading agent using the *iQuant* software we just need to:

- Specify the models in the ensemble, their training options and attributes.
- Set the parameters in the *A Posteriori Knowledge Module* to determine when a trade should be skipped or when the investment should be doubled.
- Specify the rules in the rule-based expert system.

The *iQuant* software was developed in Java. It handles the retraining of the models in the *Intuition Module* using the Weka data mining API [12]. Rules in the *A Priori Knowledge Module* are handled by the Drools engine [13]. It also uses the proprietary API of a specific broker to send trading orders directly into the market. However, it

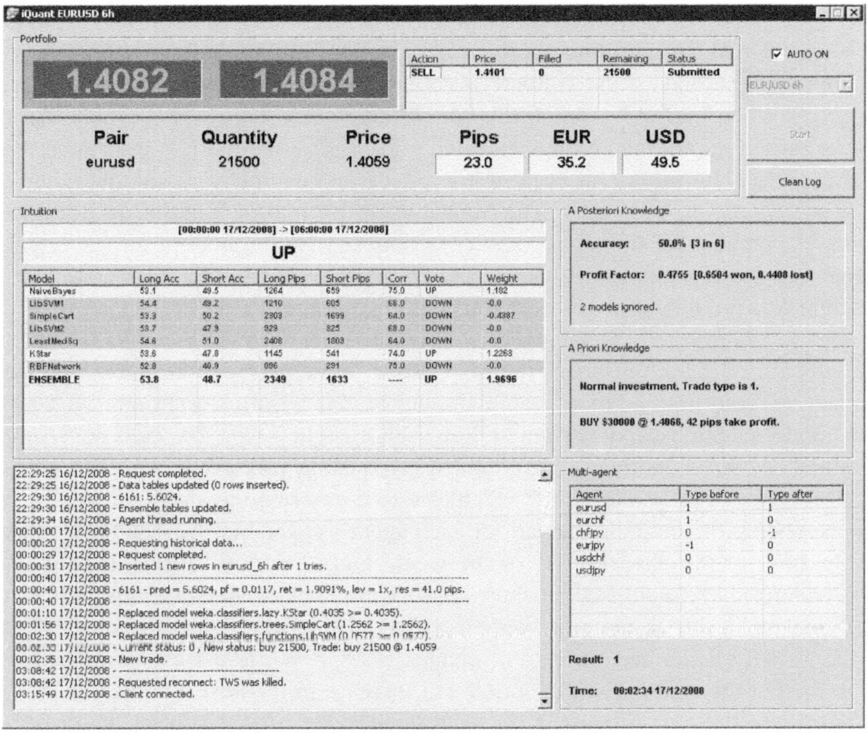

Fig. 2. The *iQuant* software

can be set to perform predictions without sending the orders, which allows us to use it as either an autonomous trader or as a tool to facilitate trading.

4 Forex Trading Agents

We used the *iQuant* software to implement six Forex trading agents, each being responsible for trading one of the following currency pairs: EUR/USD, EUR/JPY, EUR/CHF, USD/JPY, USD/CHF and CHF/JPY. Each of these agents places a trade every 6 hours, from Monday 00:00 GMT till Friday 18:00 GMT.

4.1 Agents' Intuition

The first step to implement the agents' *Intuition Modules* was to obtain historical price data that could be used to train and test the models needed for the ensembles. We were able to gather around 6 years worth of price data[1] for each currency pair, corresponding to the period from May 2003 till December 2008. After collecting the data, we had to decide which trading timeframe to use. We opted for a 6 hours

[1] The training data was obtained from www.oanda.com and www.interactivebrokers.com

timeframe, which means the agents will perform four trades per day: one at midnight, the next at 6 AM, followed by another one at midday and the last at 6 PM. If a trade is still open at the end of the 6 hours period, it will be closed before a new trade is opened. A good reason for using this timeframe is that the times at which the trades will be placed do not match the scheduled release of any important economic reports. This is extremely important, because the price can become very volatile when some of these reports are broadcast. The higher volatility makes it harder to trade, due to increased spreads and slippage. Another reason why we did not opt for a shorter timeframe was the fact that the noise in the price data increases at shorter timeframes, which makes predicting prices more difficult than it already is.

The models inserted in each agent's ensemble were selected by an automated process, which used 4,000 instances (corresponding to the period from May 2003 till December 2006) to train a considerable amount of classification and regression models, with random options and attributes. For each currency pair, the best seven models were picked, based on the profitability displayed using 100 test instances (corresponding to the month of January 2007). Tables 1 to 6 describe the attributes used in training and testing the models selected for each agent's ensemble. Among these attributes are the hour of the day, the day of the week, lagged returns (LAG) and the current class or return. We also used attributes that professional traders use regularly in technical analysis, such as moving averages (MA), the relative strength index (RSI), the Williams %R (WIL) and the rate of change (ROC). While the classification models try to predict the next class ("the price will increase in the next 6 hours" or "the price will decrease the next 6 hours"), the regression models try to predict the price return in the next 6 hours period, after which the return is converted to a class.

Table 1. EUR/USD agent's ensemble

Model	Attributes	Prediction
Naive Bayes	hour (nominal), day of week (nominal), current return	class
Support Vector Machine	hour (nominal), day of week (numeric), MA(2), RSI(11), ROC(12)	return
CART Decision Tree	hour (nominal), day of week (nominal), LAG(2), RSI(2), ROC(2), ROC(5)	class
Support Vector Machine	hour (nominal), day of week (nominal), MA(6), MA(4), MA(3), current return	return
Least Median Squared Linear Regression	hour (nominal), day of week (nominal), LAG(5), LAG(4), LAG(3), LAG(2), LAG(1), current return	return
Instance-Based K*	hour (nominal), day of week (nominal), current class	return
Gaussian Radial Basis Function Network	current return, hour (numeric), day of week (nominal), MA(12), ROC(4)	class

Table 2. EUR/JPY agent's ensemble

Model	Attributes	Prediction
Nearest-Neighbour Classifier	current class, hour (nominal), day of week (numeric), MA(8), MA(12), RSI(15)	class
Support Vector Machine	current class, hour (numeric), day of week (nominal), LAG(7), WIL(25), ROC(7)	return
CART Decision Tree	hour (numeric), day of week (numeric), MA(8)	class
PART Decision List	current class, hour (nominal), day of week (numeric), RSI(21), ROC(7)	class
Instance-Based K*	current class, hour (nominal), day of week (numeric), WIL(11), ROC(7)	return
Instance-Based K*	current class, hour (nominal), day of week (numeric), MA(8), RSI(12), RSI(20)	class
Support Vector Machine	hour (numeric), day of week (nominal), MA(7)	return

Table 3. EUR/CHF agent's ensemble

Model	Attributes	Prediction
RIPPER Rule Learner	current return, hour (numeric), day of week (nominal), WIL(24), RSI(27), RSI(39), SIGNEDRSI(8)	class
Multinomial Logistic Regression Model	current class, hour (numeric), day of week (nominal), RSI(18), RSI(28)	class
CART Decision Tree	current class, current close price, hour (numeric), day of week (numeric), LAG(5), SIGNEDRSI(7), ROC(20), ROC(21)	class
Support Vector Machine	current return, RSI(9), RSI(25)	return
Instance-Based K*	current class, MA(11), LAG(1), WIL(8), ROC(1)	return
Gaussian Radial Basis Function Network	current class, hour (numeric), WIL(17), RSI(23), ROC(3)	return
C4.5 Decision Tree	current class, hour (numeric), day of week (nominal), LAG(1), WIL(11), RSI(14)	class

Table 4. USD/JPY agent's ensemble

Model	Attributes	Prediction
Instance-Based K*	hour (nominal), day of week (nominal), MA(6), current class	class
C4.5 Decision Tree	hour (nominal), day of week (nominal), MA(6), current class	class
RIPPER Rule Learner	hour (nominal), day of week (nominal), current class	class
Naive Bayes	hour (nominal), day of week (nominal), current return	class
Logistic Decision Tree	hour (nominal), MA(6), current class	class
Instance-Based K*	hour (nominal), day of week (nominal), MA(6), current class	return
Support Vector Machine	hour (numeric), day of week (numeric), MA(10), MA(2), current return	return

Table 5. USD/CHF agent's ensemble

Model	Attributes	Prediction
Gaussian Radial Basis Function Network	hour (numeric), LAG(6), SIGNEDWIL(6), SIGNEDWIL(34)	class
K-Nearest Neighbours Classifier	hour (numeric), LAG(1), LAG(6), WIL(24)	return
Pace Regression	hour (numeric), day of week (numeric), MA(4), LAG(4)	return
Support Vector Machine	hour (nominal), LAG(4), WIL(7), WIL(23)	class
Nearest-Neighbour Classifier	current close price, hour (nominal), day of week (nominal), LAG(6), WIL(31), SIGNEDRSI(22), ROC(35)	class
CART Decision Tree	current class, hour (numeric), WIL(14)	class
Support Vector Machine	hour (nominal), day of week (numeric), MA(11), RSI(2), RSI(29)	return

Table 6. CHF/JPY agent's ensemble

Model	Attributes	Prediction
Gaussian Radial Basis Function Network	current class, hour (nominal), day of week (numeric), MA(32), LAG(6), SIGNEDWIL(7), RSI(32), ROC(32)	class
CART Decision Tree	current class, hour (numeric), MA(37), WIL(34), SIGNEDWIL(13), SIGNEDWIL(37), ROC(4), ROC(22)	class
Naive Bayes	current return, current close price, LAG(5), LAG(8), SIGNEDWIL(39), SIGNEDRSI(16)	class
Best-First Decision Tree	current class, current close price, hour (numeric), day of week (nominal), MA(20), MA(34), SIGNEDWIL(11), RSI(32)	class
Alternating Decision Tree	current return, day of week (nominal), WIL(6), SIGNEDWIL(13), RSI(7), ROC(3)	class
RIPPER Rule Learner	current close price, hour (nominal), day of week (numeric), LAG(8), WIL(4), SIGNEDWIL(17), RSI(40)	class
Support Vector Machine	hour (numeric), day of week (nominal), WIL(30)	return

As previously described, before each trade the *Intuition Module* uses a fixed size dataset to test the models and calculate their simulated profitability. Our trading agents were set to use test sets consisting of the most recent 100 instances. The decision to keep the test sets this small was based in the following reasons:

- The volatility of the price of a financial instrument is usually clustered. This means that long periods of low volatility are generally followed by short periods of high volatility. Since the weights of the models' votes are based in the profitability shown using the test instances, we need to keep the test set small enough that the weights can adapt quickly when the market enters a period of high volatility. In other words, the shorter the test set, the faster an agent can adapt to changes in the price volatility of the financial instrument it is trading.
- A new instance is available after each trade. This instance becomes a test instance, and the oldest instance in the test set becomes a training instance. This means that, as time goes by, the training set grows while the test set remains the same size and moves like a sliding window. This implies that the shorter the test set, the faster the new instances can be used for training. In other words, the shorter the test set, the faster an agent can learn new patterns.

In order to test the performance of the prediction mechanisms, we implemented the agents using only the *Intuition Modules*, as shown in Figure 3.

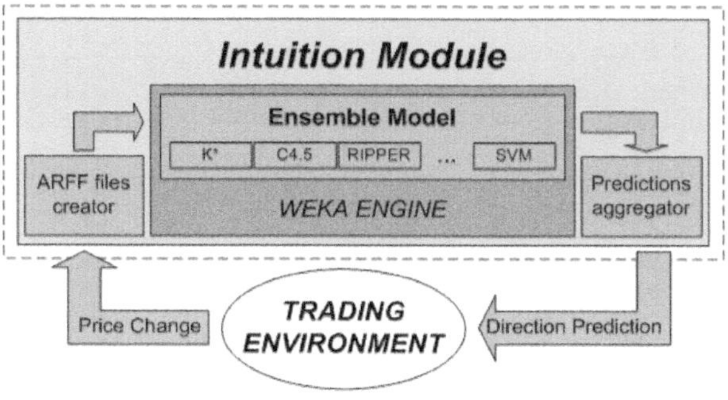

Fig. 3. Agent implementation using the *Intuition Module*

The modules executed predictions using 2,000 instances, corresponding to the period from February 2007 till December 2008, and trades were simulated accordingly: if a price increase was predicted, a buy order was simulated, and if a price drop was predicted, a short sell order was simulated. The accumulated return during this period is shown in Figure 4 and 5, and summarized in Table 7. The return results are relative to the trading capital that was available initially, considering that no leverage was used. In order to accomplish this, while at the same time maximizing the usage of the monetary resources, the size of the trades placed by each agent was made equal to the total investment money available at the beginning of the simulation.

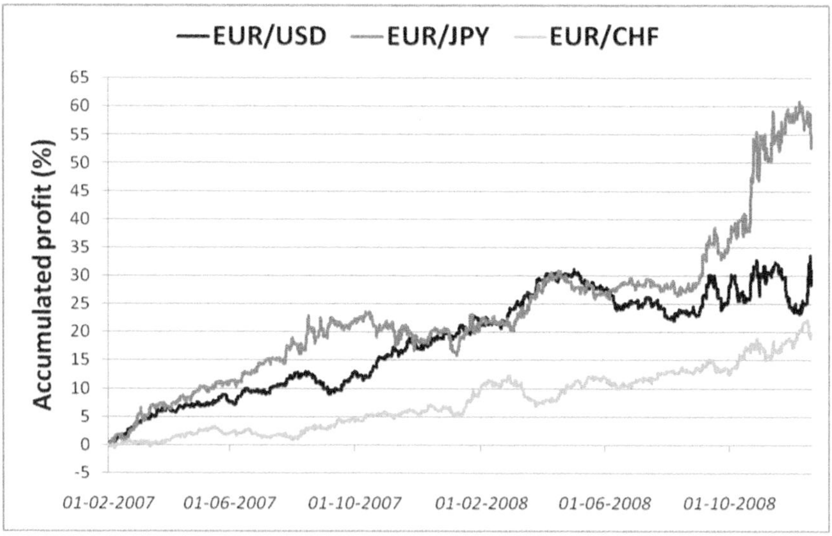

Fig. 4. Simulation results using the *Intuition Modules*

The simulation results demonstrate that the ensembles of classifiers and regression models can produce reasonably profitable predictions. Profit-wise, the best performing module was the one trading the USD/CHF currency pair. Its accuracy predicting the direction of the price of the currency pair was 54.16% in 1,861 trades. It obtained a gross profit of 56.57% in 23 months of simulated trading, with a maximum drawdown of 5.61%. The least profitable module was the one trading the EUR/CHF. Its accuracy was 53.25% in 1,818 trades, and it obtained a return of 19.83% with a maximum drawdown of 5.46%. The average return for the six simulations was 39.53% and the average maximum drawdown was 10.07%.

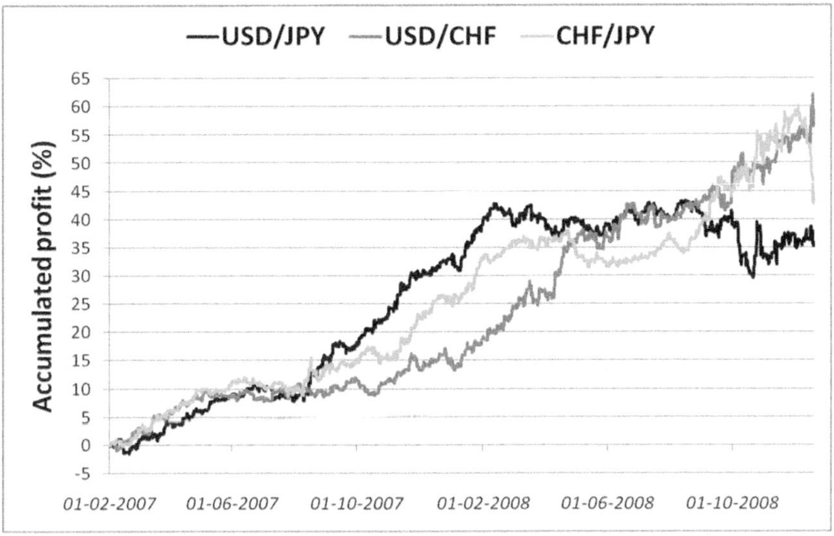

Fig. 5. Simulation results using the *Intuition Modules*

4.2 Agents' Empirical Knowledge

The agents' *A Posteriori Knowledge Modules* were implemented by setting a couple of variables that affect the way money management is handled. After several trial and error tests, we settled for the following settings:

- double the trade size if the overall profit factor of the retrieved cases is greater or equal to 1.5;
- do not place the trade if the overall profit factor of the retrieved cases is lower or equal to 0;
- require a minimum of 3 retrieved cases before making a decision.

The inclusion of the *A Posteriori Knowledge Module* in the design of the trading agents was made with two objectives in mind. First, to decrease the number of trades by avoiding trades expected to be unprofitable. Second, to decrease the agents' maximum drawdown without a proportional decrease in the profit. In order to determine if this module performed as expected, we implemented each of the six agents using a combination of the *Intuition Module* with the *A Posteriori Knowledge Module*, as shown in Figure 6.

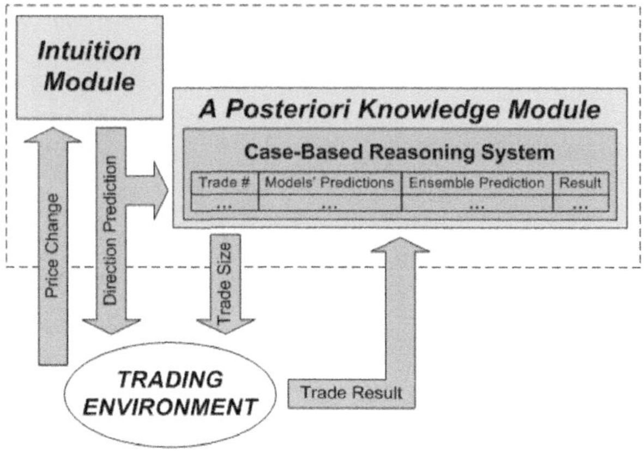

Fig. 6. Agent implementation using the *Intuition* and *A Posteriori Knowledge* modules

These module combinations were used to simulate trades for the 2,000 out-of-sample instances. The results of these simulations are shown in Figure 7 and 8, and summarized in Table 7. Once again, no leverage was used. Since the trade size could now be doubled under certain circumstances, the standard trade size had to be set to half of the capital that was initially available.

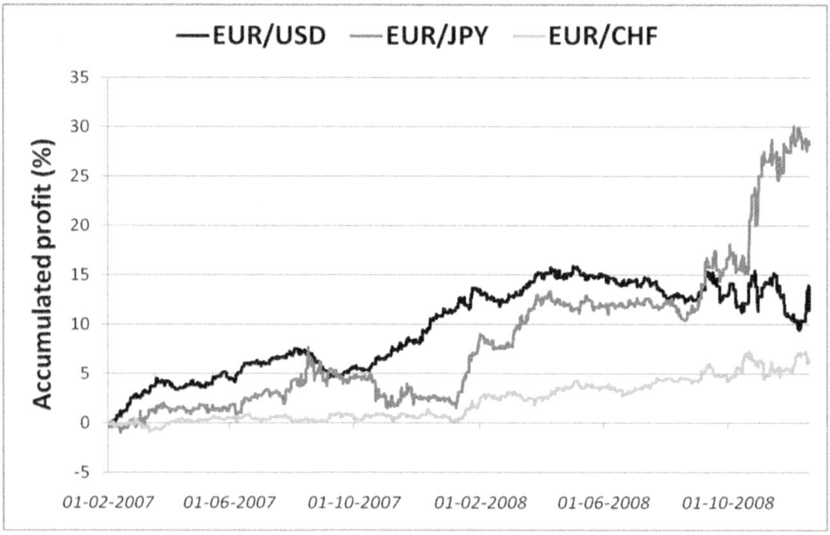

Fig. 7. Simulation results using the *Intuition* and *A Posteriori Knowledge* modules combination

From looking at the results in Table 7, we can conclude that the *A Posteriori Knowledge Modules* achieved their objectives reasonably well. The number of trades performed by each agent dropped by around 30%, compared to using the *Intuition Modules* alone. This is excellent news, because in live trading there will be a cost associated with each trade. Since we cut the trade size in half, there was also a significant decrease in the trading risk, as measured by the maximum drawdowns. However, this also had a negative impact in the profit.

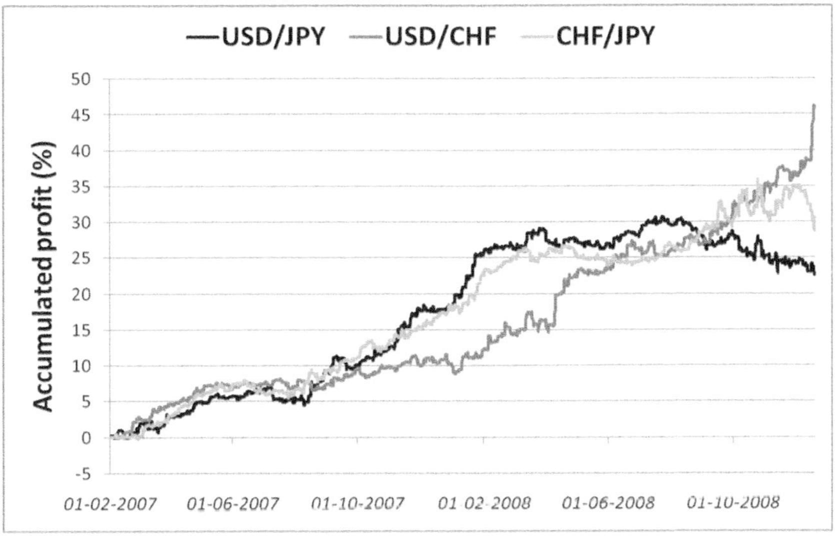

Fig. 8. Simulation results using the *Intuition* and *A Posteriori Knowledge* modules combination

The best performing combination of modules was the one responsible for trading the USD/CHF, with 55.13% accuracy in 1,159 trades. Its gross profit was 45.88% in 23 months of simulated trading, with a maximum drawdown of 2.95%. The worst performing combination was the one trading the EUR/CHF, with an accuracy of 53.74% in 1,191 trades, and a profit of 6.47% with 2.85% maximum drawdown. The average return was 24.22% and the average maximum drawdown was 5.64%. Compared to using the *Intuition Modules* alone, the average maximum drawdown decreased more than the average return, which is exactly what we wanted to achieve. The smaller number of trades also helps make up for the lower profits, if we take into account trading costs such as commissions and spreads.

4.3 Agents' Expert Knowledge

The agents' *A Priori Knowledge Modules* were implemented with two objectives in mind. First, to decrease the number of trades by stopping the agents from trading in low liquidity days and by eliminating redundant trades. Second, to decrease the trading risk, as measured by the maximum drawdown. In order to achieve these objectives, the following rules were inserted in each agent's rule-based expert system:

- Do not trade if it is Christmas Day, New Year's Day or Good Friday.
- Only open a new trade if there is not already an open trade with the same direction and size; if that is the case, just leave the old trade open.

- Close a trade if the price moves a certain percentage in the predicted direction: 0.30% for the EUR/USD, 1.00% for the EUR/JPY, 0.20% for the EUR/CHF, 0.45% for the USD/JPY, 1.30% for the USD/CHF and 0.60% for the CHF/JPY. These percentages were selected manually, by calculating the price volatility of each currency pair using the initial training data, and picking a value that worked well according to that volatility.

To test the usefulness of the *A Priori Knowledge Modules*, we used them in combination with the price prediction mechanisms to implement the agents, as shown in Figure 9.

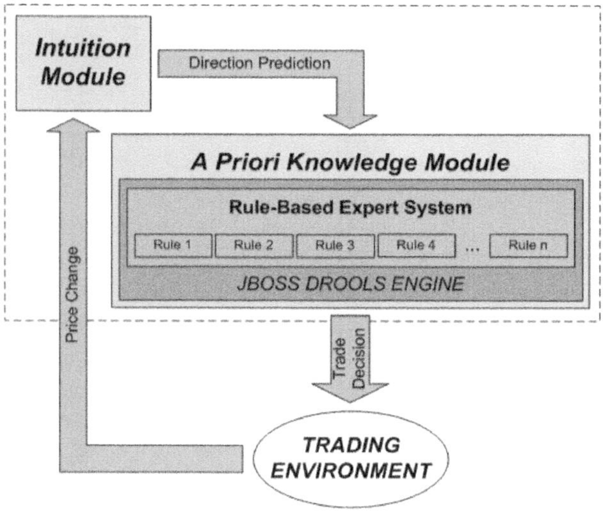

Fig. 9. Agent implementation using the *Intuition* and *A Priori Knowledge* modules

Simulation results using the 23 months of out-of-sample data are shown in Figure 10 and 11, and summarized in Table 7. Since there is no investment doubling in this simulation, we used a trade size equal to the total investment money available initially.

The results obtained with the *A Priori Knowledge Modules* were much better than those obtained with the *Intuition Models* alone. First, by eliminating redundant trades, there was a substantial decreased in the number of trades. Second, there was an increase in the profit of almost all of the agents, accompanied with smaller maximum drawdowns.

The best performing combination of modules was the one trading the EUR/JPY, with a 66.44% gross return and a maximum drawdown of 8.32%. Even though it only predicted 53.21% of the 698 trades accurately, 53.98% of these trades were successful, i.e., were closed with a profit. This was possible because the take profit rule in this agent's rule-based expert system allowed it to be profitable even if it made a wrong prediction, as long as the EUR/JPY price moved 1.00% in the predicted direction during the 6 hours period.

The worst performing modules' combination was the one trading the EUR/CHF, with a relatively high success rate of 57.06% in 1,016 trades, but finishing with a return of only 25.75% and a maximum drawdown of 4.14%. There is a reason why this pair has been associated with the worst trading results so far: it is one of the least volatile in the group of currencies that were traded, which limits the potential profits.

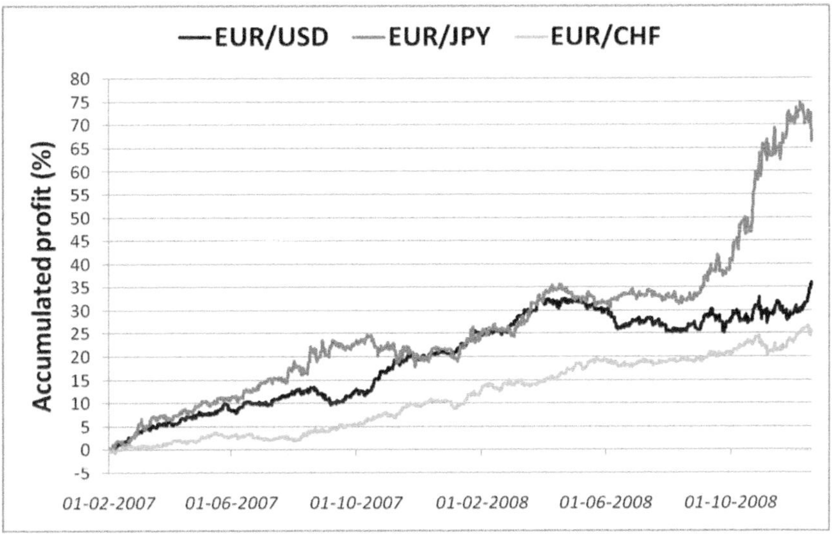

Fig. 10. Simulation results using the *Intuition* and *A Priori Knowledge* modules combination

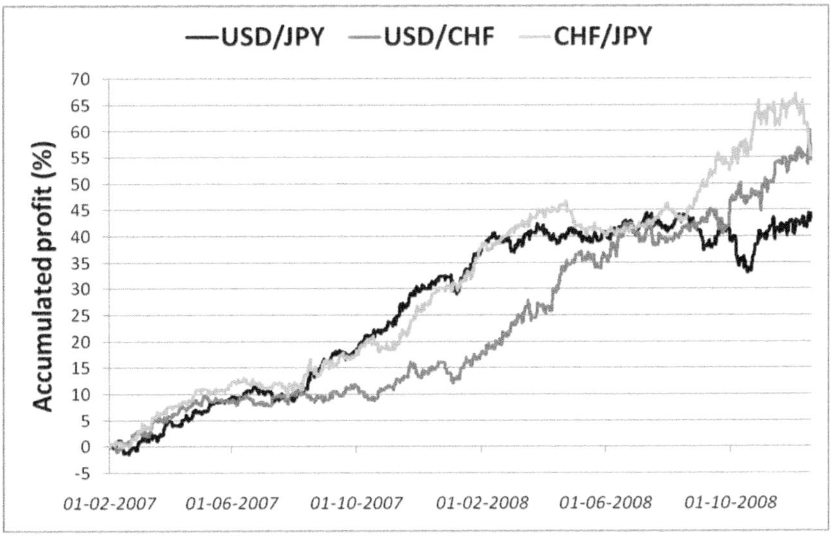

Fig. 11. Simulation results using the *Intuition* and *A Priori Knowledge* modules combination

The average return for these simulations was 47.09%, with an average maximum drawdown of 8.16%.

4.4 Agents' Simulation Results

We have demonstrated the impact of each module in the profit and maximum drawdown obtained while trading different currency pairs. It is now time to analyze the performance of the actual agents, implemented according to the architecture shown in Figure 1. The simulation results for the six agents are shown in Figure 12 and 13, and summarized in Table 7. Since the agents can double the trade size under certain circumstances, and we do not wish to use any leverage, the standard trade size was set to half of the initially available trading capital.

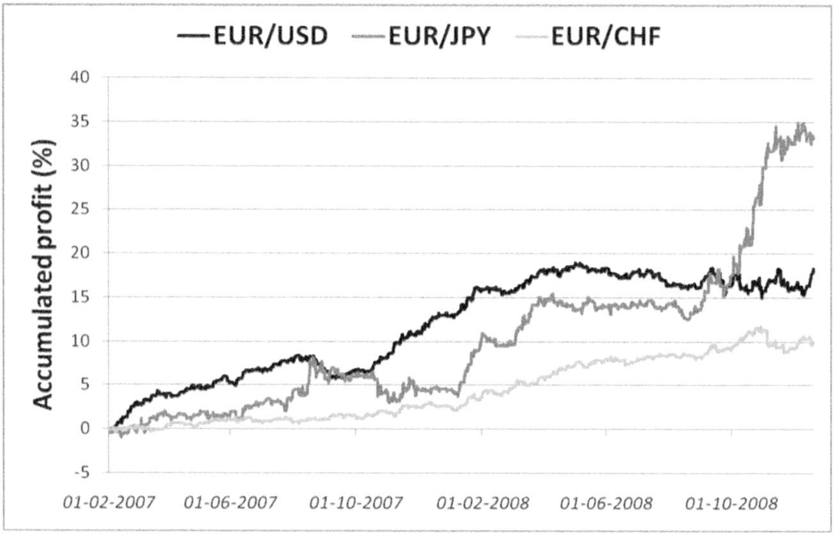

Fig. 12. Agents' simulation results

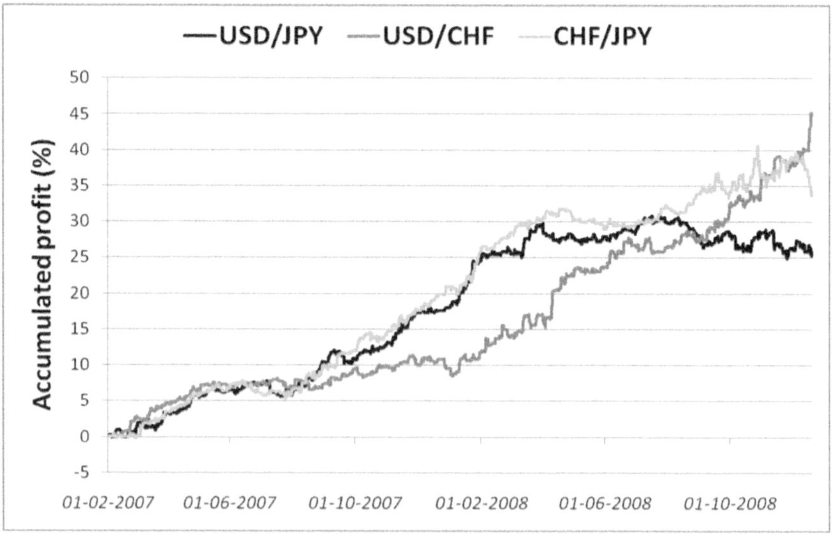

Fig. 13. Agents' simulation results

The best performing agent was the one trading the USD/CHF, with a 55.50% success rate in 915 trades, and a gross return of 44.95% with 2.99% maximum drawdown. The worst performing was, as expected, the agent trading the EUR/CHF, with a 57.06% success rate in 904 trades and a 10.06% return with 3.05% maximum drawdown. Overall, the agents behaved quite well under simulation. They all returned

Table 7. Simulation results using different module combinations

Pair	Modules	Accuracy (%)	Success (%)	Trades	Drawdown (%)	Return (%)
EUR/USD	Intuition	51.80	51.80	1,830	9.58	28.42
	Intuition + APosteriori	52.41	52.41	1,305	6.47	11.48
	Intuition + APriori	52.00	56.24	1,059	7.52	36.03
	Agent	52.57	56.53	1,045	3.95	18.37
EUR/JPY	Intuition	52.99	52.99	1,842	8.61	52.74
	Intuition + APosteriori	53.22	53.22	1,244	6.22	28.17
	Intuition + APriori	53.21	53.98	698	8.32	66.44
	Agent	53.31	53.88	901	5.16	33.15
EUR/CHF	Intuition	53.25	53.25	1,818	5.46	19.83
	Intuition + APosteriori	53.74	53.74	1,191	2.85	6.47
	Intuition + APriori	53.08	57.06	1,016	4.14	25.75
	Agent	53.46	57.06	904	3.05	10.06
USD/JPY	Intuition	53.36	53.36	1,801	13.73	35.26
	Intuition + APosteriori	53.54	53.54	1,270	8.21	22.53
	Intuition + APriori	53.22	54.98	936	11.34	44.29
	Agent	53.56	54.92	988	6.13	25.81
USD/CHF	Intuition	54.16	54.16	1,861	5.61	56.57
	Intuition + APosteriori	55.13	55.13	1,159	2.95	45.88
	Intuition + APriori	54.35	54.35	671	5.61	54.53
	Agent	55.50	55.50	915	2.99	44.95
CHF/JPY	Intuition	53.73	53.73	1,848	17.46	44.37
	Intuition + APosteriori	54.59	54.59	1,319	7.16	30.78
	Intuition + APriori	53.75	55.02	875	12.06	55.47
	Agent	54.45	55.76	988	6.97	34.34

a profit after 23 months of trading, some in excess of 17% per year, which is an excellent performance for an unleveraged strategy.

The agents' average return was 27.78% and the average maximum drawdown was 4.71%. The high ratio between this two values means that the agents should be able to trade using leverage without taking on excessive risk. As an example, if we performed the same simulation using 3:1 leverage, the performance values would triple: the average return would soar to 83.34%, while the maximum drawdown would increase to 14.13%. An impressive result by any standards, considering it would have been obtained in just 23 months of trading.

5 Diversified Investment Strategy

Given the interesting results achieved under simulation, we might feel tempted to start monetizing our work by placing the agents in the real market, and giving them real funds to trade. However, when something looks too good to be true, it usually is. The results obtained by the agents under simulation did not take into account the trading costs. Given the high number of trades that were executed, the impact of these costs in the agents' performance will be far from negligible.

In order to obtain results that better resembled the ones that would have been achieved in real life trading, we defined a fixed cost per trade and performed the same simulations again. The commissions paid will vary from broker to broker, and some offer prices as low as 0.4 pips per trade. A pip is the minimum movement in the price of a currency pair. For a pair containing the Japanese yen (JPY) that represents a movement of 0.01 in the price, while for the other pairs it represents a movement of 0.0001. But there are other costs in trading besides commissions. Slippage, spreads and execution problems will also impact the trading results. With this in mind, we decided to use a fixed cost per trade of 3 pips for the agents trading the majors (EUR/USD, USD/JPY and USD/CHF) and a 6 pips cost for the agents trading the other pairs (EUR/JPY, EUR/CHF and CHF/JPY). This difference reflects the higher liquidity of the majors, which makes it easier to execute trades at a specific price. The more realistic simulation results are shown in Figure 14 and 15, and summarized in Table 8.

While the decrease in the returns was expected, the magnitude of this decrease is an unpleasant surprise. If we take the trading costs into account, the EUR/CHF and the CHF/JPY agents would have actually lost money during the 23 months period. What is worst, as soon as these agents' accumulated return entered the negative territory, some of their trades would start requiring the use of leverage, because the capital available in the trading account would be smaller than the size of these trades. After enough losses, the agents might even become unable to trade, due to lack of funds.

Of the six agents, only the one trading the USD/CHF obtained reasonable results, with a 30.06% return and relatively small maximum drawdown of 3.75%. The agents' average return dropped to just 7.79%. Even if this percentage seems low, it would not be a major problem if the trading risk had been relatively smaller. However, the average maximum drawdown increased to 8.50%. This means it would be too dangerous to use leverage to solve the problem of the low returns, because the maximum drawdowns would increase even more than the profits. Unless we are able to find a way to decrease the maximum drawdown, the agents will have no practical use.

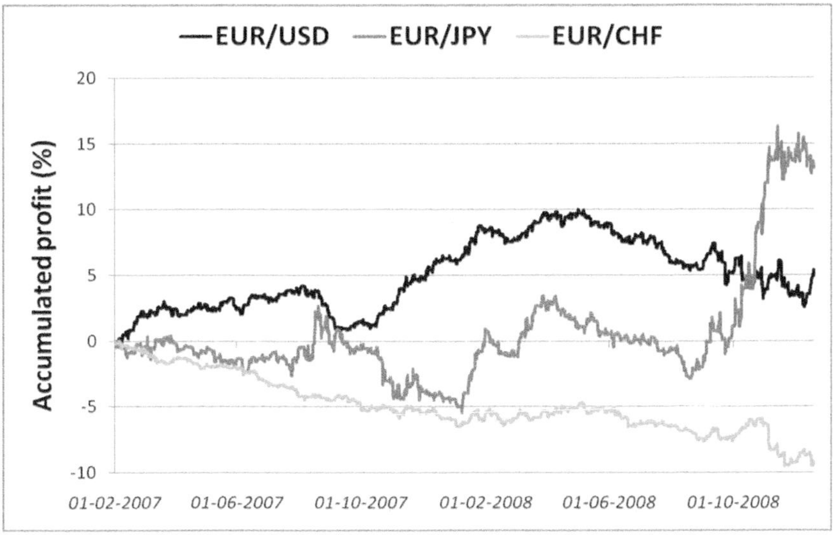

Fig. 14. Agents' simulation results, considering the trading costs

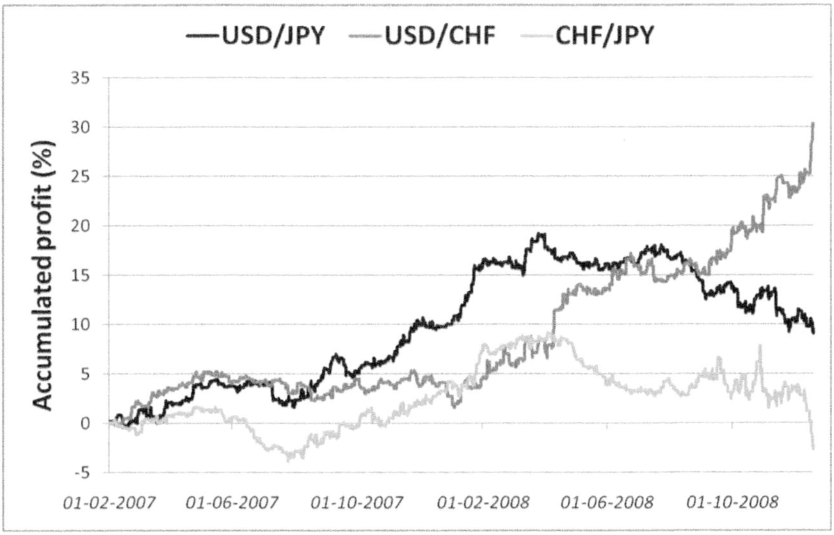

Fig. 15. Agents' simulation results, considering the trading costs

Diversification is commonly accepted as a good strategy to diminish the risk when trading financial instruments. The theory behind it is that the losses that originate from trading some financial instruments will be offset by the profits of trading other non-correlated instruments. We can implement a diversified investment strategy by making the six agents share the capital available for trading. Given that each agent can double its trade size under certain circumstances, this sharing of resources is accomplished by setting each agent's trade size to one twelfth of the initially available trading capital. As time goes by, the agents' losses and profits are credited in the same pool of capital, which should result in a smoother, less volatile profit curve. The accumulated return using this diversified investment strategy is shown in Figure 16.

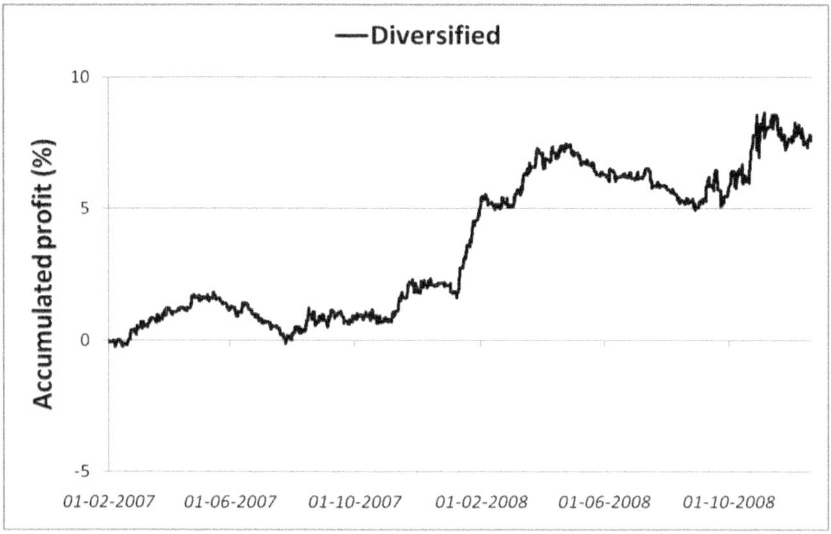

Fig. 16. Simulation results for the diversified investment strategy, considering the trading costs

The diversified strategy was a major improvement over having each agent use its own pool of capital. The net return after 23 months of trading was only 7.79%, but the maximum drawdown was comparatively low at 2.55%. As expected, diversifying the investment resulted in a much safer strategy. The lower maximum drawdown means we should be able to use leverage to increase the returns, without fearing an excessive increase in the trading risk.

Table 8. Agents' simulation results considering the trading costs

Agent	Drawdown (%)	Return (%)
EUR/USD	7.43	5.36
EUR/JPY	8.16	13.22
EUR/CHF	9.60	-9.16
USD/JPY	10.16	9.47
USD/CHF	3.75	30.06
CHF/JPY	11.91	-2.20
Diversified	2.55	7.79

6 Multi-Agent Trading System

Diversifying the investment proved to be a good option to reduce the risk associated with our agents' trading strategy. However, the profit obtained without using leverage was less than impressive. The reason for this is obvious: our agents perform too many trades, which leads to a lot of money being wasted with commissions and other costs. We developed a way to minimize the total number of trades executed, which is guaranteed to improve the agents' performance. But before going into the details, we need to explain what happens when a trade is opened in the Forex market. Whenever an agent buys a currency pair, it is actually buying the base currency and selling the other currency. When it shorts the currency pair, it is selling the base currency and buying the other one. For example, if the EUR/USD agent buys \$100.000 of the currency pair, and the EUR/USD price is 1.3990, its exposure to the market will be long €71,500 and short \$100,000. If at the same time the USD/JPY agent buys \$100,000 of the currency pair, and the USD/JPY price is 89.90, its exposure will be long \$100,000 and short ¥8,995,000. If we combine the exposures of these two agents, the result is long €71,500 and short ¥8,995,000. The exact same exposure could be obtained by simply buying \$100,000 of the EUR/JPY pair. This means that, in this case, two trades could be replaced by just one. The capital required to obtain this exposure would also be cut in half, from \$200,000 to \$100,000. Now let us imagine that the EUR/JPY agent predicts a price decrease and shorts the currency pair with a trade of \$100,000. The EUR/USD agent is expecting the price of the euro to increase in comparison to the dollar, the USD/JPY agent is expecting the price of the dollar to increase in comparison to the yen, and the EUR/JPY agent is expecting the price of the euro to decrease in comparison to the yen. This does not make any sense. If the three pairs actually moved in the predicted directions, it would constitute an obvious triangular arbitrage opportunity. However, the Forex market is efficient enough not to allow this situation to happen. If we add up the exposure of the three agents, we will notice that, even though we have three open trades, there is no exposure to the market. The three trades cancel each other, so no matter what price movements occur, the sum of the profit or loss of the three trades will always be zero. In this case, a perfect replacement for opening the three trades would be to not open any trades at all. This way, the agents would save the costs of opening the trades, and would not tie up any capital. This example demonstrates that, as is, the agents are executing a lot of redundant trades.

Each agent is capable of making five different trade decisions: buy using double the standard trade size, buy using the standard trade size, do not trade, short using the standard trade size, or short using double the standard trade size. There are 15,625 ways in which the decisions of the six agents can be combined. We get this number by calculating the number of possible permutations with repetition, or 5 to the power of 6. Some of these permutations really demonstrate the savings potential in replacing trades according to the market exposure. For example, let us consider the following six trades:

- Short \$100,000 EUR/USD,
- Buy \$100,000 EUR/JPY,

- Short \$200,000 EUR/CHF,
- Buy \$100,000 USD/JPY,
- Short \$200,000 USD/CHF,
- Short \$200,000 CHF/JPY.

The same overall exposure that results from opening these six trades could be obtained by simply shorting \$200,000 EUR/CHF. This means that six trades could be replaced by just one, which would require only \$200,000 of unleveraged capital instead of \$900,000. Most of the permutations show similar potential for decreasing the number of trades. In order to exploit this potential, we integrated the six trading agents in a multi-agent system. In this system, before opening a trade, each agent communicates its trading decision to the other five agents using XML messages. After receiving all the trading decisions, each agent calculates the minimum number of trades necessary to achieve the desired overall exposure to the market. Trades are then opened accordingly. In the previous example, even thought the six agents were supposed to open trades, after communicating with each other only the EUR/CHF agent would open a trade. By using this algorithm, the maximum number of trades open at any given point in time will be 3, which compares with 6 if the agents were not inserted in a multi-agent system. The capital requirement will also be smaller. If the agents did not communicate with each other, overall they would need 12 times the standard trade size in order to trade without leverage. The multi-agent system, on the other hand, will never have more than 8 times the standard trade size invested at any point in time. This means we can set the agents' standard trade size to one eighth of the initially available capital. This will maximize the utilization of the monetary resources without using any leverage.

The simulation results for the described multi-agent system are shown in Figure 17. As expected, this system performs much better than the strategy of investment diversifying with no inter-agent communication. Table 9 contains the comparison between these two strategies. The multi-agent system obtained a net return of 18.56% with a 2.99% maximum drawdown. Compared to using isolated agents, the total number of trades decreased 38.6%, the total volume traded decreased 19.7% and the initial capital requirement decreased 33.3%. Clearly, the multi-agent solution is much better. There are several reasons why we can consider the performance of this system to be above average:

- A return of around 9% per year, with minimal maximum drawdown, can be considered, by itself, a good return on investment.
- This return was obtained during turbulent times. The 23 months of simulation data include the end of an expansion period and a big portion of the severe recession originated by the United States subprime crisis, which caused volatility in financial markets to skyrocket to historic levels. Not surprisingly, the average return of hedge funds worldwide was negative in 2008, the year that saw the collapse of the United States investment banking industry.
- This return was obtained without using any leverage.

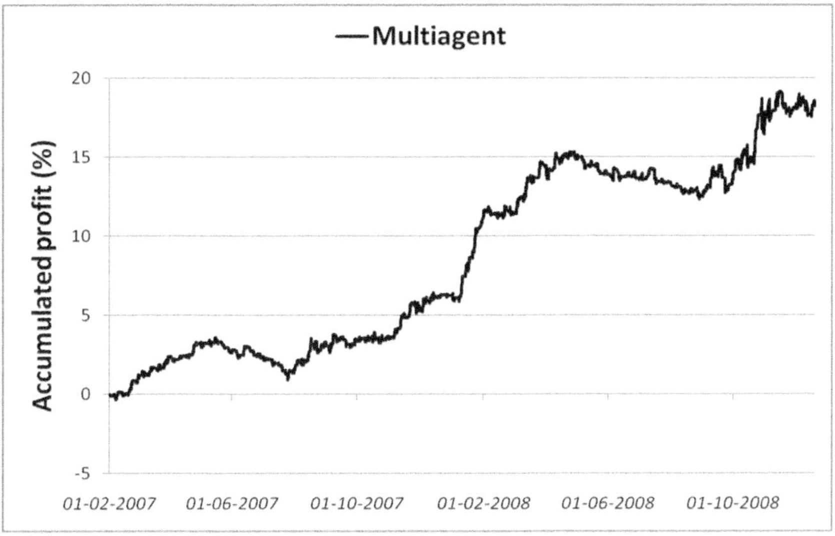

Fig. 17. Simulation results for the multi-agent system, considering the trading costs

Table 9. Comparison between using isolated agents or a multi-agent system

Strategy	Diversified without inter-agent communication	Diversified multiagent
Return	7.79%	18.56%
Maximum Drawdown	2.55%	2.99%
Total Trades	5,741	3,526
Maximum simultaneous trades	6	3
Total Volume	6,791 × trade size	5,454 × trade size
Maximum simultaneous volume	12 × trade size	8 × trade size
Initial capital requirement (no leverage)	12 × trade size	8 × trade size

The use of leverage is commonplace in the Forex market. Most brokers will offer leverage of 50:1 or more. Leverage usage can dramatically increase the trading profits. For example, let us imagine we have a trading account with \$100.000, and we allocate this capital to buy a financial instrument. If the price of the financial instrument increases 5% and we close the trade, we will make a gross profit of 5%, or \$5.000. On the other hand, if we use 50:1 leverage, then we can actually buy \$5.000.000 of the same instrument, with the \$100.000 in the account being used as collateral. In these circumstances, since the trade size is now 50 times greater, our profit will also be 50 times greater. That means the 5% increase in the price of the financial instrument will result in a 250% gross profit, or \$250.000. But using leverage is a double-edged sword. If the price of the financial instrument drops 5% instead of increasing, we will lose 250% of the capital! If that happened, not only would we lose the \$100.000 in the account, we would actually owe more than that to the broker!

Brokers protect themselves by indiscriminately closing trades when the collateral available in a trader's account drops below a certain threshold. This action is called a margin call. In the example we provided, our account would have experienced a margin call long before the instruments' price drop reached 5%, and the account would be left with little to no money. This goes to show that, while it is undeniable that using leverage can be beneficial, it will inexorably lead to greater risk. And no one is immune to this risk. Over-leveraging was one of the reasons for the bankruptcy of the Lehman Brothers investment bank, and for the worldwide financial collapse of 2008.

Table 10 shows the simulation results of the multi-agent system, when using different levels of leverage. These levels do not define the leverage used by the agents in all their trades, but instead the system's maximum initial leverage allowed. Let us assume the agents use a standard trade size of $100.000. At any given point in time, the maximum overall investment by the six agents will be 8 times that amount, or $800.000. This maximum will only be achieved under some rare circumstances. For example, when all the six agents predict a price drop with high certainty, instead of each of them shorting their pair using double the standard trade size (which would require an unleveraged amount of $1.200.000), they will communicate their predictions to each other, and the six trades will be replaced by just two: the EUR/JPY agent shorts the pair using a trade size of $600.000, and the USD/CHF agent shorts the pair using a trade size of $200.000. This is one of a very limited number of situations in which the system puts $800.000 on the line. If we are willing to use leverage of up to a maximum of 3:1, then we only need $266.666 in the account to accommodate for these situations. If the agents trade profitably, they will start adding funds to the account, thus the maximum leverage used will become even smaller over time.

As we can see from table 10, while using 10:1 leverage would have given us an astonishing return, that would have come at the expense of an unbearable maximum drawdown. Using 3:1 leverage, on the other hand, would have allowed us to obtain a yearly return of more than 27%, with a maximum drawdown under 10%. This seems to be a safe level to set the maximum leverage usage of a live trading multi-agent system. We need to be extra careful when deciding how much risk we are willing to take, specially because we are making decisions based on the past performance of the system, and past performance is not a guarantee of future returns.

Table 10. Simulation results for the multi-agent system, considering different leverages

Maximum Leverage	Maximum Drawdown (%)	Return (%)
1:1	2.99	18.56
2:1	5.99	37.12
3:1	8.98	55.68
4:1	11.98	74.24
5:1	14.97	92.98
10:1	29.95	185.61

Figure 18 shows a representation of our production multi-agent Forex trading strategy. It consists of a multithreaded distributed system that is currently trading live completely autonomously. The arrows represent the communication between the agents, and the agents' interactions with the market. The agents communicate among themselves with XML messages using the ActiveMQ message broker [14]. They interact with the Forex market using the proprietary API of an ECN Forex broker, which enables them to send orders to the market and to receive currency price updates and the results of their own trades.

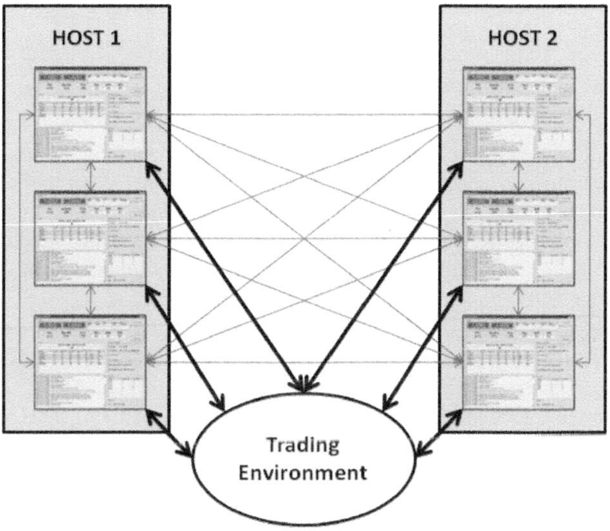

Fig. 18. Production multi-agent Forex trading system

Since the agents need to retrain the models before each prediction, we need to make sure they are able to perform this task as fast as possible. We accomplished this by spreading them over several hosts. Given that the agents' software was written in Java, they can be placed in machines with different operating systems.

The agents were implemented with autonomy in mind. However, there will be situations in which they will require human intervention, as a result of unexpected problems. For example, they cannot be expected to recover by themselves from a permanent network disconnection, or from a broker's specific failure. In order to accommodate for these potential problems, each agent has the ability to place phone calls. If an unrecoverable error occurs, it can report it by calling the system administrator's cell phone and requesting assistance.

7 Final Remarks

In this text we described the implementation of a multi-agent system that can achieve interesting results when trading in the Forex market. We demonstrated how investment diversification can be used to reduce the trading risk, and how enabling the agents to communicate among themselves can help reduce the trading costs and the capital requirements. However, there is still much room for improving this multi-agent trading system. For example, we could increase the investment diversification by adding more agents to the system, to trade other currency pairs using different timeframes. We could also use the *iQuant* software to create trading agents for other financial instruments, such as stocks or futures. This would be an even better way of further diversifying the investments, and consequently decreasing the trading risk. This improvement will be the goal of our future research, in which we intend to describe the implementation of an "intelligent" hedge fund that can operate without human intervention in the trading room.

References

1. Lee, R.: iJADE stock advisor: an intelligent agent based stock prediction system using hybrid RBF recurrent network. IEEE Transactions on Systems, Man and Cybernetics 34(3), 421–428 (2004)
2. Kimoto, T., Asakawa, K., Yoda, M., Takeoka, M.: Stock market prediction system with modular neural networks. In: 1990 International Joint Conference on Neural Networks, vol. 1, pp. 1–6 (1990)
3. Kwon, Y., Moon, B.: Daily Stock Prediction Using Neuro-genetic Hybrids. In: Cantú-Paz, E., Foster, J.A., Deb, K., Davis, L., Roy, R., O'Reilly, U.-M., Beyer, H.-G., Kendall, G., Wilson, S.W., Harman, M., Wegener, J., Dasgupta, D., Potter, M.A., Schultz, A., Dowsland, K.A., Jonoska, N., Miller, J., Standish, R.K. (eds.) GECCO 2003. LNCS, vol. 2723, pp. 2203–2214. Springer, Heidelberg (2003)
4. Franses, P., Griensven, K.: Forecasting exchange rates using neural networks for technical trading rules. Studies in Nonlinear Dynamics and Econometrics 2(4), 109–114 (1998)
5. Lu, H., Han, J., Feng, L.: Stock movement prediction and N-dimensional inter-transaction association rules. In: 1998 ACM SIGMOD Workshop on Research Issues on Data Mining and Knowledge Discovery, pp. 1–7 (1998)
6. Gençay, R.: Linear, non-linear and essential foreign exchange rate prediction with simple technical trading rules. Journal of International Economics 47, 91–107 (1999)
7. Tay, F., Cao, L.: Application of support vector machines in financial time series forecasting. International Journal of Management Science 29(4), 309–317 (2001)
8. Abraham, A.: Analysis of hybrid soft and hard computing techniques for Forex monitoring systems. In: Proceedings of the 2002 IEEE International Conference on Fuzzy Systems, pp. 1616–1622 (2002)
9. Abraham, A., Nath, B., Mahanti, P.: Hybrid Intelligent Systems for Stock Market Analysis. In: Alexandrov, V.N., Dongarra, J., Juliano, B.A., Renner, R.S., Tan, C.J.K. (eds.) ICCS-ComputSci 2001. LNCS, vol. 2073, pp. 337–345. Springer, Heidelberg (2001)
10. Barbosa, R., Belo, O.: Algorithmic Trading Using Intelligent Agents. In: Proceedings of the 2008 International Conference on Artificial Intelligence (2008)
11. Barbosa, R., Belo, O.: A Step-By-Step Implementation of a Hybrid USD/JPY Trading Agent. International Journal of Agent Technologies and Systems (2009)
12. Weka API, `http://www.cs.waikato.ac.nz/ml/weka/`
13. JBoss Drools API, `http://www.jboss.org/drools/`
14. ActiveMQ API, `http://activemq.apache.org/`

Information Coordination Using Meta-agents in Information Logistics Processes

Mats Apelkrans[1] and Anne Håkansson[2]

[1] Jönköping International Business School, Department of Informatics, Jönköping, Sweden
Mats.Apelkrans@jibs.hj.se
[2] Department of Computer and Systems Sciences, Stockholm University, Kista, Sweden
Anne.Hakansson@dsv.su.se

Abstract. To be competitive, the goal for manufacturing companies is to deliver the right products at the right time with good quality. This requires good planning, optimized purchases and well functioning distribution channels. It also requires an efficient information flow in the company. One of the problems in Business Informatics is to coordinate the information flow between the manufacturing company, its customers, and its suppliers. For example, coordinating the information flow needed in a manufacturing company to fulfil their order stock. In the production process, a major issue is the double directed information flows, one from customers to company and another from company to its suppliers. This information can be about orders, requirements and production plans. Information coordination between the actors can speed up the information exchange and, hence, optimize the production cost. However, this requires a technology that can search, combine and deliver the information needed by the manufacturing company. Our approach is to combine theories from Multi Agent Systems (MAS), meta-agents and Information Logistics (IL) in order to coordinate and deliver information at the right time and to the right place at an acceptable cost.

1 Introduction

Information logistics is a concept that becomes more used in Europe due to the fact that information itself can be thought of as a product that can be manufactured. This is a competitive manufacturing to the product manufacturing since knowledge is used instead of natural resources.

The information logistics differs from ordinary logistics and there are information systems that can support it. Rayport & Sviokla (1995) [1] exploit the concept of Virtual Value Chain, which points out the component of the logistics process that handles information as an instrument to control the physical part of the logistics process. Information is gathered, produced and refined and finally distributed in order to support the Physical Value Chain. The outcome from a logistics process can in many cases be just information products e g digital entertainment products, distributions of mail invoices. In both cases information supply is needed in order to

A. Håkansson et al. (Eds.): Agent and Multi-agent Technology, SCI 289, pp. 119–136.
springerlink.com

produce a suitable product. Information can then be processed and finally distributed on paper, CD, sent to Internet, used by an ERP system and so on. An essential part in an IL process is knowledge, knowledge of how to produce new information from older ones. This is studied in what we call the Information Logistics Process (ILP).

To enforce qualified information management and provide the right information at the right time, we apply a multi-task system. The benefit of applying agents is the ability to work with intelligent agents performing tasks in parallel, which supports handling a large number of tasks faster. We use intelligent agents to handle different tasks in the information flow. Each agent's behaviour can be thought of as an Information Logistics Process (ILP), which is handling information between several sources with given input and output. Still there is a need for coordination of the intelligent agents' performances in order to fulfil the global task of controlling a successful company. The solution is using meta-agents. The use of meta-agents on top of the information flow chart can monitor the action of the intelligent agents and be used to control the information passing between the agents. The meta-agents can provide the users with the requested information. An examination of the behaviour of an intelligent agent shows that it can perform as an ILP. Hence, there will be a number of ILP calls. The ILP is managed in overall strategy by the use of meta-agents. Each ILP has to meet the demands on e.g. time, content, and presentation.

As an example, we look at a manufacturing company receiving orders on items from customers, which need to be produced. The production line needs a continuous flow of information. In order to coordinate and deliver information in the right time and to the right place, theories from Multi-agent systems and Information Logistics are combined. The handling of this distributed information flow can be thought of as Information Logistics Processes (ILPs). Behind this there is an ERP system, which uses database technology to manage Bill of Materials (BOM), inventory situation, machine capacity and worker availability. A newly incoming order has to be calculated with BOM and matched together with earlier planning. In order to manage this production, the manufacturing company must send orders to suppliers, for example, to purchase raw material or other items. In this chapter, we suggest using meta-agents to control the behaviour of a number of intelligent agents, where the meta-agents are working with coordination of the communication that takes place in a supply chain system. The handling of this distributed information flow can be thought of as Information Logistics Processes (ILPs). We will also point out similarities between intelligent agents' behaviour and the functioning of ILPs.

The chapter is structured as follows: Section 2 gives a brief description of information logistics and information logistics process. Section 3 describes intelligent agents and meta-agents in a multi-agent system for internal use on the web and in the logistics area. Section 4 presents information flows that surround manufacturing companies and gives examples of tasks for both intelligent agents and meta-agents. Section 5 shows the multi-agent system supporting the information logistics process by having agents and meta-agents working with tasks on the intranet and extranet. Since, the work of the agents and meta-agents requires presentation of the performance to the users, views of the agents' work are provided, in section 6.

2 Information Logistics

Information logistics (IL) has appeared as a hot discipline mainly due to the Internet age. The interpretations of the concept differ from Europe to US. In US most references to IL allude to software that supports the physical logistics chain. The latter use of the IL concept can be denoted Logistics Information Management (LIM). In Europe, IL concentrates on the logistics in order to manufacture information products.

Rayport & Sviokla [1] exploit the concept of Virtual Value Chain. They point out the component of the logistics process that handles information as an instrument to control the physical part of the logistics process. Information is gathered, produced and refined and, finally, distributed in order to support the Physical Value Chain. The outcome from a logistics process is, in many cases, just information products like digital entertainment products or distribution of e-invoices. For these information products, information supply is needed in order to produce a suitable product. The information can then be processed and finally distributed, e.g., on paper, CD or published on Internet etc.

The concept of Information Logistics (IL) has been given a number of definitions. A recent definition is [2]:

"The main objective of Information Logistics is optimized information provision and information flow. This is based on demands with respect to the content, the time of delivery, the location, the presentation and the quality of information. The scope can be a single person, a target group, a machine/facility or any kind of networked organization. The research field Information Logistics explores, develops and implements concepts, methods, technologies and solutions for the above mentioned purpose"

The IL can be considered as a process that manufactures an information product and, hence, introduce ideas from the production area, like Just-In-Time (JIT). From a JIT point of view, Information Logistics can be defined as the discipline [3]:

- that will supply the right information at the right time, in the right shape and at the right place
- in a user-friendly way
- with desired quality
- to the lowest possible cost in where the final product is distributed by some kind of information carrier like paper, card, CD, Smart Card, Internet.

There are similarities between ordinary logistics and IL, e.g. efforts to deliver the product of information at the right time and to the right place. Although there are similarities, it is easy to pinpoint a number of differences between ordinary logistics and IL. The information supply in ordinary logistics is essential and often time consuming and expensive. Much effort must be put into information supply in order to handle inbound logistics, checking deliveries or out-of-stock problems or damaged materials.

Looking at Information Logistics from a process perspective, an information logistics process transforms a given input into some form of output. The input is some kind of fragmented information or knowledge description that is derived from a so-called information supplier. This input information can be handled, either manually or automatically, by the system. The process output is an information product that becomes accessible and delivered to the information receiver who can make use of the information. This workflow is called the Information Logistics Process (ILP).

The basic ingredients of the Information Logistics Process are supplying information, producing information and distributing information. ILP aims to guarantee that information with good quality is produced and delivered to the 'customer' to the right place just in time. A problem that will be addressed is how to make ILP an efficient process.

There are a lot of applications of the ILP process for example:

- Digital production and distribution of music, videos, forms
- ERP systems configuration
- Multiple Project Management
- Content Management
- Knowledge Management

ILP differs from the ordinary logistics process due to:

- Stock replenishment is no problem, just make a new copy of the information product
- The first product item has a cost to manufacture, then reproduction is more or less at no cost
- Digital assets are not consumed when delivered, but can quickly be obsolete
- Information tends to be "up to date" and get old quickly. A product, on the
- other hand, can be reused several times.

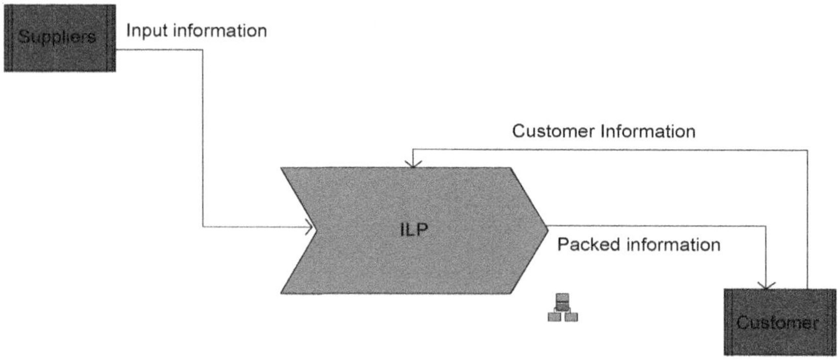

Fig. 1. The ILP process

Information suppliers deliver information to the ILP process. The consumers of the ILP deliverables are called Customers. There are applications where even the customers deliver some input to ILP. The sets of suppliers and customers can contain 0, 1 or several elements. Some ILPs need no input. They can produce the necessary information by themselves. Furthermore, some ILPs produce information, which is not delivered immediately but stored for future use and delivered in the right time.

Hence ILP can be understood as a function from incoming information I to packed output information O:

$$O = ILP(I),\ O \in Pi \text{ and } I \in Ii$$

from an input domain:

$$Ii = \{i1, i2, i3, \ldots.\}$$

to an output domain of packed information:

$$Pi = \{o1, o2o3, \ldots.\}$$

What can be included in Pi is no wrapping at all, presentation in some digital form such as PDF-files, strings of XML-code, printouts on paper, storing on a CD or Smart Card , information to another ILP or simply publishing on the Internet.

The set of suppliers can consist of both physical suppliers and more virtual ones like information from Internet or even output from a preceding ILP.

Supplier = {a Physical logistics process, customer, customers' customer, ordinary suppliers, Internet, ILP deliverables…}

In the same way a set of Customers appears as:

Customers= {'Ordinary' customer, Internet, a Physical logistics process, a subsequent ILP….}

The ILP processes will be implemented with different methods from the computer science area. To make this efficient, the computer system should be automated but there are some dependencies between users of ILP and the ILP process, which requires a dialogue with the system. The system needs knowledge to perform the required output and therefore, the ILP processes have to handle knowledge in some way.

In order to fulfil the task of delivering information products in the right time to the right place, Frauenhofer Institute [4] in Germany argues that an ILP needs in addition to Content Management (CM), Communication Management (ComM) and Time Management (TM). In [5] some examples of practical applications of this machinery are described. However, in our more general architecture CM is included in the IS and IP sub processes and ComM and TM is part of the ID process (TM can in some applications also be a part of IS).

2.1 ILP Divided into Sub Processes

Examining the ILP process in detail demonstrates that it consists of three different sub-processes: the information supply (IS), the production process (IP), and the distribution process (ID). The IS process can obtain information from suppliers, customers or create its own information. In order to enable information production IS has to extract and store knowledge. An example can be an application where IS at certain points of time produces information in order to kick-off the IP process. IS can also be the process that handles time management (TM), i.e. IS works as a trigger for the actions of IP and ID processes.

The above definition of IL from a production perspective also enables results as packed information (e. g. on a CD) hence a connection to a physical supplier must be included in the ID process (see figure†2). ID is the most appropriate sub-process to handle communication management (ComM).

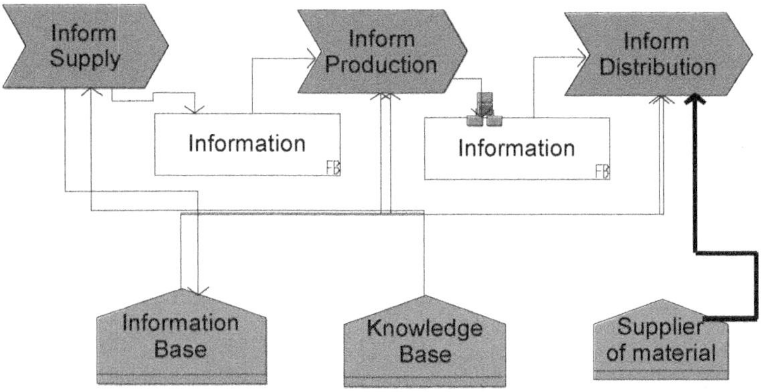

Fig. 2. ILP divided into sub processes

All sub-processes in ILP needed to handle knowledge. This knowledge is production dependent and has to be specified from case to case. Therefore, when handling the different cases, the system needs to deal with how this knowledge is elicited, stored and maintained. Also, the system needs to deal with the automation of the knowledge management.

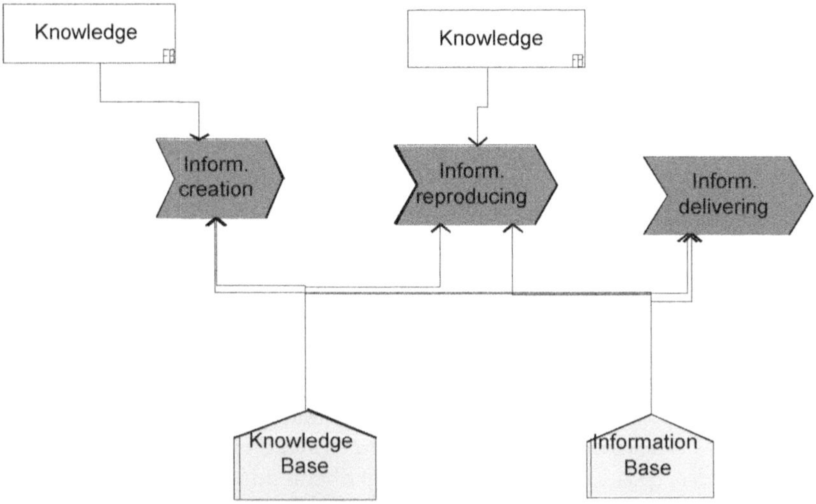

Fig. 3. A proposal for an Information Production (IP) architecture

Figure 3 illustrates a proposal on IP process architecture. It has one process for creating information, another process for reproducing already created information and, finally, a process for leveraging the produced information to the ID process. In order to handle both information and knowledge, the architecture will contain an Information Base (based on a database) and a Knowledge Base (KB). In the Information Base, the actual and/or stored information is stored. The knowledge base contains rules for use of this data.

IL is an implementation of the virtual value chain [1] where the material is much easier to handle than in the physical value chain; it is just information. But as pointed out by Apelkrans and _bom [3] one still have to look at value and quality aspects. The value of the information is dependent of the time and place, hence it should be delivered at the right time, in the right format and at the right place given to the right receiver). Therefore, a value function V is introduced:

$$V(I) = V(I(time, place))$$
$$V(O) = V(O(time, place)).$$

The desire is that ILP is a value-adding process, which means:

$$V(O) > V(I).$$

However, this is not always the case. During the ILP process, time and even place can be changed, so information can be obsolete, distributed to the wrong place or wrong receiver and so forth.

In order to solve the knowledge management problem, which is fundamental for an ILP process to be successful (see Figure 3), a number of proposals have been raised. Knowledge management problem was first studied in Apelkrans and Braf [6], where the similarities and differences between ILP and KM theories were explained. In Apelkrans and Håkansson [7], [8] visual modelling was used in order to illuminate the behaviour of an ILP process. In the first paper, [7], knowledge is applied to the ILP to support sending e-invoices between companies using external network. In the second paper, [8], knowledge is used in ILP to automate configuration of enterprise systems (ERP). The work with ERP configuration was research about choosing the right ERP implementation for a given company, hopefully, lasting for a number of years. To get the possibility to automate the decision capabilities of an ILP process, Apelkrans and Håkansson [9] applied intelligent agents with MAS and meta-agents theories.

The research presented in this chapter solves the problem with information exchange between the manufacturing company and its customers. The manufacturing company receives orders on items from customers, which then need to be produced.The input to the ILP comes from the user via an application and the receiver can be either a new ILP or a database or system user. From the input, the ILP communicates with the knowledge base to produce the desired output, see Figure 4. A request to the system can be to send output information as messages.

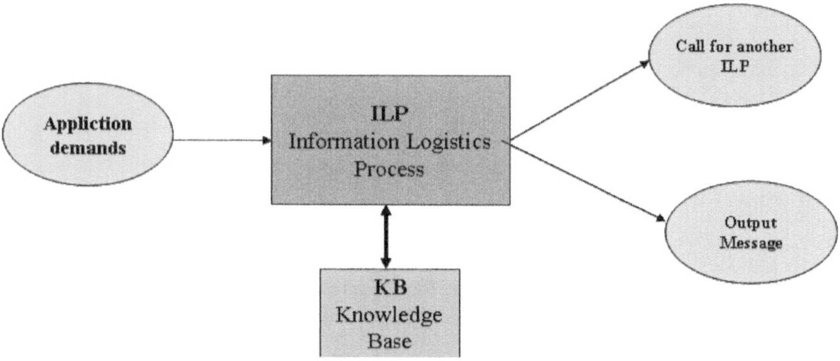

Fig. 4. The Information Logistic Process with a knowledge base

Right decisions are likely to be made in a short time (dynamical) perspective. Incoming orders may be rejected, renegotiated and/or purchase orders may be sent quickly to suppliers. Looking at the business process perspective, many papers have been written in order to optimize the work-flow in a manufacturing situation [10; 11]. However, these contributions do not include ILP. The research in this chapter is more on the decision side, i.e., choosing the right solution to supplier problems at the best cost and at right time. For this, a technology of using self-propelled software like agents for searching information, provides the ILP with information and provides the manufacturing company with information. A benefit of using agents is that, there are similarities between an ILP and the behaviour of an intelligent agent. Both concepts have the ability to communicate with others, produce messages in an intelligent way, and send messages at right time. The difference is that ILP usually does not work externally at the web to find information, which agents can do. Hence, the use of intelligent agents offers a solution to the knowledge management tasks of an ILP.

The example company studied for the research in this chapter, produces goods that are assembled from a number of items, some of them manufactured at place and some of products are bought from different suppliers to be assembled with the product produced by the manufacturing company, see Figure 5.

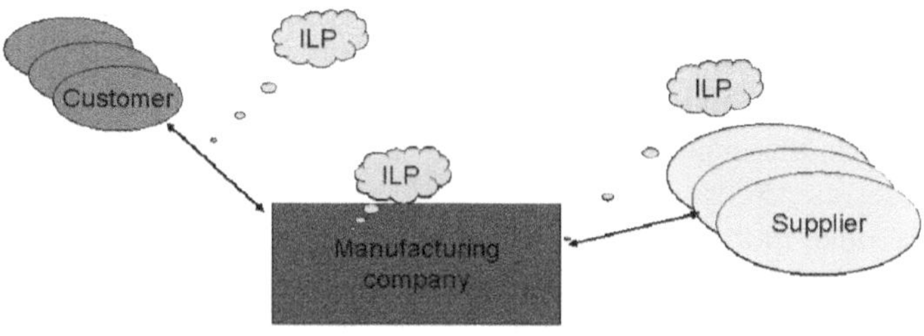

Fig. 5. The example company and its surroundings; ILP is Information Logistics Processes

As mentioned above, a lot of the information flow in the system consists of messages, some of them produced by an ERP system supporting the physical value chain and some of them are manually delivered. A message can be directed as input to an ILP, which answers with a new message. This message can either be input to another ILP or directed towards a final goal. Some examples of messages are:

- "Buy 25 pieces of item #1234 to best possible price with delivery not later then xx-xx-xx"
- "Take care of a new order of 30 pieces of item #2345 which should be delivered yy-yy-yy"
- "Check inventory for pieces of item#3456"
- "Tell customer ABC that his order is delayed 4 days"

We need a multi-agent system that can take care of these kinds of messages.

3 Intelligent Agents and Meta agents in Multi Agent Systems

Intelligent agents in multi-agent systems (MAS) can be used to perform tasks in complex environments. The tasks can be searching and retrieving information from sources on the web as well as accessing and executing these services. Some of the agents can efficiently work with, e.g., commodity products, usually offering a list of products based on some criteria. Some of these agents can take actions on production process requirements and support assembling important information from several sources.

In logistics area, a multi-agent environment can handle the tasks that are too complex to be accomplished by a single software agent. To our best knowledge, not many of the logistic agents work on the intranet and extranet, simultaneously, supporting information logistics process by supplying the process with external information. The intranet includes the MAS and the ILP system to which the agents need to provide input information. The extranet is extended to users outside the company and can support doing business with a pre-approved set of other companies over the network.

The agents must be capable of autonomous actions when situated in different environments [12]. Software agents are autonomous and can perform actions without a directive. Moreover, some of the software agents are intelligent and, thus, adaptable to the environment. The environment, of the web, has some characteristics that the agents must observe and act upon. The web is a vast network with over 100 million web sites where 74% are commercial sites or other sites operating at the .com. The agents must find the right requested product in the environment that is partially observable, stochastic, dynamic, continuous, and episodic.

In a partially observable environment only some of the information is known and, therefore, can be a limitation. Environments such as the intranet and extranet are considered to be partially observable [13] because of its nature of continuously growth. However, the agents do not have to maintain their internal state to find satisfactory information. The agents, in our work, are working with the ILP system to solve tasks, but also use the intranet and the extranet for finding information. Even though we consider the environment to be partially observable, the agents must still able to find the significant information needed for solutions. Finding the necessary information corresponds to fully observable environment, where the agents can obtain complete, accurate and up-to-date information about the environment state [13]. The agents do not need to maintain any internal state to track the world and can easily achieve the task based on the information in the environment.

In a partially observable environment, the environment can appear to be stochastic [10]. The agent cannot predict the behaviour of the environment, as in real-world cases, since the state, which will result from performing an action, is unclear. Although the environment is deterministic, there can be stochastic elements that randomly appear. A partially observable environment with stochastic elements is what will be expected for our agents. The agents have a task of searching and combining data but the information and combinations will vary with time, i.e., when actions these are performed.

In a dynamic environment [13], the environment changes while agents are deliberating its contents [10], which is beyond the agents' control. In these dynamic environments, the agents need to interact with the environment to continuously check the surroundings to be able to act properly. This environment's characterises the intranet and the extranet since these change on a daily bases. However, the environment might be static over smaller time intervals, thus, remaining unchanged while the agents consider their course of action between states. Nonetheless, we will consider continuous changes, which will affect the agents and make them monitor changes during each task.

In continuous environments, there might be an uncountable number of states, arising from the continuous time problem [13]. Continuous time can be a problem for the agents at the intranet and the extranet because of the number of states and possible actions. This requires special treatment of the agents using an execution suspension to control the agents' performance. A common suspension of execution occurs when the agents have found information and returned with result. This limits the possibility of finding several solutions. A better solution in our system is an execution suspension for a short time interval followed by a resumption of the search.

In an episodic task environment [10] the choice of action in each episode depends on the episode itself. The agents, in our work, perform one task at the time while moving between commercial sites to find information. The task itself can be more complex, but before it is assigned to agents, the task is divided into smaller, single tasks, where each task is applied to an agent. In multi-agent systems for information logistics, a lot of intelligent agents will be used to perform a task, which in complex task environments will require a vast number of the agents. Thus, one task is assigned to many agents to perform the task since it speed up the searching and expands the number of results. However, it can be difficult to keep track of all the agents. In these cases, meta-agents can be used to monitor the agents and their results produced from accomplishing tasks within acceptable time-range.

The concept of meta-agents is based on the idea of meta-reasoning. Meta-reasoning is a technique that supports the system to reason about its own operation. This reasoning can be used to reconstruct the agents' behaviour [14], but also to help in the interaction among agents. The meta-reasoning can be applied in the implementation strategies or plans to respond to requests [15].

Besides reasoning, the meta-level agents (so called meta-agents) can plan actions and maintain individual agents' state information but also support future behaviour by controlling and direction the agents, as well as, classify conflicts and resolve these conflicts [15]. Another benefit with meta-agents is that they can work with intelligent agents in network to control and guide their performance [16] but also calculate the optimal decisions based on the intelligent agents' performances in the time-dependent characteristics in the environment [17]. For our work, it is interesting to use meta-agents to combine different information pieces and calculate and select the best option for company on the basis of pre-defined features. Thus, evaluation is important since the results from the intelligent agents must be compared to find the best result that is most useful for the company.

Moreover, using strategies and plans is important to either look for information at the intranet or the extranet. There might be a mix of the searching, which arises when the information is not found at the intranet. Then, the agents need to redirect the work to the extranet. The features of the meta-agents are applied in our system to support information flow in the information logistic process.

4 Information Flow around a Manufacturing Company

Specialized intelligent agents that fulfil special tasks at right time can cover the behaviour of the ILPs. Thus, looking at them as simple information searching engines travelling between parts in the information flow systems, where the intelligent agents (IA) are solving the ILP information problems see fig.6.

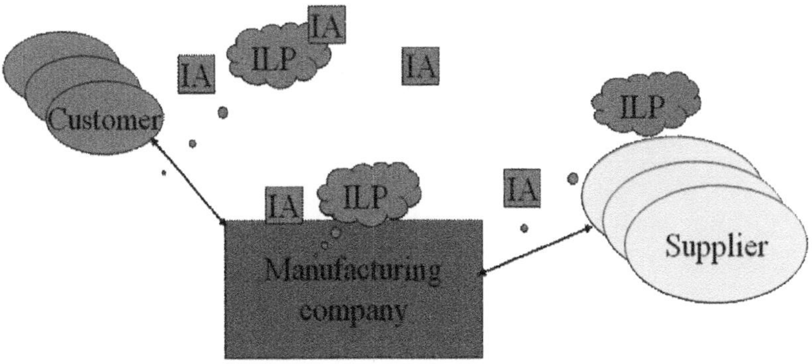

Fig. 6. A number of IAs try to solve the ILP information problems

In this multi-agent system, intelligent agents are individual performers, in order to get useful solutions; their global effects have to be controlled by a number of meta-agents, which have the overall strategy for a controlled solution. Much of the manufacturing process is conducted by operating on the information stored in a database, hence the ILP have to communicate with a number of registers like:

Customer records
Supplier records
Load Planning
Returns Management
Product description with drawings
Price lists

The messages examples given above are simplified in their task nature. In fact the intelligent agents have to penetrate much harder tasks, such as:

- Find out how to manufacture a product ordered from a customer
- Plan the master schedule
- Calculate Bills of material
- Produce Buying lists

For these different tasks both intelligent agents and meta-agents are applied. The intelligent agents work with information between customers, the manufacturing company and suppliers. The meta-agents work with monitoring and controlling the intelligent agents, and also accepting and combing information for optimal purchases for the manufacturing company.

5 The Multi-Agent System

In the multi-agent system, the intelligent agents and the meta-agents have several different tasks that all concern production. The intelligent agents' tasks are searching for suppliers of material and finding material according to pre-decided attributes for purchasing the raw material suiting optimal conditions. Additionally, the intelligent agents are keeping track of the material and the attributes for the material and the product. The meta-agents' tasks, on the other hand, are combining information based on the results of the intelligent agents. Moreover, the meta-agents are planning manufacturing, calculating the total cost of the manufactured product but also producing lists and plans. Furthermore, the meta-agents keep information about time delivery and records about the customers.

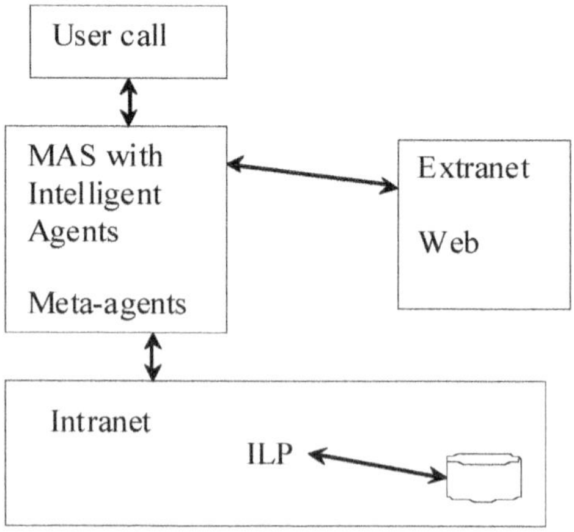

Fig. 7. Architecture of MAS

The process starts with customers ordering manufactured products. The orders are handled by the system, where each order becomes a planned action. The action launches the production line, which will begin to look for the parts and amount of pieces used in manufacturing. Each time a part is needed, the multi-agent system is invoked and the intelligent agents and meta-agents are activated. As an example, let us look at an order from customer ABC on 500 pieces of end-item #1. An intelligent agent finds the corresponding Bill of materials (BOM), which describe the "parts list" of components needed to complete a saleable end-item. The result can be 1500 pieces of item #1234, 1000 pieces of #2345 and finally 500 pieces of items #3456 and #4567 each. This performance is monitored by the meta-agents.

In the case of finding information from suppliers, the intelligent agents search for information at the intranet or the extranet depending on the requested part. The intelligent agents move between the manufacturing company and the different suppliers that are offering the parts for sale, and ask for information about the particular part. If these suppliers are subsidiaries, the agents are looking for information at the intranet, otherwise, at the extranet. Intranet is used when the agents need information from the warehouse of the manufacturing company.

When the intelligent agents identified all the suppliers offering the raw material, they need to find out the price for the pieces. Moreover, the agents need to know the amount of parts to order and find out if the suppliers have the parts in their stock or need to order these, in their turn. Also the price can change with the quantity and the agents must derive the price for the entire stock.

Finally, the delivery date is important. It is the supplier that sets this date but the agents need to consider the date and keep it stored. Moreover, if several customers want to buy the same product from the manufacturing company, the intelligent agents need to find enough raw materials for the product to cover the orders from all customers. This requires buying extra material from the suppliers and, moreover, might require buying the material from several different suppliers.

The results from the intelligent agents are compared to find the best options for the products. Moreover, commonly several raw material parts are used to produce a product and all these pieces are important to assemble the final product and to calculate the delivery time to the customer. Since many intelligent agents are involved in the process, it is too complex to let them compare all the information about the raw material collected at the intranet and the extranet. Therefore, the meta-agents are applied to the intelligent agents.

The meta-agents correspond to goals of the intelligent agents. The intelligent agents work to find information on the parts and report to the meta-agents. Since many agents are involved, a strong part of the meta-agent needs to be a "reasoner" that evaluated alternative strategies. As the information reaches the meta-agents, they need to select the best solutions based on the user-given attributes such as price, time, quantity, and quality. Hence, the user describes the attributes that the raw material needs to meet and the meta-agents have to combine these attributes to select the best option. For the combination, calculation is used.

The calculation for the combination is complex. To evaluate the different attributes, they are given values within a range between 1-5, with the linguistic from of extremely significant to total irrelevant. The formula for the costs is:

Combination(c) -> Attribute (x) X Attribute (y)

Where x,y ={1-5}

The combination is then used together with the cost of the raw material to meet the conditions to the best possible price. It has to be a balance of the cost, i.e., the material versus the sacrifice of other aspects. For example, a cheap product may not be the best in quality and vice versa. Quality versus price raises questions, such as, is the quality more important than the price and if so, what is the acceptable price for the material. Also time can be a cost for the company, which is why a slightly more expensive raw material can be acceptable for a particular production. Another aspect is quantity. If the company can buy a larger quantity, which might be more than needed for the manufacturing, they may get an offer that is price-worthy. However, this will require storage, which also is a cost. If it is a seldom used product, the manufacturing company should not buy a larger quantity. The agents need to be aware of the products utility.

In additional to keeping track of the amount of raw material, the meta-agents also need to keep track of the inventory of the pieces in the warehouses. The parts in the warehouse together with new material should barely cover the orders from the customers. Especially, for the companies that consider the importance of keeping the efficiency of lean production and reduce the stock in warehouses.

Up till now, we have only presented the process of finding information about suppliers for one part of raw material. However, in manufacturing, usually several parts are involved. The meta-agent needs to be the controlling agent, which holds information about all the parts that are being assembled. Thus, they need to direct the tasks of finding different suppliers for these parts to the intelligent agents. Then, from the results of intelligent agents, the meta-agents need to combine all parts to assemble the product and calculate the total cost and delivery time for the manufacturing. This combination is also used for planning the manufacturing. To accomplish the search of all parts needed for manufacturing the product, information about the parts is essential. There are many ways to represent the information about, e.g., bills of material, frames and rules. For example, when the goal is to build a widget, we need to have a part-whole relation to describe the parts of the widget. Note that a widget can be either information or a physical thing and that some widgets may have multiple possible parts. Either, it is possible to buy a set of parts to build a subassembly or buy the finished assembly.

Another important role of meta-agents is to keep record of the customers. The meta-agents keep track of the information about the quantity of the manufactured products, the customer ordered. They also need to be aware of the information about requested delivery time of each product. Finding the raw material for all products, planning the manufacturing and calculate the delivery is a scheduling task for the meta-agents.

In the case of changes, such as the manufacturing, e.g., a broken machine that needs repairing or the customer have made changes in the order, the meta-agents need to react. The meta-agents have to take necessary steps to handle the situation, and depending on the problem, there will be different options. If they are machine problems, the meta-agents need to recalculate the delivery time. In the case of changed orders, they may redo the complete process from searching for the parts needed to planning manufacturing and scheduling the delivery. Another task for the meta-agents is to check the producing capacity. If a customer, for example, extends the order with 50 more pieces, the meta-agents have to check if the producer has sufficient capacity for the additional quantity. Capacity concerns both personnel and machines, and the suggested solution to the customer may be to delay delivery time a number of days.

6 Presenting the Information

An important part of the multi-agent system is to present the result to the users. For this we use similar diagrams of UML, more precisely sequence diagrams and collaboration diagrams [8; 16].

The purpose of using sequence diagrams is that it suits well when illustrating static information of knowledge bases [18], which can also be used for production systems and show strategies [19]. Collaboration diagrams on the other hand are good at illustrating dynamic phenomena, as inserted data can affect executing order [18].

The sequence diagrams are used to show the parts found in the manufactured product. These diagrams are static in the sense that it is always the same parts assembled to constitute the product.

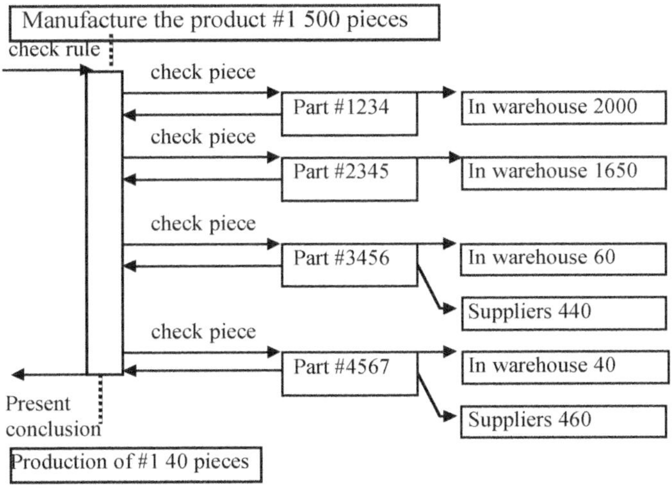

Fig. 8. Sequence diagram illustrating the manufacturing chain

The sequence diagram is used illustrate the manufacturing chain by showing the raw material and time aspect. The sequence diagram in Figure 8, illustrates the parts that are needed to manufacture the product #1. Moreover, the diagram shows where the parts can be found, that is either in the warehouse or at suppliers. In this case, the part #1234 and the part #2345 are in the stock already. The part #3456 and the part #4567 are not in the warehouse and need to be supplied for. The amount of pieces to produce is presented in the diagram. The order is 500 pieces, "Manufacture the product #1 500 pieces. Then, to the left in the figure, the pieces found in the warehouse are presented but also how many must be purchased. Finally, the conclusion is presented, which is the amount of parts that can be produced directly from the pieces found in the warehouse.

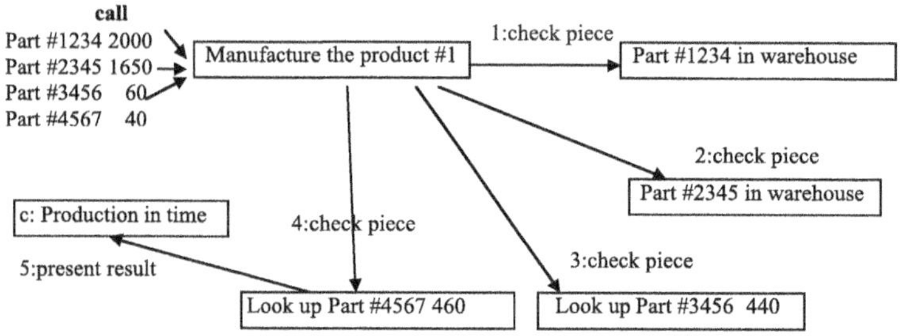

Fig. 9. Collaboration diagram illustrating the manufacturing chain

Also the collaboration diagram is used illustrate the manufacturing chain by showing the raw material and time aspect. The collaboration diagram in Figure 9 illustrates how the manufacturing chain looks for parts to be used for manufacturing the product. If all the pieces are present, the product can be delivered in time, which is also the result presented in the diagram. The collaboration diagram is a simplified diagram but, yet, it illustrates what is needed to assembly the product and what is found in the warehouses. The call to the diagram is two different parts provided by external suppliers. The square with look up pieces is to search for the pieces in the warehouse.

7 Conclusions

In this chapter, we have presented a combination of theories of MAS, including agents and meta-agents, and information logistics in order to search, coordinate and deliver information at the right time and to the right place at an acceptable cost. The information flows between the manufacturing company, its customers, and its suppliers in the form of orders, requirements and production plans, which can speed up the information exchange and optimize the production cost.

Currently, the MAS send agents and receive information. From the performance of these agents, meta-agents are built. However, neither the agents nor the meta-agents are tested on the business area, demonstrating the extraction of information from the suppliers and delivering information the to customers. That is the next step in the research. The problem is that we do not know what kind of systems and ontologies, the different manufacturing companies use today. We have found a range of ontologies but there are no standards for the structure and the contents of these ontologies. These ontologies need to be investigated further. Moreover, we need to investigate what other systems present information at the companies' web sites. The majority use SAP or other similar system, which prevents us from scanning the information on the web sites. These systems give our current system problems.

Acknowledgement

This work is supported by Sweden-Korea Research Cooperation Programme funded by STINT, The Swedish Foundation for International Cooperation in Research and Higher Education. http://www.stint.se/

References

1. Rayport, J., Sviokla, J.: Exploiting the virtual value chain. Harvard Business Review 73(6), 75–85 (1995)
2. Russell, S.J., Norvig, P.: Artificial Intelligence: A Modern Approach. Prentice Hall, Upper Saddle River (2003)
3. Apelkrans, M., Åbom, C.: Information logistics in a process view. In: Proceedings to Hawaiian International conference on business (2002)
4. Frauenhofer Institute, `http://www.isst.fhg.de/EN/research_focus/ilog/ilog/index.jsp`
5. Costantini, S.: Meta-reasoning: a survey. In: Kakas, A., Sadri, F. (eds.) Computational Logic: From Logic Programming into the Future: Special volume in honour of Bob Kowalski (in print), Springer, Berlin (2002)
6. Apelkrans, M., Braf, E.: Content Management: An Integrated Approach of Information Logistics and Knowledge Management. In: Proceedings of the Is One World conference, Las Vegas (2004)
7. Apelkrans, M., Håkansson, A.: Visual knowledge modeling of an Information Logistics Process - A case study. In: ICICKM-2005, 2nd International Conference on Intellectual Capital, Knowledge Management and Organisational Learning Dubai, Förenta Arab Emiraten (2005)
8. Apelkrans, M., Håkansson, A.: Enterprise systems Configuration as an Information Logistics Process - A Case Study. In: Cardoso, J., et al. (eds.) Proceedings of 9th International Conference on Enterprise Information Systems, ICEIS 2007, INSTICC, Portugal, pp. 212–220 (2007)
9. Apelkrans, M., Håkansson, A.: Information coordination using Meta-Agents in Information Logistics Processes - A Study. In: Lovrek, I., Howlett, R.J., Jain, L.C. (eds.) KES 2008, Part III. LNCS (LNAI), vol. 5179, pp. 788–798. Springer, Heidelberg (2008)
10. Reijers, H., Limam Mansar, S.: Best practices in business process redesign: An overview and qualitative evaluation of successful redesign heuristics. Omega: The International Journal of Management Science 33(4), 283–306 (2005)
11. Minglun, R., Weidong, Z., Shanlin, Y.: Data oriented analysis of workflow optimization. In: Proceedings of the 3rd World Congress on Intelligent Control and Automation, vol. 4, pp. 2564–2566 (2000)
12. Sandkuhl, K.: Information logistics in networked Organisations Issues, Concepts and Applications. In: Proceedings of 9th International Conference on Enterprise Information Systems, ICEIS 2007, INSTICC, Portugal, pp. 23–30 (2007)
13. Wooldridge, M.: An Introduction to MultiAgent Systems. John Wiley & Sons Ltd., Chichester (2002)
14. Pechoucek, M., Stepánková, O., Marík, V., Bárta, J.: Abstract Architecture for Meta-reasoning in Multi-agent Systems. In: Mařík, V., Müller, J.P., Pěchouček, M. (eds.) CEEMAS 2003. LNCS (LNAI), vol. 2691, p. 84. Springer, Heidelberg (2003)

15. Chelberg, D., Welch, L., Lakshmikumar, A., Gillen, M., Zhou, Q.: Meta-Reasoning For a Distributed Agent Architecture, http://zen.ece.ohiou.edu/~robocup/papers/HTML/SSST/SSST.html
16. Håkansson, A., Hartung, R.: Calculating optimal decision using Meta-level agents for Multi-Agents in Networks. In: Apolloni, B., Howlett, R.J., Jain, L. (eds.) KES 2007, Part I. LNCS (LNAI), vol. 4692, pp. 180–188. Springer, Heidelberg (2007)
17. Håkansson, A., Hartung, R.: Using Meta-Agents for Multi-Agents in Networks. In: Arabnia, H., et al. (eds.) Proceedings of The 2007 International Conference on Artificial Intelligence, ICAI 2007, vol. II, pp. 561–567. CSREA Press, U.S.A. (2007), http://www.world-academy-of-science.org
18. Håkansson, A.: UML as an approach to Modelling Knowledge in Rule-based Systems. In: ES 2001: The Twenty-first SGES International Conference on Knowledge Based Systems and Applied Artificial Intelligence, December 10-12, Peterhouse College, Cambridge (2001)
19. Håkansson, A.: Transferring Problem Solving Strategies from the Expert to the End Users - Supporting understanding. In: Proceedings of 7th International Conference on Enterprise Information Systems, ICEIS-2005, INSTICC, Portugal, vol. II, pp. 3–10 (2005)

Social Network Analysis Based Approach to Trust Modeling for Autonomous Multi-agent Systems

Grzegorz Kołaczek

Wroclaw University of Technology,
27 Wyb.Wyspianskiego, 54-210, Wroclaw, Poland
Grzegorz.Kolaczek@pwr.wroc.pl

Abstract. The chapter presents a novel concept of trust modeling and analysis method for autonomous multi-agent systems. The method is based on assumption that multi-agent system constitutes social network and so it is an instantiation of a complex network. Two well-known and much studied classes of complex networks are scale-free networks and small-world networks. Small-world and scale free networks for example can be characterized by the clustering coefficient and the average network distance. The clustering coefficient is the probability that any two nodes are connected to each other, given that they are both connected to a common node. The average network distance measures the average minimal number of links connecting any two nodes in the network. When we consider multi-agent system all these parameters describing small-world or scale-free networks can be also used to analyze the position and relations among agents within their society.

1 Introduction

One of the most important factors in human interaction and communication is trust. Trust is also very important feature for all autonomous multi-agent systems. Performing their activities agents collaborate with other agents this means that they obtain and process data provided by them. So, the final decision or action performed by the agent strongly depends on the quality of the previously obtained data. The evaluation of the risk related to the agent's decisions in a context of uncertainty about data quality provided by other agents is one of the most important problems and so interesting research area in multi-agent systems.

Trust and reputation measure gives an idea of the confidence one can have on the quality of an agent's responses [32]. The disadvantage of typical trust and recommendation evaluation mechanisms is that they need the explicit and frequent involvement of users that issue ratings. This implies that a good reputation calculation and maintenance depends on the involvement of users and continued contribution of ratings. Less intrusive and less demanding in terms of users involvement methods are more interesting. The problem is how reputation can be measured in the absence of user feedback for agent's responses.

The chapter presents a novel concept of trust modeling and analysis method for autonomous multi-agent systems which increases automation and quality of trust level evaluation and evolution. The proposed method is based on such an assumption that

A. Håkansson et al. (Eds.): Agent and Multi-agent Technology, SCI 289, pp. 137–156.
springerlink.com

multi-agent system constitutes social network and so it is also an instantiation of the complex network. There are two main types of complex networks the small world and the other one is scale-free networks [2,23]. Small-world networks can be characterized by the clustering coefficient and the average network distance. The clustering coefficient is the probability that any two nodes are connected to each other, given that they are both connected to a common node. The average network distance measures the average minimal number of links connecting any two nodes in the network. Many regular networks have high clustering coefficients and large network distances. Random networks, on the other hand, have small network distances and low clustering coefficients [4,29]. Small-world networks are somewhere in between these two extremes as they have large clustering coefficients and small average network distances [14,28]. The scale-free property of the complex network is defined by an algebraic behaviour in the probability distribution P(k) of the number k of links at a node.

When we consider multi-agent system all these parameters characteristic for small-world or scale-free networks can be used to analyze the position and relations among agents within their society. The other important attribute of the method is application of Subjective Logic formal description to the agent's opinions expression and so to evaluation of trust level [8]. Another assumption that has been made is that the trust has subjective nature and so that it is possible to consider three elementary components that influences each agent's opinion:

- Belief – the world around us will follow our intuitions,
- Disbelief – the world around us will not follow our intuitions,
- Uncertainty – the world around us is undefined, unknown and we are reluctant to make any categorical statement about its future behavior.

One of the consequences of application Subjective Logic to reasoning about trust are such that agent's A trust level in an agent's B statements can be obtained as the Subjective Logic consensus of modified trust levels of all agents that trust the agent B. The trust level modification is performed in relation to the current position of the agent in the network. The precise description of the method and corresponding algorithms are presented in the next sections of this chapter.

In the next section of the chapter some basic concepts of trust modelling in information systems are presented. After that, related works are described and the basic notions and concepts related to social networks metrics has been analyzed, especially in relation to the trust modelling task. Subjective Logic as the one of the important elements of the novel trust modelling system has been introduced in the subsequent section. The next section presents the main contribution of the chapter, algorithms and examples of trust level evaluation using selected complex networks parameters. After the presentation of algorithms there has been introduced another new proposition – the trust evolution modeling. The last section summarizes the chapter results.

2 Trust and Information Systems Security

There are two main reasons why the trust is so important in contemporary information systems security. The first one is that there is no information system we could believe that is 100% secure. The consequence of this is that we cannot be completely sure about the subject identity and its true intensions. The second is that many of the information systems are a part of open networks. These networks allow subjects to communicate without any prior arrangements like for example organizational membership and so this also makes information authenticity difficult to verify [21].

The level of an autonomous agent uncertainty about other agents intensions, behavior, etc. may be expressed and evaluated by trust level. However there are a few problems on the way to define the notion of trust formally [30]. The first one is that there are some differences in defining of the most fundamental nature of trust. Second difficulty arises when a set of attributes which values decide about a final trust level must be calculated. The third and probably the most important one is that trust will always have some subjective aspects that always create problems in formal usage. This idea was expressed by A.Jøsang in the following way: "we claim that there can be no other measure for security and authenticity than subjective trust" [7,8,9,10,11,12].

Apart from many controversial points about trust formal modeling there are also some elements that are commonly accepted. These aspects are as follows [5,15,16]:

- Trust is subjective
  Trust is the social knowledge that is derived from personal observation and serves for future personal decision-making.
- Trust is affected by some actions that cannot be monitored
- Trust is context dependant
  Trust can have different meaning dependently from the context. For example, one can trust TV news when it has been broadcast by X but not when it has been broadcast by Y.
- Trust level strictly depends on how our actions are affected by the others.
- Trust is directed.
  Trust relation is always composed by two entities: trustor and trustee, so the relation is asymmetric. A trusts B does not implies that B trusts A.
- Trust is measurable belief.
  Trust is usually associated with a value that represents the strength or degree of trust. A can trust B more than it trusts C.
- Trust is conditionally transitive.
  When we consider trust as the recommendation factor it is partially transitive.
- Trust has a temporal dimension.
  Trust is derived from the past observations and trust values evolve with a new observations and experience. The behavior of B may force the A to recalculate its trust
- Trust between collectives does not distribute automatically to trust between their members.
  If a state A trusts the state B it does not imply that all state members trust state B.

The other important idea related to trust is reputation. The most typical real life context of reputation is when someone needs to work with a new, unknown person. Then, the typical practice is to ask people with whom he/she already has relationship for the information about such person. Based on the obtained information one can form an opinion about the reputation of the new person. Further, a positive reputation leads to confidence or trust in the person while negative reputations leads to lower confidence or social status. How it could be observed in real life societies the reputation propagation works well and it is true even in cases of large societies. The main source of this effect is related to the structure of societies where members tend to be highly interconnected and the topological distance between society members is small. This propriety is know as the Small World effect and it has been proved to be true for a many different real and virtual communities.

The formal representation of trust should create possibility to use this notion in a context of information systems in a similar way as it is being used in reality. If only we could have a good, functional and working model of trust for virtual societies it would be possible for agents to use trust in a similar way as in real societies and so also to perform their tasks more effectively.

The most general idea about representing trust for security system purposes states that trust level must be estimated according to [13,14]:

- agent's own experience,
- recommendation from other agents,
- agent's context (history, time, location, etc.).

The presented in subsequent sections algorithms assume that all three above mentioned components are important for achieving the intended effects.

3 Related Work

Trust is very important feature of security engineering. In this context it is related to risk management, surveillance, auditing and communication. Extensive knowledge on security engineering has already been collected and analyzed by Taipale [24] and has been studied in the Trusted Systems [22] project, which is a part of Global Information Society Project [22] lead by the World Policy Institute. It investigates systems in which some conditional prediction about the behavior of people or objects within a system that has been determined prior to authorizing access to system resources.

Another research area related to this topic is the concept of "web of trust" systems. This concept emerged in applications related to cryptography and is an element of such technologies like PGP, OpenPGP-compatible or public key infrastructure. They have been defined to support the trust endorsement of the PKI generated certificate authority (CA)-signed certificates.

Finally, the most popular research area in the context of trust is called a trust metric. The aim of it is to propose a measure of how a member of a group is trusted by other members. An exemplary overview of such metrics has been prepared on the Internet community at TrustMetrics Wiki and it presents a brief classification and provides many examples.

There are many different metrics that are diverse in many aspects. For example TrustMail or FilmTrust [5] propose to take advantage of a Semantic Web-based social network. Other different approaches are based on graph walking or use subjective logic to express trust level [8]. In the PeerTrust Project [33] a decentralized Peer-to-Peer electronic community has been considered and a proposed trust model considers only three factors: the amount of satisfaction established during peer interaction, the number of iterations between peers and a balance factor for trust. At the other hand, the EigenTrust [25] algorithm has similar ideas to PageRank [20] but has been used in the context of file-sharing systems. This method computes global trust for peers, where the value is based on the history of uploads. It is dedicated to support the system to choose the peers with a history of reliable downloads and to exclude the malicious peers from the network.

The trust and reputation of subjects has been typically assessed as a function of the quality of their response to requests coming from other members in the community. This approach is used in some organizational learning systems as, for example, Answer Garden [1] or knowledge communities [26]. Discussion about the different ratings that can be obtained by analyzing response quality is to be found in [17,19]. These systems rely on ratings provided as a feedback from the subject receiving the response to a previous demand. Subsequently these ratings are combined and finally each subject calculates its own trust value. Trust and reputation measure gives an idea of the confidence one can have on the quality of a subject's responses. The disadvantage of this type of mechanism is that it needs the explicit and frequent involvement of users that issue ratings. This implies that a good reputation calculation and maintenance depends on the involvement of users and continued contribution of ratings. Less intrusive and less demanding in terms of users involvement methods are more interesting. The problem is how reputation can be measured in the absence of any user feedback for subject's responses.

Although much of the previous related work presented in this chapter is related to trust metrics, the proposed approach differs from them with regard to several aspects. The described in this chapter trust level evaluation method for virtual communities takes advantage of the capabilities of the social network analysis. The presented method enables the autonomous agent to combine recommendation data obtained from other agents and data related to the agent's context. As the area of possible contexts of agent's interactions is very diverse we will focus on agent's position in agent's society described by its location in the social network.

4 Social Network Metrics

The basic idea about social networks is very simple. It could be understood as a social structure made of actors which can be represented as network nodes (which are generally individuals or organizations) that are tied by one or more specific types of interdependency, such as values, visions, idea, financial exchange, friends, kinship, dislike, conflict, trade, web links, etc. The resulting structures are often very complex [5]. Social relationships in terms of nodes and ties among them could be used in various types of analysis. A number of academic researches have shown that dependences form social fields play a critical role also in many other fields and could be used in determining the way problems could be solved.

Better understanding of social networks requires a complete and rigorous description of a pattern of social relationships as a necessary starting point for analysis. The most convenient situation is when we have complete knowledge about all of the relationships between each pair of actors in the population. To manage all pieces of information related to social network the mathematical and graphical techniques have been used. This formal apparatus allows us to represent the description of networks compactly and systematically. In this context, social network analysts use two kinds of tools from mathematics to represent information about patterns of ties among social actors: graphs and matrices.

Network analysis uses one kind of graphic display that consists of nodes to represent community members and edges to represent ties or relations. There are two general types of situation when there are a single type of relations among the community members and more than one kind of relation. The first one can be represented by simplex graph while in the second case we use multiplex graphs. Additionally, each social tie or relation represented by graph may be directed or undirected (tie that represents cooccurrence, co-presence, or a bonded-tie between the pair of community members). Another important feature related to the social networks and their graph representation is the strength of ties among community members. In a graph it may be one of the following types: nominal or binary (represents presence or absence of a tie); signed (represents a negative tie, a positive tie, or no tie); ordinal (represents whether the tie is the strongest, next strongest, etc.); or valued (measured on an interval or ratio level).

Other basic social network proprieties that can be formally described and so can constitute a good background for analysis of community dynamics and also trust modelling are as follows [23]:

- Connections
  - the number of immediate connections may be critical in explaining how community members view the world, and how the world views them, so it could be also important factor while modeling trust relations within community
  - the number and kinds of ties are a basis for similarity or dissimilarity to other community members
  - the direction of connections may be helpful to describe the role of the community member in the society, it can be "a source" of ties , "a sink", or both.
- Size
  - the size of a network is indexed simply by counting the number of nodes,
  - critical element for the structure of social relations because of the limited resources and capacities that each community member has for building and maintaining ties
  - in a bigger groups more likely it is that differentiated and partitioned subgroups will emerge
  - the size of a network also influences trust relations while in bigger group it is easier to preserve anonymity and its more difficult to evaluate trust values
- Density
  - the proportion of all ties that could be present to that actually are
  - the number of logically possible relationships then grows exponentially as the number of actors increases linearly

- in communities with greater value of density parameter it should be easier to maintain trust relations as we get more information about the other community members

- Degree
  - it tells us how many connections a community member has
  - out-degree - the sum of the connections from the community member to others
  - in-degree - the sum of the connections to the particular community member from others
  - community members that receive information from many sources may also be more powerful
  - community members that receive a lot of information could also suffer from "information overload" or "noise and interference" due to contradictory messages from different sources
  - impact for the trust relation is similar to that described in a case of density, dependently from in/out-degree when an individual has more or less information about its neighborhood

- Reachability
  - a community member is "reachable" by another if there exists any set of connections by which we can find link from the source to the target entity, regardless of how many others fall between them
  - if some community members in a network cannot reach others, there is the potential of a division of the network
  - disconnected community members could have more problems to evaluate trust value

- Transitivity
  - the transitivity principle holds that, if A is tied to B, and B is tied to C, then A should be tied to C.
  - the triadic relationships (where there are ties among the actors) should tend toward transitivity as an equilibrium condition
  - the trust relation is not strictly transitive and so this propriety not necessarily influences trust evaluation process

- Distance
  - an aspect of how individuals are embedded in networks,
  - two actors are adjacent when the distance between them is one
  - how many community members are at various distances from each other can be important for understanding the differences among community members in the constraints and opportunities they have as a result of their network location
  - community members located more far apart from each other in the community have more problems with trust evaluation than the members which are close

- Geodesic distance
  - geodesic distance is defined as the number of relations in the shortest possible walk from one community member to another

- many algorithms in network analysis assume that community members will use the geodesic path when communicating with each other
- it could be useful to find the exact trust value between two community members

- Diameter
  - the diameter of a network is the largest geodesic distance in the connected network
  - tells us how "big" the community is, in one sense quantity in that it can be used to set an upper bound on the lengths of connections that we study
  - the diameter value could be an important coefficient while calculating trust value
- Cohesion
  - the strength of all links as defining the connection
  - count the total connections between actors
  - more strong connection between community members should determine grater trust values
- Centrality, Power, Betweenness
  - centrality , closeness, betweenness describe the locations of individuals in terms of how close they are to the "center" of the action in a network – though there are a few different definition of what it means to be at the center
  - the more important community member is, the more important its opinions should be, it should be also reflected while establishing new trust relations
- Eigenvector of the geodesic distances
  - an effort to find the most central community members in terms of the "global" or "overall" structure of the network, and to pay less attention to patterns that are more "local"
- Cliques
  - a sub-set of community members who are more closely tied to each other than they are to community members who are not part of the group.
  - idea of cliques within a network is a powerful tool for understanding social structure and the embeddedness of individuals
  - cliques reflects the groups of community members with strong relationship so in natural way it also influences trust relations between them

As it was depicted here above most of the typical parameters used in social networks description has more or less strict relation with trust modelling problem. The basic assumption about trust model presented in this work is also that some of these typical for social networks parameters can be used to improve trust modelling.

5 Subjective Logic

Subjective logic was proposed by Josang as a model for reasoning about trust propagation in secure information systems. It is compatible with Dempster-Shafer's theory of evidence [8] and binary logic [7]. Subjective logic includes standard logic operators and additionally two special operators for combining beliefs – *consensus* and *recommendation*. The basic definitions of subjective logic given in this section come from [10,11].

Subjective logic has been used to express so-called *opinions* (see below) about facts with assumptions that we do not require the knowledge of how these facts were grounded or inferred. We may also have an opinion about some subject (source of information). These features make Subjective Logic a promising formalism for the use within trust level evaluation architecture.

When expressing belief about a statement (predicate) it is assumed that it is either true or false, but we're not certain about it – we are only able to have an opinion about it (because of our imperfect knowledge). Let's denote *belief*, *disbelief* and *uncertainty* as b, d and u respectively.

Definition 1. A tuple $\omega = \langle b,d, u\rangle$ where $\langle b,d, u\rangle \in [0,1]^3$ and $b + d + u = 1$ is called an opinion.

As we see, the three components of an opinion (*belief*, *disbelief* and *uncertainty*) are dependent and one of them is redundant. In spite of this it is useful to use all three of them, because in this way we may obtain relatively simple expressions when defining subjective logic operators. Opinions with uncertainty component equaling to 0 are called *dogmatic opinions*. They correspond to traditional frequentist probabilities and represent situations without uncertainty. Opinions with either $b = 0$ or $d = 0$ are referred to as *absolute opinions*, and correspond to TRUE and FALSE propositions of binary logic.

From Definition 1 we may also graphically express an opinion as a point belonging to so-called *opinion triangle* (Figure 1. - point ω marks opinion $\langle 0.8, 0.1, 0.1\rangle$).

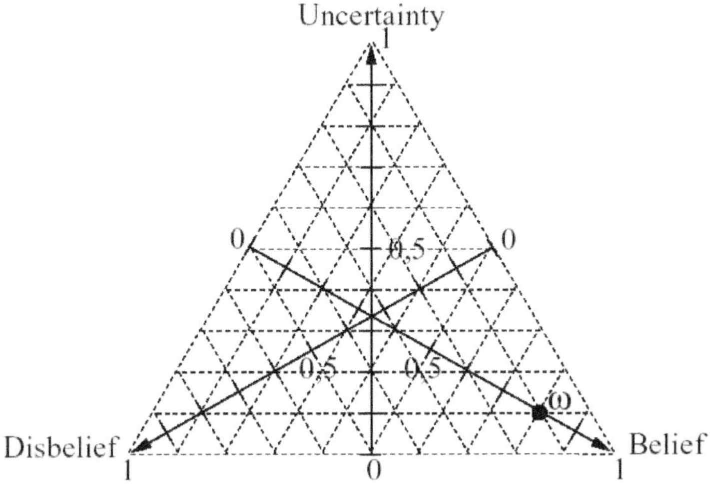

Fig. 1. The opinion triangle

Opinions have always assigned membership (are expressed by certain agents) and are not inherent qualities of objects but *judgments* about them (that feature will be used in the next section to define opinion about concepts correspondence on the basis of the results of similarity assessments).

For any opinions $\omega_p = \langle b_p, d_p, u_p\rangle$ and $\omega_q = \langle b_q, d_q, u_q\rangle$ about logical value of predicates p and q the following operators (equivalent to well-known operators of binary logic) may be defined (proofs and in-depth discussion were presented by Josang (2001):

Definition 2. (Conjunction)

$$\omega_{p\wedge q} = \left\langle b_p b_q, d_p + d_q - d_p d_q, b_p u_q + u_p b_q + u_p u_q \right\rangle \tag{1}$$

Definition 3. (Disjunction)

$$\omega_{p\vee q} = \left\langle b_p + b_q - b_p b_q, d_p d_q, d_p u_q + u_p d_q + u_p u_q \right\rangle \tag{2}$$

Definition 4. (Negation)

$$\omega_{\neg p} = \left\langle b_p, d_p, u_p \right\rangle \tag{3}$$

For absolute opinions both conjunction and disjunction operators give the same results as AND and OR of binary logic. For dogmatic opinions they return the results of product and co-product of probabilities respectively.

Now assume two agents, A and B, where A has an opinion about B. Opinion expressed by A about other agent B is interpreted as opinion about proposition "*B's opinion is reliable*". We'll denote opinion expressed by agent B about given predicate p and agent's A opinion about B as ω_p^B and ω_B^A respectively. Assuming that ω_p^B and ω_B^A are known, the opinion of agent A about p is given by *discounting operator* (also known as *reputation* operator):

Definition 5. (Recommendation, denoted by ⊗)

$$\omega_p^{AB} = \omega_B^A \otimes \omega_p^B = \left\langle b_B^A b_p^B, b_B^A d_p^B, d_B^A + u_B^A + b_B^A u_p^B \right\rangle \tag{4}$$

It may be proved that recommendation operator is associative but not commutative. This implies that the order of opinions in recommendation chains is significant. The joint opinion of two agents A and B about given predicate is computed by *consensus* operator:

Definition 6. (Consensus, denoted by ⊕)

$$\omega_p^{AB} = \omega_B^A \oplus \omega_p^B = \left\langle \left(b_p^A u_p^B + b_p^B u_p^A\right)/k, \left(d_p^A u_p^B + d_p^B u_p^A\right)/k, u_p^A u_p^B / k \right\rangle \tag{5}$$

where $k = u_p^A + u_p^B - u_p^A u_p^B$

Consensus operator is commutative and associative thus allowing to combine more opinions. Note that ⊕ is undefined for dogmatic opinions, reflecting that there is no joint opinion, if agents are certain about given fact. However, in this case the limits do exist and may be computed for all three components of resulting opinion. Recommendation and Consensus require also assumption that elementary opinions used are independent (i.e. we cannot take into account opinion of any of the agents more than once).

Opinions about binary events can be projected onto a 1-dimensional probability space resulting in *probability expectation* $E(\omega_p)$ value for a given opinion:

Definition 7. (Probability expectation)

$$E(\omega_p) = E(\langle b, d, u \rangle) = b + \frac{u}{2} \tag{6}$$

From (6) we see that probability expectation function removes information contained in an opinion – there may be infinitely many opinions with the same values of probability expectation functions E.

The opinions may also be ordered. For any two opinions ω_1 and ω_2 about predicate p, we say that ω_1 is *greater* than ω_2 (denoted by $\omega_1 > \omega_2$) if ω_1 expresses stronger belief that p is true. Thus, when ordering opinions the following rules (listed by priority) hold:

1. The opinion with the greatest probability expectation E is the greatest.
2. The opinion with the smallest uncertainty is the greatest.

For instance: $\langle 0.5, 0, 0.5 \rangle > \langle 0.4, 0.2, 0.4 \rangle > \langle 0.2, 0, 0.8 \rangle$.

6 Model of Trust in Complex Networks

The term "complex network" refers to a network that has certain non-trivial topological features that do not occur in simple networks. Most social and technological networks can be considered complex by virtue of non-trivial topological structure (e.g., social network, computer network). Among these non-trivial features are: a heavy-tail in the degree distribution; a high clustering coefficient; assortativity or disassortativity among vertices; community structure at many scales; and evidence of a hierarchical structure [8].

Some of these features can be very useful in a case of trust modelling and management. Two parameters has been selected that are frequently used to characterise complex network structure: clustering coefficient and centrality.

The clustering coefficient of a node in a network quantifies how close the node and its neighbours are from being a clique. This measure has been introduced by Duncan J. Watts and Steven Strogatz [27,28] to determine whether a graph is a small-world network. At the other hand, the clique is defined as an exclusive group of people who share common interests, views, purposes, or patterns of behaviour [8,12]. A clique is a subset of individuals from a larger group, who are more closely identified with one another than the remaining members of the group, and who exchange something among themselves, such as friendship, affection, or information [13].

In a context of trust analysis existence of cliques in a trust network or a fact that a particular node is a member of the clique can be interpreted in two completely different ways. Positive interpretation is that a member of the clique has deeper knowledge about other clique members so its recommendation is more trustworthy. However in negative scenario we may assume that the clique members can collaborate to cheat other agents.

This ambiguity is the reason why in the proposed trust evaluation method the clique membership is an element of *uncertainty* component of the agent's opinion.

The numerical value of clustering coefficient can be calculated using for example following formula:

$$C_i = \frac{\left|\{e_{jk}\}\right|}{k_i(k_i-1)} : v_i, v_k \in N_i, e_{jk} \in E \qquad (7)$$

where:

- v_i , - nodes,
- N_i – neighbourhood of the node v_i ,
- e_{jk} –link between v_i and v_k,
- k_i – degree of the node v_i

Let us define a *CLIQUE* as the value of the clustering coefficient for a particular network node. In further discussion the parameter *CLIQUE* will be used in evaluation of agent's opinion *uncertainty* component. This parameter should be understood as the normalised value of clustering coefficient calculated using selected method (e.g. formula (4)) for a given node.

The second parameter - centrality determine the relative importance of a vertex within the graph (e.g. how important an agent is within a social network). There are four main measures of centrality that are used in network analysis:

- degree centrality,
- betweenness,
- closeness,
- eigenvector centrality.

We propose to use one of the well known centrality measures (e.g. Google's PageRank which is a variant of the Eigenvector centrality measure [20]) to describe the level of confidence in agent's opinion. We assume that value of centrality measure is a kind of recommendation which originate from the network structure and so the opinions of the "more central" nodes are much worthy to be believed in.

In the next section the parameter *CENTRALITY* is used to refer to the normalised value of a certain centrality measure for a particular network node.

6.1 Trust Level Evaluation Algorithm

Let us consider a simple multi-agent network with trust relations as it is shown in the Figure 2.

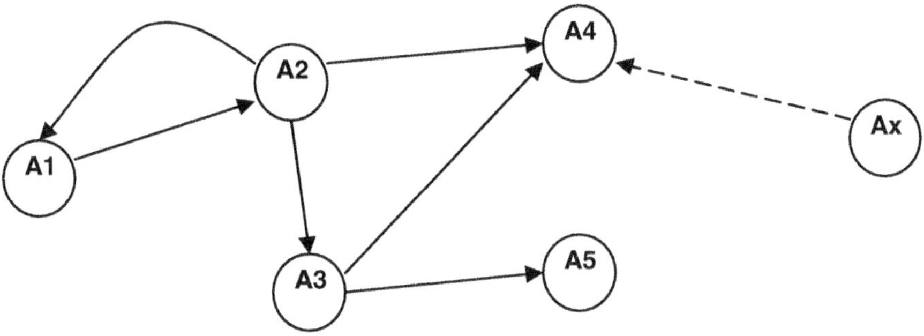

Fig. 2. Simple trust network in multi-agent system

where:

- A1,A2, - autonomous agents with established trust relations
- Ax – a new agent which tries to derive trust of agent A4

At the early stage of each multi-agent system there is a moment where we have several agents and no trust relations. There is no historical records of interaction between agents so the agents can't use their own experience or recommendation to establish trust relations and to calculate trust levels. The only possible source of information that could be used is the agent's context. We assume that in such cases each agent will propose arbitrary some trust level with high values of uncertainty component and this level will be updated according to the further interactions among agents.

The second case, described in the Figure 3. is when a new agent joins the network. The new agent can use the existing relations between agents and their recommendations to elaborate their own opinions. In such situations agent uses some data coming from the structure of existing trust network.

We propose the following algorithm which will enable the new agent to calculate its own opinion about any other agent in the network:

Given:

- An agent *A* who calculate its trust level to the agent *B*,
- trust network – a set of agents' opinions about other agents

Result:

- ϖ_B^A - the agent's *A* opinion about the agent *B*

BEGIN

1. Get all opinions about the agent *B*. Let **C**={C_1, C_2,…, C_k} be a set of all agents that have opinion about *B*.
2. Measure the network context for each agent from the set **C** and present it in a form of Subjective Logic opinion $\varpi_{C_i}^{net} = (b_{netC_i}, d_{netC_i}, u_{netC_i})$.

Where $\varpi^{net}_{C_i}$ is a recommendation and it reflects the agent's C_i position in the network and is calculated as follows:

$$b_{net,C_i} = CENTRALITY \tag{8}$$

$$d_{net,C_i} = 1 - b_{C_iB} - u_{C_iB} \tag{9}$$

$$u_{net,C_i} = \min(1 - CENTRALITY, CLIQUE) \tag{10}$$

CENTRALITY and *CLIQUE* are normalised values calculated using selected complex network metrics as it was discussed at the beginning of the this section.

3. Calculate the modified opinion for each agent from the set **C** using context from Step 2. and Subjective Logic recommendation operation.

$$\varpi'^{C_i}_{B} = \varpi^{net}_{C_i} \otimes \varpi^{C_i}_{B} \tag{11}$$

Where opinion $\varpi^{C_i}_{B}$ is the original opinion of the agent C_i about the agent B

4. Calculate the agent A opinion about the agent B as the Subjective Logic consensus of all modified in the step 3 opinions of agents from the set **C**.

$$\varpi^{A}_{B} = \varpi'^{C_1}_{B} \oplus \varpi'^{C_3}_{B} \oplus \ldots \oplus \varpi'^{C_k}_{B} \tag{12}$$

END

6.2 Trust Level Update Algorithm

The other type of situation that happens during trust network life period is when an existing node tries to establish a new relation with another node.

Let us consider the situation presented in the Figure 3. The agent A3 has established trust relation with the agents A2 and A5 and wants to establish a new relation with the agent A4.

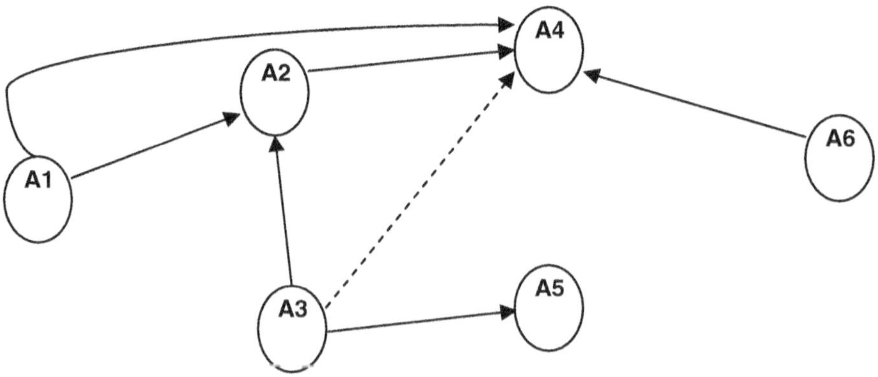

Fig. 3. Trust network update

To evaluate the trust level between A3 and A4 we use the similar parameters as we used in the previous situation and some additional knowledge coming from the trust relations included in the existing trust network. The evaluation process is described by the following algorithm:

Given:

- An agent *A* who calculate its trust level to the agent *B*,
- trust network – a set of agents' opinions about other agents

Result:

- ϖ_{B}^{A} - the agent's *A* opinion about the agent *B*

BEGIN

1. Find in the trust network all paths leading from the agent *A* to the agent *B*.
2. Let **P**={P_1, P_2,..., P_j} be a set of all paths form *A* to *B* where P_i=<A,p_1, p_2,..., p_z,B>
3. Let **D**={D_1, D_2,..., D_k} be a set of all agents that are present in the paths form *A* do *B*.
4. Let **C**={C_1, C_2,..., C_l} be a set of all agents that have opinion about *B* and are not the elements of the set **D**.
5. Measure the network context for each agent from the set **C** and present it in a form of Subjective Logic opinion $\varpi_{C_i}^{net} = (b_{netC_i}, d_{netC_i}, u_{netC_i})$. Where $\varpi_{C_i}^{net}$ is a recommendation and it reflects the agent's C_i position in the network and is calculated as follows:

$$b_{net,C_i} = CENTRALITY \tag{13}$$

$$d_{net,C_i} = 1 - b_{C_iB} - u_{C_iB} \tag{14}$$

$$u_{net,C_i} = \min(1 - CENTRALITY, CLIQUE) \tag{15}$$

   CENTRALITY and *CLIQUE* are normalised values calculated using selected complex network metrics as it was discussed at the beginning of this section.
6. Calculate the modified opinion for each agent from the set **C** using context from Step 5. and Subjective Logic recommendation operation.

$$\varpi'^{C_i}_{B} = \varpi_{C_i}^{net} \otimes \varpi_{B}^{C_i} \tag{16}$$

   Where opinion $\varpi_{B}^{C_i}$ is the original opinion of the agent C_i about the agent *B*

7. Calculate the opinion for each path from the set **P** using the trust network and Subjective Logic recommendation operation.

$$\varpi'^{P_i}_{B} = \varpi^{A}_{p_1} \otimes \varpi^{p_1}_{p_2} \otimes \varpi^{p_2}_{p_3} \ldots \otimes \varpi^{p_z}_{B} \tag{17}$$

Where opinion $\varpi^{p_i}_{p_j}$ is the opinion of the agent p_i about the agent p_j which are the elements of the given path leading from A to B.

8. Calculate the agent A opinion about the agent B as the Subjective Logic consensus of all opinions evaluated in steps 6 and 7

$$\varpi^{A}_{B} = \varpi'^{C_1}_{B} \oplus \varpi'^{C_3}_{B} \oplus \ldots \oplus \varpi'^{C_k}_{B} \oplus \varpi'^{P_1}_{B} \oplus \varpi'^{P_2}_{B} \ldots \oplus \varpi'^{P_l}_{B} \tag{18}$$

END

7 Simulation of Trust Dynamics in Social Networks

This section describes some general ideas how to develop model that can represent evolution and dynamic of the trust relation between agents of multiagent system. This model could be useful in evaluation process of the algorithms presented in earlier sections of this chapter.

Social structure emerges from the interaction and information exchange between members of a given population [3]. Graph model, where nodes represents agents and edges represent communication pathways, can be used to simulate individual influence and the communication of ideas within given population. The interscientific field that investigate behaviour of social groups using simulation tools related to modern physics is called sociophysics. Sociophysics emphasises the underlying similarities between all systems. It's based on random simulations and tries to give answers concerning to societies, such like: how mass-media or "missionaries" influence our beliefs, how various factors determines social hierarchy, what are the chances to establish consensus or when we are threatened with escalation of disagreement. All this make sociophysics a wide-range applicable science [18].

There are several simulation approaches that are being used in social science nowadays. One type of simulation techniques have been derived from mathematics (represented by points on the field with gray background on the figure below), the other techniques (points on the white background), come from computer science, especially from artificial intelligence and automata theory. However, the dividing line between both groups is not that sharp – "cellular automata may be considered as an alternative (to differential equations) and in some respects complementary basis for mathematical models of nature." [17]

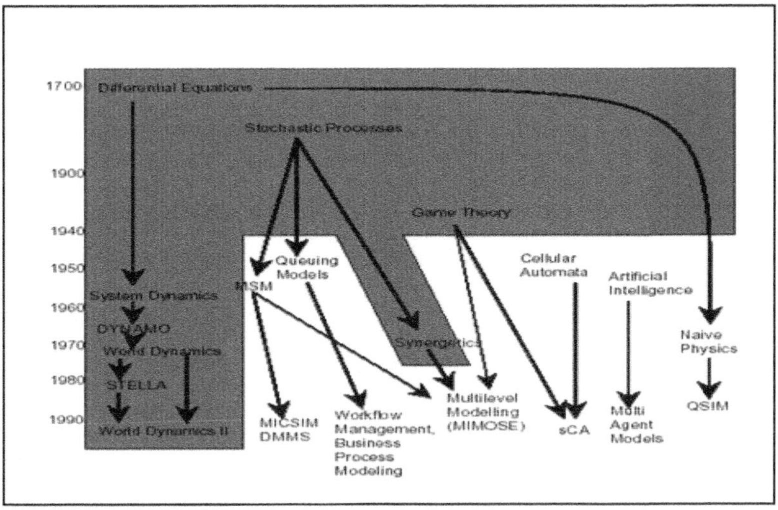

Fig. 4. Historical development of contemporary simulation approaches [18]

Simulation is a type of a thought experiment which is carried out with the help of a computer, but without any direct interface to the target system. Simulation helps to find out answers to several questions from two main categories [31]:

1. Qualitative prediction:
   - will the system stabilize or lock (and in which of several stable states will it do so), will it go into more or less complicated cycles, will it develop chaotic behavior (such that longtime quantitative predictions are impossible)?
   - will the system display some emergent structures like stratification, polarization or clustering?
2. Quantitative prediction:
   - which kinds of behavior can be expected under arbitrarily given parameter combinations and initial conditions?
   - which kind of behavior will a given target system (whose parameters and previous states may or may not have been precisely measured) display in the near future?
   - which state will the target system reach in the near future, again given parameters and previous states which may or may not have been precisely measured?

In a case that is being considered a simulation process focus on investigation of trust formation and flow in network system.

The fundamental level of the model is a graph where agents are represented by network nodes. Each node can be in a number of states. In a presented proposition states are opinions in a sense defined in Subjective Logic and represented by triple {b,d,u}. Edges between the nodes represent the pathways through which information exchange can occur. There must be also a few additional rules to describe the behaviour of agents and topological rules to describe connectivity patterns. According to a scheme proposed in [8] in a context of simulation models of social network structures proposed rules are as follows:

1. General rules – each individual is represented as a threshold automaton.
   - Each node's state (opinion) influences the state of neighbouring nodes to which it is directly connected. Similarly each, node's state is influenced by the state of the neighbouring nodes that are directly connected to it.
   - Each node is initially randomly seeded with a starting state which is a vector of its opinions about trustworthiness of all other nodes. This values represent recommendation as it has been defined in Subjective Logic.
   - Each node is initially randomly seeded with an opinion vector that describes the trust of a given subject to values of some attributes related with other agents. It may be elements like subject's public key, application, service, data, etc.
2. Topologic rules.
   - Connections between nodes are established randomly as the edges through which communication can take place.
   - The average number of connections per node (expressed as a percentage of the population) can be varied to investigate its effect on opinion (trust level).
3. The model provides each node with the ability to change state.
   - States are being changed in a synchronous manner. This means that all nodes recalculate appropriate values or none of them do so.
   - New values of opinions for nodes are calculated according to the conjunction, recommendation and consensus operation defined in Subjective Logic as it was described in earlier sections.

8 Conclusions

The paper presents a general concept about trust modeling and management using network based analysis of relations between autonomous agents. The method is based on assumption that multi-agent system constitutes social network and so the network specific parameters can be used to evaluate the trust levels. Proposition describes how metrics like centrality and clustering coefficient used in description of this type of networks can be used to model trust relation between agents. Within the presented model the agent's A opinion about agent's B statements is obtained as the Subjective Logic consensus of modified trust levels of all agents that trust the agent B. The trust level modification is performed in relation to the current position of the agent in the network.

The research related to social network metrics based trust modeling is still under continuous development. The following topics are considered as the promising directions for the next step of research:

- experimental verification which network measure fits the best our expectation about trust modeling,

- experimental verification how much presented method is vulnerable to intentional manipulation of collaborating agents,
- simulations that could help to find the answer to the question how fast opinions (trust levels) can be spread within a network,
- verification if the model is convergent to the real world observation that shows that connectivity level influences propagation of opinions among nodes,
- discovering type of the relation between connectivity level and opinion propagation speed,
- discovery of the relation describing how much network structure determine possibilities of opinion change,
- discovery of the relation describing how distribution of values among opinion's components: believe, disbelieve and certainty, influence opinion change dynamics.

References

1. Ackerman, M.S., McDonald, D.W.: Answer Garden 2: Merging organizational memory with collaborative help. In: Computer Supported Cooperative Work, pp. 97–105 (1996)
2. Barabasi, A.L., Albert, R.: Emergence of scaling in random networks. Science 286, 509–512 (1999)
3. Bollobas, B.: Random Graphs. Academic Press, New York (1985)
4. Garcia, R.: Extensió col.laborativa del servei de localització d'expertesa. Master's thesis. Technical University of Catalonia (2001)
5. Golbeck, J., Hendler, J.: Inferring trust relationships in web-based social networks. ACM Transactions on Internet Technology, 145–165 (2005)
6. Haheswaran, M., Ghunaim, A.: Towards a Gravity-Based Trust Model for Social Networking Systems. In: Distributed Computing Systems Workshops, pp. 240–248 (2007)
7. Jøsang, A.: Trust and Reputation Systems. In: Aldini, A., Gorrieri, R. (eds.) FOSAD 2006/2007. LNCS, vol. 4677, pp. 209–245. Springer, Heidelberg (2007)
8. Jøsang, A.: A Subjective Metric of Authentication. In: Quisquater, J., et al. (eds.) ESORICS 1998. LNCS, vol. 1485, pp. 329–344. Springer, Heidelberg (1998)
9. Jøsang, A.: An Algebra for Assessing Trust in Certification Chains. In: Kochmar, J. (ed.) Proceedings of the Network and Distributed Systems Security Symposium (NDSS 1999). The Internet Society (1999)
10. Jøsang, A.: Artificial reasoning with subjective logic. In: Nayak, A., Pagnucco, M. (eds.) Proceedings of the 2nd Australian Workshop on Commonsense Reasoning, December 1997. Australian Computer Society, Perth (1997)
11. Jøsang, A., Knapskog, A.: A Metric for Trusted Systems (full paper). In: Proceedings of the 21st National Information Systems Security Conference, NSA (October 1998)
12. Josang, A.: An Algebra for Assessing Trust in Certification Chains. In: Kochmar, J. (ed.) Proceedings of the Network and Distributed Systems Security Symposium, pp. 124–140 (1999)
13. Kołaczek, G.: A new concept of trust modeling and management in complex networks. In: Lovrek, I., Howlett, R.J., Jain, L.C. (eds.) KES 2008, Part I. LNCS (LNAI), vol. 5177, pp. 237–244. Springer, Heidelberg (2008)
14. Kołaczek, G.: Dynamika relacji zaufania i jej modelowanie w systemach teleinformatycznych. Internet w społeczeństwie informacyjnym, Zakopane (2004)

15. Kołaczek, G.: Evaluation of trust in multiagent systems using social network analysis metrics. In: Józefczyk, J., Thomas, W., Turowska, M. (eds.) W: 14th International Congress of Cybernetics and Systems of WOSC. 14th ICCS. Proceedings, pp. 595–603. Oficyna Wydaw. PWroc., Wrocław (2008)
16. Kołaczek, G.: Trust management in E-commerce. In: Wachowicz, J. (ed.) W: Electronic commerce. Theory and applications, pp. 53–59. Wydział Zarządzania i Ekonomii PGdań., Gdańsk (2005)
17. Kołaczek, G.: Trust networks and cellular automata. In: Grzech, A., Wilimowska, Z. (eds.) W: Information Systems Architecture and Technology ISAT 2004. Proceedings of the 25th international scientific school. Information models, concepts, tools and applications, pp. 272–278. Wrocław Oficyna Wydaw. PWroc (2004)
18. Kruczkowska, M., Ciepiela, E.: Sociophysics simulations, `http://student.uci.agh.edu.pl/~eciepiel/sociophysics/ss`
19. Maheswaran, M., Tang, H.C., Ghunaim, A.: Towards a Gravity-Based Trust Model for Social Networking Systems. In: Distributed Computing Systems Workshops, pp. 240–248 (2007)
20. Page, L., Brin, S., Motwani, R., Winograd, T.: The PageRank Citation Ranking: Bringing order to the Web. Technical Report, Computer Science Department, Stanford University (1998)
21. Pipkin, D.L.: Information Security: Protecting the Global Enterprise. HP Professional Series (2000)
22. Schillo, M., Funk, P., Rovatsos, M.: Who can you Trust: Dealing with Deception. In: Castelfranchi, C., Tan, Y., Falcone, R., Firozabadi, B.S. (eds.) Proceedings of the Workshop "Deception, Fraud and Trust in Agent Societies" of the Autonomous Agents Conference, pp. 293–320 (1999)
23. Scott, J.: Social Network Analysis: A Handbook, 2nd edn. Sage, Newberry Park (2000)
24. Stocker, R., Green, D.: Newth D.Consensus and cohesion in simulated social networks. Journal of Artificial Societies and Social Simulation 4(4), `http://www.soc.surrey.ac.uk`
25. Teacy, W.T., et al.: Travos: Trust and reputation in the context of inaccurate information sources. Autonomous Agents and Multi-Agent Systems 12(2) (2006)
26. Troitzsch, K.: Approaching Agent-Based Simulation FIRMA meeting (June 2000)
27. Wang, Y., Vassileva, J.: A Review on Trust and Reputation for Web Service Selection. In: First International Workshop on Trust and Reputation Management in Massively Distributed Computing Systems (TRAM 2007), pp. 322–340 (2007)
28. Watts, D.J., Strogatz, S.H.: Collective dynamics of 'small-world' networks. Nature 393, 432–440 (1998)
29. Watts, D.J.: Small Worlds: The Dynamics of Networks between Order and Randomness. Princeton University Press, Princeton (1999)
30. Webster's Revised Unabridged Dictionary (1913)
31. Wolfram, S.: Preface to Cellular Automata. In: Proceedings of an Interdisciplinary Workshop. North-Holland Physics Publishing, Amsterdam (1984)
32. Xiong, L., Liu, L.: PeerTrust: supporting reputation-based trust for peer-to-peer electronic communities. IEEE Transactions on Knowledge and Data Engineering 16(7), 843–857 (2004)
33. Zacharias, G., Maes, P.: Trust Management Through Reputation Mechanisms. Applied Artificial Intelligence 14, 881–907 (2000)

Logical Analysis of Multi-agent Kripke Structures

Sergey Babenyshev and Vladimir Rybakov

Department of Computing and Mathematics,
Manchester Metropolitan University,
John Dalton Building, Chester Street, Manchester M1 5GD, U.K.
and
Institute of Mathematics, Siberian Federal University,
79 Svobodny Prospect, Krasnoyarsk, 660041, Russia
Sergei.Babyonyshev@gmail.com, V.Rybakov@mmu.ac.uk

The last three decades witnessed a gradual shift from standalone-computer focused computations towards more and more distributed computational environment. This ever quickening shift brings new technological and theoretical challenges. Instead of seeing the computational system as completely controllable, which if malfunctioned should and can be immediately fixed, the contemporary user has to accept, sometimes grudgingly, the fact that the major part of her everyday computational environment is completely out of control or any reach. This feature brings up the question of how such environments can be organized to function efficiently. The metaphor of multi-agent system is one of the answers for these conceptual challenges. In this chapter we discuss several interdependent theoretical tools that modern mathematical logic and computer science can suggest for formalizing the informal notion of multi-agent system. We start by explaining how the modal propositional language may be used to specify the behavior of the processes, then we introduce formally the Kripke semantics for modal languages. Further we discuss different techniques of combining simple modal languages into complex ones, more suitable for practical purposes, and conclude with an extended example, illustrating the major points of the abovementioned techniques.

1 Modal Formulae as Specifications

Specification languages for multi-agent systems neither intend to provide a toolbox for constructing individual software agents nor strive to predict agents' individual behavior and possible results of their interaction. From the current state of science and technology these tasks seem to be too ambitious. On the one hand, the behavior of agents can depend on the behavior of their human hosts or can intentionally employ some random combinations of strategies. On the other hand, as we know, systems built from even very simple elements can demonstrate an unpredictable behavior, as illustrated by Turing machines and the cell automata (as in the game of "Life").

A. Håkansson et al. (Eds.): Agent and Multi-agent Technology, SCI 289, pp. 157–180.
springerlink.com

Thus the main focus of possible specification languages should be to provide an outer observer with adequate tools for monitoring important qualities of the multi-agent system as a whole, qualities like consistency of the shared data, observance of access rights and so on. Since we have no way of controlling or even knowing the specifics of the computational environment in question, the only reasonable point of view is the black-box view of the system's behavior.

To see the major distinctions between the structural and behavioral approaches toward equivalence of object, compare two specification of the same datatype — stack. (The example is taken from [21].)

```
%%%% Algebraic specification of the stack  %%%%%%
speck STACK
constructors
  new : * -> stack
  push: {0,1} # stack -> stack
operations
  <top,pop> : stack -> {0,1} # stack + *
axioms
  <top(new),pop(new)> = *
  <top(push(b,s)),pop(push(b,s))>=<b,s>

%%%% End of algebraic specification %%%%%%%%%%%%%
```

and

```
%%%% Coalgebraic specification of the stack  %%%%%%
speck STACK
operations
  new : * -> stack
  push: {0,1} # stack -> stack
observers
  <top,pop> : stack -> {0,1} # stack + *
axioms
  <top(new),pop(new)> = *
  <top(push(b,s)),pop(push(b,s))>=<b,s>

%%%% End of coalgebraic specification %%%%%%%%%%%%%
```

Although both specifications seem to be very similar (the only differences is the words "operations" replaced with "observers" and "constructors" with "operations"), this small change makes a big difference in interpretation of the respective data structures. Algebraic stacks are thus essentially identified with the terms constructed from operations `new` and `push`. In particular, two stacks are different if they are presented by different terms (there is no *confusion* in the datatype [21]). The second implication, is that there are no other stacks besides constructed as terms in `new` and `push` operations (there is no *junk* in the datatype [21]). On the contrary, seeing `new` and `push` as mere operators means that they are not essential for generating stacks. What is essential for being stack, under the coalgebraic specification, is being able to apply observers `top` and `pop` to a data structure.

Both approaches have their advantages and disadvantages. We just mention that coalgebraic description does not see distinction between finite and infinite stacks and, secondly, one may note that object-oriented programming have some features of the coalgebraic approach.

We summarize the differences between the algebraic and coalgebraic specifications in the following list:

- Algebraic specification
  - only data structures that are constructed by `new` and `push` are considered to be stacks,
  - whenever two stacks are constructed in two different ways, they are different.
- Coalgebraic specification
  - any data structure that can be observed with `top` and `pop` is considered to be a stack,
  - two stacks with the same behavior are considered to be equal.

The black-box view abandons the considerations of inner structures and focuses only on behavioral aspects of the systems. In the next example (Fig. 1), two noticeably distinct non-deterministic automata nevertheless demonstrate the same behavior (for the outer observer, which is not able to see the inner states). This can be shown by using a variant of Ehrenfeucht-Fraisse games [11] (see [22] for computer-science oriented introduction). On the figure, the inner states are presented by circles, labeled by Latin letters inside. Binary outputs are marked by symbols 0 and 1 next to the respective state.

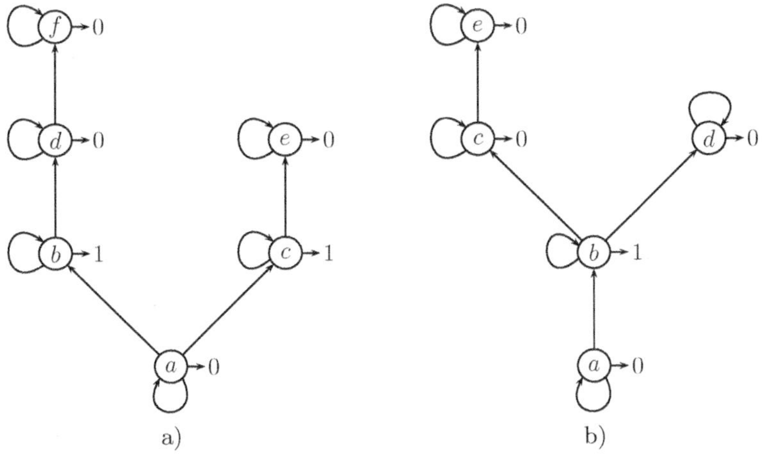

Fig. 1. An example of two different automata that are behaviorally undistinguishable

To be able to describe the behavior of the automata presented on Fig. 1, let us introduce two operators $\Box$ and $\Diamond$ with the following meaning (p takes binary values):

$\Box p$ is true at some state w of the automaton if for every state u reachable from w the output will be p;

$\Diamond p$ is true at some state w of the automaton if there is a state u reachable from w, where the output will be p.

These operators are known as *modalities* ("it is necessary" and "it is possible", respectively). Therefore, a sentence that is built from boolean connectives and modalities is called a *modal* formula. From the results of van Benthem [33,35], it follows that modal formulas express exactly the (first-order) statements about behavior, that are preserved by behavioral equivalence.

Returning to multi-agent systems, note that by means of an appropriate coding, we can present a state of a typical multi-agent system (excluding inner states of the agents) by a finite word in a certain language. Therefore, from a behavioral point of view, a multi-agent system can be seen as a non-deterministic automaton (with the empty input alphabet, because we assume that all impelling impulses come from inside the system) generating the sequences of such words.

Thus the focus on theoretical and specificational aspects of multi-agent systems bring us to the following conclusion:

> *Modal languages allow to specify the evolution of a system only up to behavioral equivalence. This can be seen as a viable solution for specifying the behavior of multi-agent systems, partially because it is not practically feasible to control every aspect of the underlying system.*

In the next section, we will put our informal discussion on the firmer ground, by presenting the *Kripke* semantics for modal languages. The discussed thus far modal language with one modality can be used to describe the behavior of a system with one simple prototypical agent. But this language is too weak to be useful for specifying situations of practical complexity and importance. The major problem is how to expand the system and the language and still retain the manageability and adequacy of the description. In the section "Fibring of Logics" we will present some mechanisms used for expansion of modal languages (and their corresponding logics), that are in a sense *modular*, i.e. allow to build more complex, *multi-modal* languages and logics out of simpler, uni-modal ones.

2 Kripke Semantics

In this section we will formally present the notions of *modal language, semantics* and *logic*, that we discussed in the previous section. For more detailed introduction and history of the topic we recommend [7,4,27]. When it concerns the Kripke semantics, we mainly follow the exposition of [7].

2.1 Modal Language

A *propositional language* is just a set of logical connectives taken with their *arities.* As an example, consider the language $\mathcal{L} := \{\wedge^2, \vee^2, \rightarrow^2, \neg^1\}$ of the classical propositional logic CPL. The arity is shown as an upper index of the corresponding connective. The arity of a connective shows how many formulas of the same language the connective can bind (or connect, therefore the term "connective"). There is also a chosen countable set $Var = \{p_1, p_2, \dots\}$ of so called *propositional variables.* The propositional variables are the simplest formulas of the language.

As for more complex expressions, the arities dictate which ones are considered *well-formed.* They are called *well-formed formulas of the language* $\mathcal{L}$ (wff) or $\mathcal{L}$*-formulas*, and the formula formation rules are as follows

1. $p_1, p_2, \dots$ are wff's,
2. if $\lambda^n \in \mathcal{L}$ and $\alpha_1, \dots, \alpha_n$ are wff's, then $\lambda(\alpha_1, \dots, \alpha_n)$ is wff.

The set of all $\mathcal{L}$-formulas is denoted by $For\mathcal{L}$. So, in the case of the classical logic CPL, we have that

$$\alpha \wedge \beta,\ \alpha \vee \beta,\ \alpha \rightarrow \beta \text{ and } \neg\alpha \text{ are } \mathcal{L}\text{-formulas, whenever } \alpha \text{ and } \beta \text{ are,}$$

where as usually the connectives $\wedge, \vee, \rightarrow, \neg$ stand for *and, or, implies* and *it is not*, respectively.

The *propositional modal language* $\mathcal{ML}$ is obtained from the language $\mathcal{L} = \{\wedge^2, \vee^2, \rightarrow^2, \neg^1\}$ of the classical propositional logic by enriching it with the new unary connective $\Box^1$. Therefore we have to add the following formula-formation rule

$$\text{if } \alpha \text{ is an } \mathcal{ML}\text{-formula, then } \Box\alpha \text{ is also an } \mathcal{ML}\text{-formula.}$$

We also define the connective $\Diamond^1$ as dual to $\Box$, i.e., by assigning

$$\Diamond\alpha := \neg\Box\neg\alpha, \text{ for all } \alpha \in For\mathcal{ML}.$$

The connectives $\Box$ and $\Diamond$ are usually read as "it is necessary" and "it is possible" and called the *necessity* and *possibility* operators.

There is a great number of modal formulas, that express important properties of *Kripke frames* (the latter will be introduced in the next section). In the following table we present some of them for illustration.

Modal formula	The property of Kripke frames
$\Box(\alpha \rightarrow \beta) \rightarrow (\Box\alpha \rightarrow \Box\beta)$	holds in all Kripke frames
$\Box\alpha \rightarrow \Box\Box\alpha$	transitivity
$\Box(\Box\alpha \rightarrow \beta) \vee \Box(\Box\beta \rightarrow \alpha)$	linearity
$\Box\alpha \rightarrow \alpha$	reflexivity
$\Box\alpha \rightarrow \Diamond\alpha$	seriality
$\alpha \rightarrow \Box\Diamond\alpha$	symmetry

2.2 Modal Frames and Models

We have provided thus far only pure syntactical description of the modal language $\mathcal{ML}$. According to this description we can tell by just looking at the expression in the language $\mathcal{ML}$, whether it is correctly built (wff) or nonsensical. To tell whether a statement represented by a well-formed formula is *true* or *false* we need the notion of *semantics.* Out of several viable alternatives, we will be working mainly with the *Kripke* or *possible worlds* semantics, sometimes also called *relational.*

A *modal frame* $\mathcal{F}$ *of the Kripke signature* $\{\Box^1\}$ is a pair $\langle W, R\rangle$, where W is a non-empty set of *states*, and $R \subseteq W \times W$ is an *accessibility relation* between states. If aRb then b is considered to be *accessible* or *reachable* from a.

Other names that can be used for an element of W roughly reflect the history of modal logic, originally developed for the analysis of philosophical arguments: *a possible world, a state of affairs, a state* or, more neutrally, *a point.*

A *valuation* V of the set of variables $P \subseteq Var$ in a frame $F = \langle W, R\rangle$ is a mapping $V : P \to 2^W$ associating with every variable $p \in P$ a set $V(p) \subseteq W$ of states in W, $V(p)$ is understood as the set of states where p is *true.*

A *Kripke model* $\mathcal{M}$ is a pair $\langle \mathcal{F}, V\rangle$, where $\mathcal{F} = \langle W, R\rangle$ is a frame and $V : P \to 2^W$ is a valuation in $\mathcal{F}$. Given a model $\mathcal{M} = \langle W, R, V\rangle$ we can define a *truth-relation* $(\mathcal{M}, w) \models \varphi$, "$\varphi$ is true at the state w in the model $\mathcal{M}$", by induction on the construction of a formula φ, built over variables P:

$$(\mathcal{M}, a) \models_V p \Longleftrightarrow a \in V(p);$$

$$(\mathcal{M}, a) \models_V \varphi \wedge \psi \Longleftrightarrow (\mathcal{M}, a) \models_V \varphi \text{ and } (\mathcal{M}, a) \models_V \psi;$$

$$(\mathcal{M}, a) \models_V \varphi \vee \psi \Longleftrightarrow (\mathcal{M}, a) \models_V \varphi \text{ or } (\mathcal{M}, a) \models_V \psi;$$

$$(\mathcal{M}, a) \models_V \varphi \to \psi \Longleftrightarrow (\mathcal{M}, a) \not\models_V \varphi \text{ or } (\mathcal{M}, a) \models_V \psi;$$

$$(\mathcal{M}, a) \models_V \neg\varphi \Longleftrightarrow (\mathcal{M}, a) \not\models_V \varphi;$$

$$(\mathcal{M}, a) \models_V \Box\varphi \Longleftrightarrow \forall b \in \mathcal{M}\,[(aRb) \text{ implies } (\mathcal{M}, b) \models_V \varphi];$$

$$(\mathcal{M}, a) \models_V \Diamond\varphi \Longleftrightarrow \exists b \in \mathcal{M}\,[(aRb) \text{ and } (\mathcal{M}, b) \models_V \varphi].$$

A formula φ is *satisfied* in a model $\mathcal{M}$ if there is a state w of $\mathcal{M}$, such that $(\mathcal{M}, w) \models \varphi$. On the contrary, a formula φ is *refuted* in a model $\mathcal{M}$ if there is a state w of $\mathcal{M}$, such that $(\mathcal{M}, w) \models \neg\varphi$. In this case we also say that $\mathcal{M}$ is a *countermodel* for φ. A formula φ is *true* in a model $\mathcal{M}$, in symbols $\mathcal{M} \models \varphi$, if $(\mathcal{M}, w) \models \varphi$ for every state w of $\mathcal{M}$.

A formula φ is *satisfied* in a frame $\mathcal{F}$, if it is satisfied in some model based on $\mathcal{F}$, i.e., if there is a valuation V and a state w of $\mathcal{F}$, such that $(\langle \mathcal{F}, V\rangle, w) \models \varphi$. A formula φ is *valid* in a frame $\mathcal{F}$, in symbols $\mathcal{F} \models \varphi$, if φ cannot be refuted in any model based on $\mathcal{F}$. Suppose $\mathcal{K}$ is a class of Kripke frames. Let $\mathcal{ML}$ be the respective modal language. We denote by $Log(\mathcal{K})$ the set of formulas

$$\{\alpha \in For\mathcal{ML} \mid \mathcal{K} \models \alpha\}.$$

An element α of $Log(\mathcal{K})$ is traditionally called a *theorem*. $Log(\mathcal{K})$ is indeed a modal normal logic, because it is closed under uniform substitutions, modus ponens and necessitation rules for all $\Box$-modalities. On the other hand, if L is a modal logic, let $Fr(L)$ be the class of all frames $\mathcal{F}$ of the respective Kripke signature, such that all theorems of L are valid on $\mathcal{F}$. A frame from $Fr(L)$ we will call *adequate* for L or an *L-frame*. Whenever $L = Log(Fr(L))$, the logic L is said to be *Kripke complete*.

The operators $\Box$ and $\Diamond$ can be interpreted not just as logical modalities "it is necessary" and "it is possible", but in a variety of other ways, depending on the intended meaning of the language. The major types, besides the already mentioned case of logical modalities, include

- *epistemic* modalities: $\Box$ means "it is known";
- *provability* modalities: $\Box$ means "it is proven";
- *deontic* interpretation: $\Box$ means "it is obligatory" and $\Diamond$ is "it is permitted";
- *tense* modalities: $\Box$ means "it is true and will be always true in the future" and $\Diamond$ reads "it is true or will be true sometimes in the future".

Depending on the intended interpretation of the language, the laws that govern modalities change, which leads in its turn to different classes of admissible Kripke frames. Later in this chapter, we will be working only with tense and a variant of epistemic modalities.

2.3 An Example: Modal Logic $S5$

We will define logic $S5$ *semantically*, i.e. as a set of modal formulas valid on a certain class of Kripke frames. (The logic LPD (Logic of Plausibility of Discovery), which is the main topic of this chapter, will be also defined semantically.)

Definition 1. *An $S5$-frame is a Kripke frame $\langle W, R\rangle$ such that the accessibility relation R is*

1. reflexive, *i.e.,* $\forall x(xRx)$,
2. transitive, *i.e.,* $\forall xyz(xRy \wedge yRz \rightarrow xRz)$,
3. symmetric, *i.e.,* $\forall xy(xRy \rightarrow yRx)$.

Thus, in $S5$-frame, R is just an *equivalence relation* on W, thereby partitioning W into the family of non-intersecting connected subsets, known as *clusters*.

For illustration, consider a simple example of an $S5$-frame. Suppose $F = \langle W, R\rangle$, where $W = \{a_1, a_2, a_3, a_4, a_5, a_6, a_7\}$ and

$$R = id_W \cup \{a_2, a_3, a_4, a_7\} \times \{a_2, a_3, a_4, a_7\} \cup \{(a_5, a_6), (a_6, a_5)\}$$

(see Fig. 2.3(a)). This $S5$-frame consists of 3 clusters.

Logic $S5$ is determined by the class of all $S5$-frames:

$$S5 := \{\alpha \in For\mathcal{ML} \mid \text{ for all } S5\text{-frames } \mathcal{F}\text{: } \mathcal{F} \models \alpha\}.$$

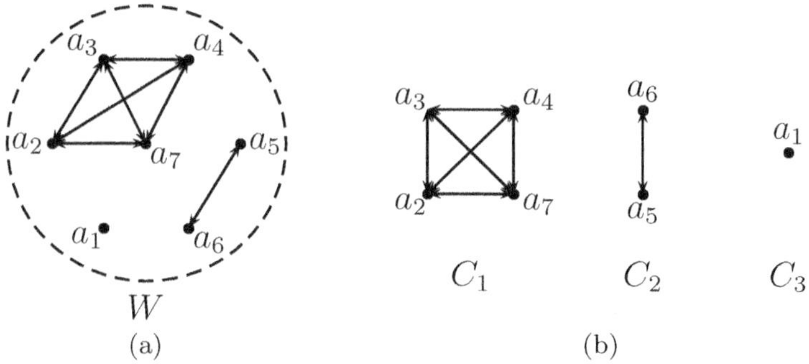

Fig. 2. An example of a finite $S5$-frame. (a) The frame $\langle W, R\rangle$, where R is an equivalence relation on W. (b) The same frame shown as a family of clusters.

This way of defining logics through semantics historically was not first. At the early era of modal logic researches (logicians, philosophers and mathematicians alike) generally preferred to use *axiomatic* method. In a nutshell, it means that bearing in mind the desired behavioral properties we must guess and outline a set (preferably finite) of formulas that explicitly express the desired properties and can serve as axioms (i.e. must be true together with their substitutional variants). The semantical way has the advantage that we can directly postulate the behavioral properties, without guessing what formulas express it. But a good axiomatization (usually it means finite), if exists, has the advantage that we have a constructive way of proving the theorems of the logic. In the case of $S5$ it is possible and easy to provide a finite axiomatization.

Proposition 1. *Logic $S5$ is the smallest subset of $For\mathcal{ML}$, with the properties: (i) $CPL \subseteq S5$; (ii) $S5$ is closed under rules of* modus ponens: $\alpha, \alpha \to \beta \vdash \beta$ *and* necessitation: $\alpha \vdash \Box\alpha$; *and (iii) for all $\alpha, \beta \in For\mathcal{ML}$*

1. $\Box(\alpha \to \beta) \to (\Box\alpha \to \Box\beta) \in S5$,
2. $\Box\alpha \to \Box\Box\alpha \in S5$,
3. $\Box\alpha \to \alpha \in S5$,
4. $\alpha \to \Box\Diamond\alpha \in S5$.

Because of the given axiomatization, any formula $\alpha \in S5$ can be obtained from the axioms by a deduction-like process, therefore formulas of $S5$ (and, in fact, of any logic) are traditionally called *theorems.*

An explicit finite or even only recursively enumerable axiomatization is a very desired property for any logic, both from the theoretical and computational point of view. Such axiomatization would allow to generate the theorems of the logic by a simple procedure of the syntactical derivation. Unfortunately (as we will witness later), it is not always easy to find an explicit axiomatization for a semantically defined logic.

Our choice for $S5$ to illustrate modal logics is in part motivated by the fact, that we will use the multi-modal variants of $S5$-frames to model the interaction of agents in a multi-agent system at certain time moment.

3 Fibring of Logics

"Fibring" in the logical parlance means "combining". Why would one need to combine different logics? An obvious reason in our case is that different agents in a multi-agent system might have different visions of the same situation, i.e. different name-spaces or different interpretations and assigned types for the same names. But these differences are not logical in nature, because distinct agents in a system might be using the same system of reasoning about their data, therefore employing the same *logic*. In fact we will assume that all agents in a given system use the same *Boolean* logic for reasoning about their intrinsic knowledge.

More fundamental problem is connected with the necessity to consider *evolutions* of a multi-agent system. For computability reasons, we want to present this evolution as a discreet sequence of states of the given system (provided that we know how to represent formally a state of a multi-agent system). Furthermore, to account for our incomplete data about the system and non-determinacy in general, the sequence of states might become a tree (so called, branching time model). The logic describing the time flow and the logic describing interaction of agents, in general, are fundamentally different. For instance, two agents can communicate through a bi-directional channel, which can be represented by an undirected (or, equivalently, two-way) edge, but the time flow cannot reverse its course and return to a state in the past.

There were numerous approaches tried for combining propositional logics. Further we will discuss pro and contra of some of them. The major two, that assume full combination of signatures, are *product* and *fusion* (cf. [6]). Basic fusions do not allow for interaction of modalities, while products of logics are undecidable even in cases of simple constituent logics, like $S5$ [15]. A powerful unifying mechanism of "fibred" logics was proposed by D. Gabbay [12,13,14], as a generalization of these methods.

We will use the metaphor of "*continents and bridges*" from [3] to describe the general framework for "fibring" of logics.

Let $\mathcal{L}$ and $\mathcal{R}$ be two classes of structures representing two different ontologies we intend to combine, and let Z be a family of relations between elements of $\mathcal{L}$ and $\mathcal{R}$. A *trio* is the triple $\langle \mathcal{L}, Z, \mathcal{R} \rangle$, where $\mathcal{L}$ is called the *left continent* of the trio, $\mathcal{R}$ is called the *right continent* of the trio, and Z is called the *bridge* of the trio. Relations R from Z represent the combined structures. What shall be the language of the combined structures? According to [3], we will need two predicates "left" and "right", to be able to determine at which continent we are on at the moment of statement. Secondly, we will need a special modality $\langle z \rangle$—"zooming in", to be able to invoke a statement in the language of $\mathcal{R}$ from within the statements in the language of $\mathcal{L}$, and, symmetrically, the "zooming out" modality $\langle z^{-1} \rangle$ is needed.

Later we will use this metaphor to "place" our approach inside a bigger picture of "fibred" logics.

3.1 Fusion

Fusion or *independent join* is the simplest way of combining languages.

The *fusion* $\mathcal{F}_1 \otimes \mathcal{F}_2$ of two frames $\mathcal{F}_1 = \langle W, R_1 \rangle$ and $\mathcal{F}_2 = \langle W, R_2 \rangle$ with the same universe is the frame $\langle W, R_1, R_2 \rangle$.

Suppose we have two Kripke-complete modal logics L_1 and L_2. Let $\mathcal{L}$ be the language $\{\wedge, \vee, \rightarrow, \neg, \Diamond_1, \Diamond_2\}$. The *fusion* $L_1 \otimes L_2$ of L_1 and L_2 is the logic over the language $\mathcal{L}$ defined by

$$Log\{\mathcal{F}_1 \otimes \mathcal{F}_2 \mid \mathcal{F}_1 \in Fr(L_1), \mathcal{F}_2 \in Fr(L_2)\}.$$

We can easily iterate the fusion obtaining multi-modal logics with any finite number of modalities. For illustration, let us look at multi-modal logic $S5_n$. By definition, a $S5_n$-frame is the Kripke frame $\langle W, R_1, \ldots, R_n \rangle$, where each R_i is an equivalence relation on W independent from the rest of R_j's.

3.2 Product

The *product* $\mathcal{F}_1 \times \mathcal{F}_2$ of two frames $\mathcal{F}_1 = \langle W_1, R_1 \rangle$ and $\mathcal{F}_2 = \langle W_2, R_2 \rangle$ (see Fig. 3) is the frame $\langle W_1 \times W_2, R_h, R_v \rangle$, where for all $u, u' \in W_1$, $v, v' \in W_2$

$$\langle u, v \rangle R_h \langle u', v' \rangle \iff uR_1u' \text{ and } v = v';$$

$$\langle u, v \rangle R_v \langle u', v' \rangle \iff vR_2v' \text{ and } u = u'.$$

Product logics are defined semantically: they are sets of formulas in the combined language, that are valid on all product frames. More formally, let $\mathcal{K}_1$, $\mathcal{K}_2$ be two classes of Kripke frames with their respective Kripke signatures, and suppose that $L_1 = Log(\mathcal{K}_1)$, $L_2 = Log(\mathcal{K}_2)$. Then the product $L_1 \times L_2$ of logics L_1, L_2 is

$$L_1 \times L_2 := Log\{\mathcal{F}_1 \times \mathcal{F}_2 \mid \mathcal{F}_1 \in \mathcal{K}_1, \mathcal{F}_2 \in \mathcal{K}_2\}.$$

It follows from the definition that the product is defined only for Kripke complete logics.

3.3 Refinement

Let us fix a class of Kripke frames $\mathcal{K}$. Suppose $\mathcal{F} = \langle W, R \rangle$ is a transitive Kripke frame (not necessarily from $\mathcal{K}$) and consider a family

$$\mathcal{X} = \{\mathcal{G}_C \in \mathcal{K} \mid \ C \text{ is a cluster of } \mathcal{F}\},$$

of frames from $\mathcal{K}$ indexed by clusters of $\mathcal{F}$, where each $\mathcal{G}_C$ is based on the cluster $C \subseteq \mathcal{F}$ (i.e., $\mathcal{G}_C = \langle C, R_C \rangle \in \mathcal{K}$, where $R_C \subseteq C \times C$). A *refinement of* $\mathcal{F}$ *by* $\mathcal{X}$ is the frame $\langle W, R, \bigcup_C R_C \rangle$ (see Fig. 4(b)).

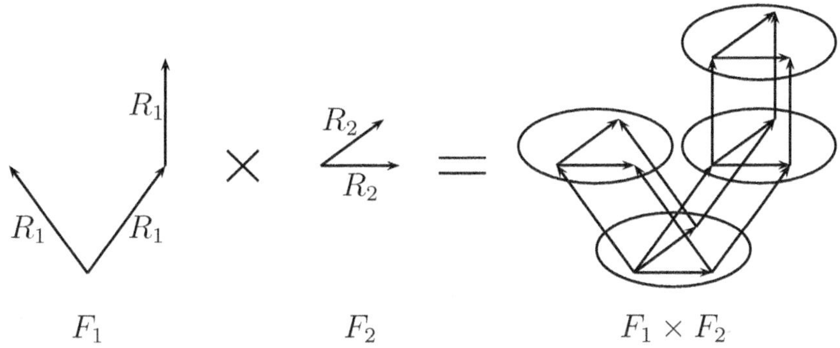

Fig. 3. A product of two frames (schematically)

Thus, in a refinement, the Kripke structure of frames from the indexed family $\mathcal{X}$ is imposed on the clusters of worlds of a frame $\mathcal{F}$, in addition to the existing relation R. It is important, for technical reasons, that all elements of $\mathcal{G}_C$ are universally accessible from each other by R.

We denote by $Ref(\mathcal{K}, \mathcal{J})$ the class of all refinements of frames from the class $\mathcal{K}$ by the frames of the class $\mathcal{J}$. The frames from the class $\mathcal{K}$ we will call *basic*, and the frames from $\mathcal{J}$ — *secondary*.

Suppose we have two Kripke-complete modal logics $\mathcal{B}$ and $\mathcal{S}$, (that we call *basic* and *secondary* logic, respectively) of Kripke signatures $\Diamond_R$ and $\Diamond_Q$ and let $\mathcal{B}$ be transitive. Let $\mathcal{L}$ be the language $\{\wedge, \vee, \rightarrow, \neg, \Diamond_R, \Diamond_Q\}$. We define the *refinement of* $\mathcal{B}$ *by* $\mathcal{S}$, denoted by $\mathcal{B}_\mathcal{S}$ or $Ref(\mathcal{B}, \mathcal{S})$, to be the logic

$$Log(Ref(Fr(\mathcal{B}), Fr(\mathcal{S}))).$$

It is easy to see from the definition that so defined set of formulas is indeed a multi-modal logic, i.e., it is closed under uniform substitutions and modus ponens.

To see how refinements fit in the bigger picture of fibred logics, recall the metaphor of "bridges and continents". The classes of structures $\mathcal{L}$ and $\mathcal{R}$, representing two different ontologies, are presented by classes of Kripke frames for two Kripke-complete modal propositional logics. Members $\mathcal{F} \in \mathcal{L}$ of the "left continent" represent our *basic* structures, while the "right continent" $\mathcal{R}$ represents the class of secondary structures. The "bridge", in our case, consists of the relations of the form

$$\{\langle \mathcal{F}, \mathcal{G}_C \rangle \mid C \text{ is a cluster of } \mathcal{F}\}.$$

Since universes of $\mathcal{G}_C$ are embedded in the universe of $\mathcal{F}$, the elements of $\mathcal{R}$-structures $\mathcal{G}_C$ can be *identified* with the elements of the basic $\mathcal{L}$-structure $\mathcal{F}$, hence we do not need to consider special "*zooming in*", "*zooming out*" modalities to combine the languages of $\mathcal{L}$ and $\mathcal{R}$. Also there is no use for "*left*" and "*right*" predicates, that distinguish "*states*" of the left and right continents.

It is the refinement that we will use to define the logic LPD later.

3.4 Products vs. Refinements

To see the major differences between products and refinements of frames take a look at Fig. 4. According to the definition of the product of frames (see Fig. 4(a)), the product of a linear frame, isomorphic to $\mathbb{N}$, and an $S5$-cluster C consists of countably many isomorphic copies of C, indexed by natural numbers. The situation in the refinement is different. Each state in a cluster has access to every state of the next cluster (and of any successive cluster, since we assume that time-relation R is transitive). The clusters, which describe particular frame configurations in various moments of times, are generally not isomorphic.

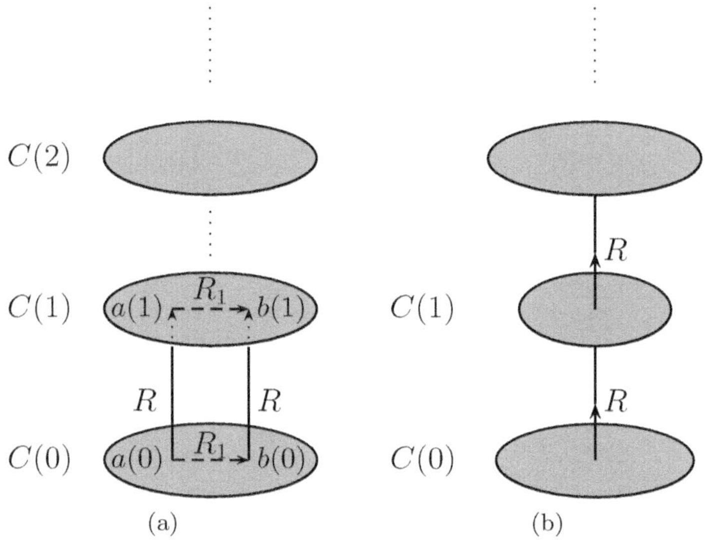

Fig. 4. Products vs. refinements. Example: a) Product of frames. Note that $b(1)$ is accessible from $a(0)$ only through composition of R with R_1. All clusters are isomorphic. b) Refinement of frames. Every element of $C(i+1)$ is accessible from every element of $C(i)$ via time relation R. Clusters are in general non-isomorphic.

4 Augmenting the Modal Language

The standard modal operators, that we have considered so far, work at the level of an individual agent. For the purposes of specification, monitoring and control of the behavior of a typical multi-agent system, and also for the goal of describing agents' interaction, additional operators are needed. There were several attempts to augment the modal and multi-modal propositional languages with the more expressive operators. A well known example (cf. [10]) is, so called, the "common knowledge" operator $\Box_K$. This operator cannot be expressed by finitary means in terms of "elementary" agent-knowledge operators (although it can be defined through itself by a fixed-point method [10]). In fact, $\Box_K\alpha$ is equivalent to an infinite disjunction

$$\Box_K\alpha = \bigvee_{\bar{n}\in\{1,\ldots,n\}^+} \Box_{\bar{n}}\,\alpha,$$

where $\Box_{\langle i_1,\ldots,i_t\rangle}\alpha := \Box_{i_1}\ldots\Box_{i_t}\alpha$.
The dual form $\Diamond_K$ of the common knowledge operator $\Box_K$ was used in Rybakov [29,30] to model the interaction of agents in the framework of Linear Temporal Logic (LTL). The operator $\Diamond_K$, also known as "knowledge through interaction of agents", has a simple meaning: $\Diamond_K\alpha$ is true at a state w if there is a sequence of states $w = w_0, w_1, w_2, \ldots, w_n$, where each successor is accessible from the predecessor by a communication channel available to one of the agents, and α is true at w_n. Symbolically,

$$w \models \Diamond_K\alpha \iff \exists w_1 w_2 \ldots w_n : wR_{i_1}w_1R_{i_2}w_2\ldots R_{i_{n-1}}w_n \text{ and } w_n \models \alpha.$$

Unfortunately, currently there is no an established general theory of what kind of operators can be added to the language without breaking good logical and semantical properties of the presumed logic.

Below, we will describe an ad hoc technique which goes back to the "renaming method", sometimes attributed to M. Wajsberg. The main idea is to transform a formula containing additional operators to a special equivalent reduced form. This can be done, according to Lemma 1, if the additional operators satisfy the following condition

$$\alpha_1 \leftrightarrow \beta_1, \ldots, \alpha_n \leftrightarrow \beta_n \vdash \lambda(\alpha_1, \ldots, \alpha_n) \leftrightarrow \lambda(\beta_1, \ldots, \beta_n), \text{ for } \lambda^n \in \mathcal{EML} - \mathcal{ML},$$

where $\mathcal{EML}$ is the extended modal language. There is another complication though. This technique was developed originally (cf. [27]) for the treatment of inference rules, not formulas. So we will have to work with presentation of formulas by inference rules.

An *inference rule* is an expression

$$r := \frac{\varphi_1(p_1, \ldots, p_n), \ldots, \varphi_k(p_1, \ldots, p_n)}{\psi(p_1, \ldots, p_n)},$$

where $\varphi_1(p_1, \ldots, p_n), \ldots, \varphi_k(p_1, \ldots, p_n)$ and $\psi(p_1, \ldots, p_n)$ are some formulas built over the variables $p_1, \ldots, p_n$. The variables $p_1, \ldots, p_n$ are called the *variables of the rule* r, and the corresponding set is denoted by $Var(r)$. It is common to write $\frac{\varphi_1(\bar{p}),\ldots,\varphi_k(\bar{p})}{\psi(\bar{p})}$ as $\varphi_1(\bar{p}), \ldots, \varphi_k(\bar{p})/\psi(\bar{p})$.

Definition 2. *A rule r is said to be* valid *in the Kripke model $\mathcal{M}$ with the valuation V (we will use notation $\mathcal{M} \models_V r$) if*

$$\forall w \left((\mathcal{M}, w) \models_V \bigwedge_{1\leq i\leq k} \varphi_i \right) \Rightarrow \forall w\,((\mathcal{M}, w) \models_V \psi).$$

Otherwise, r is refuted *in $\mathcal{M}$, or* refuted in $\mathcal{M}$ by V, *and we write $\mathcal{M} \not\models_V r$.*

Since conjunction $\wedge$ will be in any considered in this chapter modal language, we can assume that our rules have only one premise.

Definition 3. *A rule r is* valid *in a frame $\mathcal{F}$ (symbolically $\mathcal{F} \models r$) if, for every valuation V of $Var(r)$, $\mathcal{F} \models_V r$.*

We will use the easy verifiable fact, that a formula φ is valid in a frame $\mathcal{F}$ iff the rule $x \to x/\varphi$ is valid in $\mathcal{F}$, and, consequently, φ is satisfiable in a frame $\mathcal{F}$ if and only if the rule $x \to x/\neg\varphi$ is refuted in $\mathcal{F}$.

Lemma 1. *Any rule r can be transformed in exponential time to an equivalent rule r_{nf} in the reduced normal form.*

Proof. We shall specify the general algorithm, described in Lemma 3.1.3 and Theorem 3.1.11 [27]. We assume for simplicity that the language in question has only unary non-boolean connectives.

Suppose we have a rule

$$r = \frac{\varphi_1(x_1, ..., x_n), ..., \varphi_m(x_1, ..., x_n)}{\psi(x_1, ..., x_n)}$$

It is obvious that r is equivalent to the rule

$$r_0 = \frac{\bigwedge_{1 \le j \le m} \varphi_j(x_1, ..., x_n) \wedge (x_c \leftrightarrow \psi(x_1, ..., x_n))}{x_c}$$

where x_c is a new variable.

Therefore we can consider only rules in the form $\varphi(x_1, ..., x_n)/x_c$.

A variable or a unary logical operation applied to a variable we will call a *terminal* formula. If $\varphi = \alpha \circ \beta$, where $\circ$ is a binary logical operation and both formulas α and β are not terminal, take two new variables x_α and x_β and the rule

$$r_1 := (x_\alpha \circ x_\beta) \wedge (x_\alpha \leftrightarrow \alpha) \wedge (x_\beta \leftrightarrow \beta)/x_c.$$

If one of the formulas α or β is terminal and another is not, we apply this transformation again but only to a non-terminal formula. It is clear that r and r_1 are equivalent w.r.t. validity in frames.

If $\varphi = *\alpha$, where $*$ is a unary logical operation and α is not a variable, take a new variable x_α and the rule

$$r_1 := *x_\alpha \wedge (x_\alpha \leftrightarrow \alpha)/x_c.$$

Again r and r_1 are equivalent. We continue to apply similar transformation to successive rules:

$$r_k = \frac{\gamma \wedge \bigwedge_{i \in I_k} (x_{\alpha_i} \leftrightarrow \alpha_i)}{x_c}$$

until all formulas α_i in the premise become terminal. The resulted compound transformation is polynomial in time. Next, we transform the premise of the final rule to the perfect disjunctive normal form and obtain another equivalent rule r_2. The latter transformation, as well as all known ones for reduction of boolean formulas to perfect disjunctive forms, is time-exponential. As the result the final rule $r_{\mathrm{nf}} := r_2$ has the required form. □

5 Example: Logic of Plausibility of Discovery

5.1 Historical Background

The area of non-classical mathematical logic have provided up to date a multitude of techniques with applications to artificial intelligence and computer science. Significant part of such applications is connected with multi-agent logic oriented to reasoning about knowledge, time and computation (see for instance, Goldblatt [16], van Benthem [34]). Such logics were often defined by semantic objects based on Kripke/Hinttikka models and temporal algebras (cf. Thomason [32], Goldblatt [17]). The fundamental reason for such applicability, as we already mentioned, is that modal logics arise naturally as description and verification languages for processes.

We will illustrate the techniques presented in the previous sections by providing a non trivial example: Logic of Plausibility of Discovery (LPD). The language of this logic augments standard multi-modal language with the operations for *chance of discovery, time and knowledge accessibility* and *plausibility of statements*. The choice of operators owes to the topic of chance discovery. Chance discovery (cf. Ohsawa and McBurney [23]) is a modern area in artificial intelligence and computer science which analyzes events with uncertain information, incomplete past data, chance events, which are typically rare or hard to find. The logic of discovery has a solid prehistory, possibly starting with the monograph *Logic of Discovery and Logic of Discourse* by Jaakko Hintikka and Fernand Vandamme [20]. This logic is of interdisciplinary nature and was influenced by (and had influence on, in its turn) by various ideas coming from researchers with diverse background. In particular, the modeling of environmental decision-support systems has been undertaken (cf. Cortés, Sánchez-Marré and Ceccaroni [8], Avouris [2]), applications of the fuzzy logic were discovered (cf. Bugarin and Barro [5], Dubois and Prade [9]), tools for semantic web and multi-agent systems has been developed (cf. Harmelem and Horrocks [18], Hendler [19], Arisha et al [1]).

5.2 Logic of Plausibility of Discovery

Our basic idea for this section is to combine logical operations for *agents' knowledge*, with *time, discovery* and *plausibility operations*. To impose the desired properties on these operations, we introduce a logic LPD that combines these operations in one propositional language. Note that the use of *plausibility operation* seems to be quite new in this context.

We introduce LPD semantically (as $S5$ before), i.e. as a set of formulas of a given language that are valid on all Kripke frames from a certain class. More precisely, let

$$\mathcal{N}_C := \langle \bigcup_{i \in N} C(i), R, R_1, \dots R_m \rangle,$$

where N is the set of all integer numbers, all $C(i)$ are some disjoint nonempty sets representing time clusters, R and $R_1, \dots, R_m$ are binary accessibility relations, such that R is linear and $R_1, \ldots, R_m$ are equivalence relations when restricted to each of $C(i)$ (R is imitating linear time flow, R_i's are agents' accessibility relations inside clusters).

We introduce a propositional language to reason about agents' knowledge, time, discovery and plausibility of statements. It contains agent's knowledge operations K_i, $1 \leq i \leq m$, standard modal operation $\Diamond$ — *"is possible"*, unary decision operations $\Diamond_{D,l}$, $\Diamond_{D,g}$ for local and global possibility to discover information, and finally the plausibility operation $\mathrm{pl}(A, q)$, where $q \in Q$ (q is a rational number from [0,1]) with the meaning: "the plausibility of the statement A is q". We discuss the language and rules for computing truth values of formulas in the suggested language in models based on frames $\mathcal{N}_C$. These rules allow to formulate satisfiability problem for our logic LPD — Logic of Plausibility for Discovery. The main question we are dealing with is construction of an algorithm for checking satisfiability of formulas in LPD. The presented algorithm is based on reduction of formulas to rules and converting rules to reduced normal form, and, then, on checking validity of such rules in finite models of effectively bounded size. The technique which we use is borrowed from our research (cf. [25,28,31,26]) devoted to study of inference rules in non-classical logics, in particular, it worked well for Logic of Discovery in uncertain situations [24]. At the end we discuss some open problems and briefly comment on the essence and features of the proposed algorithm.

5.3 Notation

We start with introduction of semantic models which motivate our choice for the language to model multi-agent reasoning about chances to discover necessary facts, knowledge, information, etc. The following Kripke frames form the basis for our models.

$$\mathcal{N}_C := \langle \bigcup_{i \in N} C(i), R, R_1, \ldots R_m \rangle,$$

where N is the set of integers, all $C(i)$ are nonempty sets, R and $R_1, \ldots, R_m$ are binary accessibility relations. For all elements a and b from $\bigcup_{i \in N} C(i)$,

$$aRb \Longleftrightarrow [a \in C(i) \text{ and } b \in C(j) \text{ and } i \leq j];$$

any R_j is a reflexive, transitive and symmetric relation (S5-relation), and

$$\forall a, b \in \bigcup_{i \in N} C(i),\ aR_j b \Rightarrow [a, b \in C(i) \text{ for some } i\ ].$$

Intuition behind this formal definition is as follows:

The numbers $i \in N$ serve to model linear discrete flow of time (like steps in a computational process, etc), any $C(i)$ is the time cluster of all states at the current time-moment i, any R_j is the accessibility relation for the j-th agent between states in any time cluster $C(i)$.

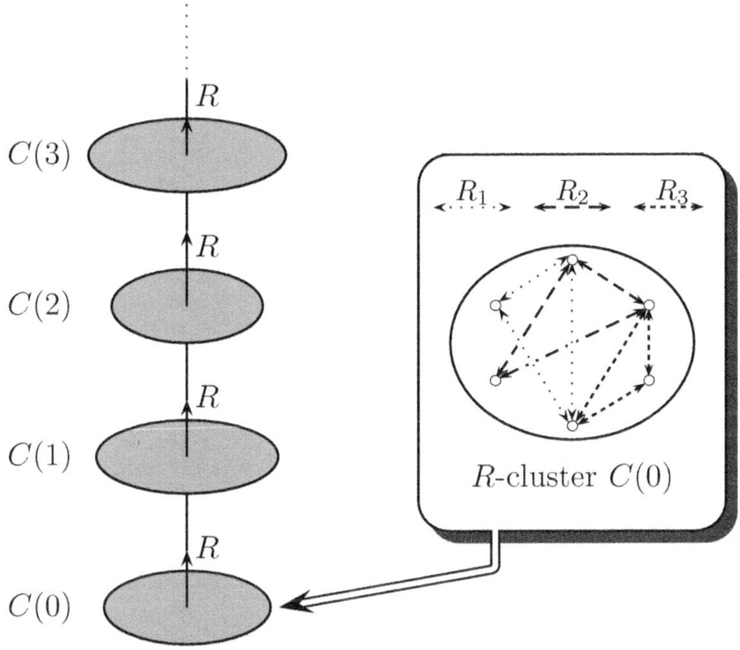

Fig. 5. A typical frame $\mathcal{N}_C$

To model multi-agent reasoning about properties of propositions in models based at $\mathcal{N}_C$, their mutual relationships and their plausibility to discover some information, we introduce the following logical language $\mathcal{L}$: $\mathcal{L}$ contains a countably infinite set of propositional variables Var; the logical operations include usual boolean operations and usual unary agent-knowledge operations K_i, $1 \leq i \leq m$, as well as the modal operation $\Diamond$ — "*is possible*". We extend the language by taking unary decision operations $\Diamond_{D,l}$, $\Diamond_{D,g}$ for local and global possibility to discover information, and an additional binary operation $\mathrm{pl}(\varphi, q)$ with arguments to be: first — an arbitrary formula, and second — any rational number from [0,1]. This operation is meant to say that φ from the viewpoint of agents is q-plausible. The formation rules for formulas are standard. In particular, for any formula φ and $q \in [0,1] \subseteq Q$ (i.e., q is a rational number from [0,1]), the expression $\mathrm{pl}(\varphi, q)$ is a well-formed formula of our language. Intuitive meaning of introduced operations is as follows:

- all boolean connectives have standard meaning;
- $K_i\varphi$ can be read: agent i knows φ in the current state (informational node) of the current time cluster;
- $\Diamond_{D,l}\varphi$ has meaning: a state were φ is true is locally discoverable (there is a state in current time cluster where φ is true).
- $\Diamond_{D,g}\varphi$ is meant to say that φ globally discoverable; in any future time cluster there is a state, where φ is true.

- $\Diamond\varphi$ says that φ there is a future time cluster and a state within this cluster where φ is true;
- for any formula φ and $q \in [0,1] \subseteq Q$ (q is a rational number from [0,1]) $\mathrm{pl}(\varphi, q)$ says that the plausibility (in accordance with summarized agents' knowledge) of φ to be true is equal to q.

Now we define rules for computing truth values of formulas in models $\mathcal{N}_C$ with valuations V of propositional letters $P \subseteq Var$. Given a frame $\mathcal{N}_C$, and a set of propositional letters P, a valuation V of P in $\mathcal{N}_C$ is a mapping of P into the set of all subsets of the set $\bigcup_{i \in N} C(i)$, in symbols,

$$\forall p \in P : V(p) \subseteq \bigcup_{i \in N} C(i).$$

If, for an element $a \in \bigcup_{i \in N} C(i)$, we have that $a \in V(p)$, then we say *the fact p is true in the state a*. In the notation below $(\mathcal{N}_C, a) \models_V \varphi$ is meant to say that the formula φ is true at the state a in the model $\mathcal{N}_C$ with respect to the valuation V. The truth-values for classical connectives are defined the same way as in Section 2, for additional operators the rules are listed below:

$$(\mathcal{N}_C, a) \models_V K_i\varphi \iff \forall b \in \mathcal{N}_C[(aR_ib) \Rightarrow (\mathcal{N}_C, b) \models_V \varphi];$$

$$(\mathcal{N}_C, a) \models_V \Diamond_{D,l}\varphi \iff a \in C(i) \,\&\, \exists b \in C(i)\ (\mathcal{N}_C, b) \models_V \varphi;$$

$$(\mathcal{N}_C, a) \models_V \Diamond_{D,g}\varphi \iff a \in C(i)\&\forall j \geq i \exists b \in C(j)(\mathcal{N}_C, b) \models_V \varphi;$$

$$\forall a \in \mathcal{N}, \quad \mathrm{pr}(\varphi, a, V) := \frac{||\{i \mid (\mathcal{N}_C, a) \models_V K_i\varphi\}||}{m};$$

$$\forall a \in \mathcal{N}, \forall q \in Q : (\mathcal{N}_C, a) \models_V \mathrm{pl}(\varphi, q) \iff \mathrm{pr}(\varphi, a, V) = q.$$

Under this approach, there is no need to "localize" an agent to any particular informational node. (In fact an instance of each agent is present at each node, by reflexivity.) Instead we can identify the ith agent with the scope of its "awareness", i.e. with the set of nodes, that are related through relation R_i (excluding isolated R_i-reflexive points).

Example. As we already mentioned, the operation $\mathrm{pl}(\varphi, q)$ seems to be new in such framework. To illustrate it, assume that $m = 3$ (i.e., there are three independent agents) and in some model $\langle \mathcal{N}_C, V \rangle$ with a valuation V, there is an R-cluster $C(i) = \{a, b, c, d\}$ (see Fig. 6). Suppose that the valuation V is such that for a formula φ

$$(\mathcal{N}_C, a) \models_V \varphi, \ (\mathcal{N}_C, b) \models_V \varphi, \ (\mathcal{N}_C, c) \not\models_V \varphi, \ (\mathcal{N}_C, d) \not\models_V \varphi.$$

Then

$$(\mathcal{N}_C, a) \models_V K_1\varphi, \ (\mathcal{N}_C, a) \models_V K_2\varphi, \ (\mathcal{N}_C, a) \not\models_V K_3\varphi,$$

In particular, $\mathrm{pr}(\varphi, a, V) = 2/3$, and $(\mathcal{N}_C, a) \models_V \mathrm{pl}(\varphi, 2/3)$, thus the formula $\mathrm{pl}(\varphi, 2/3)$ says that, in this model, from the summarized viewpoint of all agents the plausibility of the formula φ to be true is 2/3.

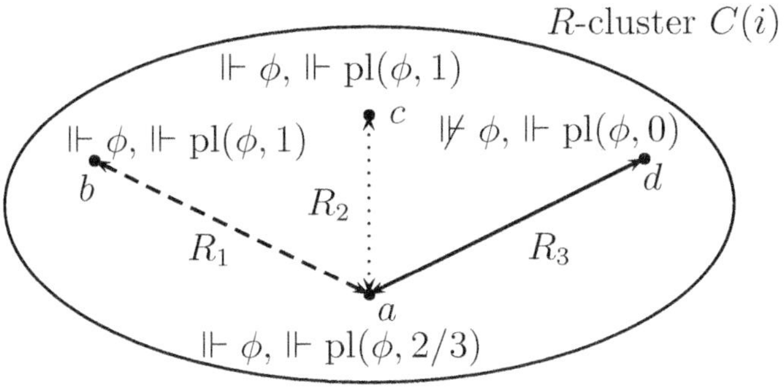

Fig. 6. Example: truth-value for the plausibility operator

For the other states we have

$$(\mathcal{N}_C, b) \models_V \mathrm{pl}(\varphi, 1), \ (\mathcal{N}_C, c) \models_V \mathrm{pl}(\varphi, 1), \ (\mathcal{N}_C, d) \models_V \mathrm{pl}(\varphi, 0).$$

We cannot define the logic with such semantics in the standard way, as set of all formulas which are true in any state of any frame $\mathcal{N}_C$ w.r.t. any valuation, because plausibility operations employ rational numbers, and truth values of plausibility operations vary in the same frame under different valuations. Nevertheless, the satisfiability problem for our logic LPD has standard formulation.

Definition 4. *A formula φ is satisfiable in the logic LPD if and only if there is a frame $\mathcal{N}_C$ and a valuation V at $\mathcal{N}_C$ such that $\mathcal{N}_C \models_V \varphi$ (i.e., φ is true at any state of $\mathcal{N}_C$ w.r.t. the valuation V).*

For $\Box := \neg\Diamond\neg$, the pure $\Box$-fragment of LPD (formulas with no other modalities and which hold in each frame $\mathcal{N}_C$) is well-known modal logic $S4.3$ (this follows directly from the finite model property of $S4.3$). Also the following holds:

Formulas $\Diamond p \to \Diamond_{D,l} p$ and $\Diamond p \to \Diamond_{D,g} p$, are not logical laws of LPD, they may be refuted in some frames $\mathcal{N}_C$ by certain valuations. Formulas $\Diamond_{D,g} p \to \Diamond_{D,l} p$, and $\Diamond_{D,g} p \to \Diamond p$, on the contrary, are logical laws, they are true in all frames $\mathcal{N}_C$ w.r.t. all valuations. So, the meaning of the operations well suits their names.

Obviously the language of LPD is much more expressive comparing with the standard languages combining multi-agent knowledge and time. It uses additional operations for local and global discovery together with evaluation operations for plausibility of statements. For most logics investigated in the literature the prime questions are problems of satisfiability and decidability, and in the next section we will address them.

5.4 Decidability

We aim to provide an algorithm which would decide by a formula whether it is satisfiable in LPD or not. To construct such an algorithm we will use the technique borrowed from our previous research focused on applying inference rules to logics originating in artificial intelligence and computer science (cf. [26,27,24]). In particular, this approach has been successfully implemented for Logic of Discovery in [24]. To apply this technique we need to customize to our language, the representation of formulas by rules, and conversion of rules to reduced normal forms, as it was discussed in the section "Augmenting the Modal Language".

A rule r is said to be in the *reduced normal form* if $r = \varepsilon/x_1$, where

$$\varepsilon := \bigvee_{1\leq j\leq m}\left(\bigwedge_{1\leq i\leq n}\left[x_i^{t(j,i,0)} \wedge (\Diamond x_i)^{t(j,i,1)} \wedge (\Diamond_{D,l} x_i)^{t(j,i,2)} \wedge \right.\right.$$

$$\left.\left.(\Diamond_{D,g} x_i)^{t(j,i,3)} \wedge \bigwedge_{s=1}^{m} (\neg K_s \neg x_i)^{t(j,i,4,s)} \wedge \bigwedge_{s=1}^{m} \mathrm{pl}(x_i, s/m)^{t(j,i,5,s)}\right]\right),$$

and x_i's are variables, $t(j,i,z)$, $t(j,i,k,z)$ are some functions with values in $\{0,1\}$, and, for any formula α above, $\alpha^1 := \alpha$, $\alpha^0 := \neg\alpha$.

For any formula φ in the language of LPD, we can convert it into the rule $x \to x/\varphi$ and employ technique of reduced normal forms as follows:

Definition 5. *Given a rule r_{nf} in the reduced normal form, r_{nf} is said to be the* normal reduced form for a rule r *iff, for any frame $\mathcal{N}_C$,*

$$\mathcal{N}_C \models r \Longleftrightarrow \mathcal{N}_C \models r_{\mathrm{nf}}.$$

Based on proofs of Lemma 3.1.3 and Theorem 3.1.11 from [27], as was discussed in the previous section, we obtain

Theorem 1. *There exists an algorithm running in exponential time, which, for any given rule r, constructs its normal reduced form r_{nf}.*

Recall, that a formula φ is valid in a frame $\mathcal{N}_C$ iff the rule $x \to x/\varphi$ is valid in $\mathcal{N}_C$, and, consequently, φ is satisfiable in LPD iff the rule $x \to x/\neg\varphi$ is refuted in some $\mathcal{N}_C$ frame. Therefore, from Theorem 1 we obtain

Proposition 2. *A formula φ is satisfiable in LPD iff the rule $(x \to x/\neg\varphi)_{\mathrm{nf}}$ is refuted in some frame $\mathcal{N}_C$ by a valuation V.*

Thus, for checking satisfiability in LPD, it is sufficient to find an algorithm recognizing rules in reduced normal form, which may be refuted in frames $\mathcal{N}_C$. For this we need the following special finite Kripke models.

Take any frame $\mathcal{N}_C$ and some numbers n_1, m_1, where $m_1 > n_1 \geq 3$. The frame $\mathcal{N}_C(n_1, m_1)$ has the following structure:

$$\mathcal{N}_C(n_1, m_1) := \left\langle \bigcup_{1\leq i\leq m_1} C(i), R, R_1, \ldots R_m \right\rangle,$$

where R is the accessibility relation from $\mathcal{N}_C$ extended by pairs (x, y), where $x \in C(m_1)$, $y \in C(n_1)$. For any given valuation V of letters from a formula φ in $\mathcal{N}_C(n_1, m_1)$, the truth values of φ can be defined at the elements of $\mathcal{N}_C(n_1, m_1)$

literary by the same rules as for the frames $\mathcal{N}_C$ above, even though the new frames are finite. To illustrate this similarity, below we provide formal definitions for some of logical operations on the new frames:

$$(\mathcal{N}_C(n_1, m_1), a) \models_V \Diamond\varphi \iff \exists b \in \mathcal{N}[(aRb) \text{ and } (\mathcal{N}_C(n_1, m_1), b) \models_V \varphi];$$
$$(\mathcal{N}_C(n_1, m_1), a) \models_V \Diamond_{D,l}\varphi \iff a \in C(i) \,\&\, \exists b \in C(i) : (\mathcal{N}_C, b) \models_V \varphi;$$
$$(\mathcal{N}_C(n_1, m_1), a) \models_V \Diamond_{D,g}\varphi$$
$$\iff [a \in C(i) \,\&\, b \in C(j) \,\&\, aRb \Rightarrow \exists c \in C(j)\, (\mathcal{N}_C, c) \models_V \varphi];$$
$$\forall a \in \mathcal{N}_C(n_1, m_1), \forall q \in Q : (\mathcal{N}_C(n_1, m_1), a) \models_V \mathrm{pl}(\varphi, q) \iff \mathrm{pr}(\varphi, a, V) = q.$$

Based on this modified Kripke structures $\mathcal{N}_C(n_1, m_1)$ we can prove:

Lemma 2. *A rule r_{nf} in the reduced normal form is refuted in a frame $\mathcal{N}_C$ w.r.t. a valuation V if and only if r_{nf} is refuted in a frame $\mathcal{N}_C(n_1, m_1)$ by a special valuation U, where*

1. *the size of any cluster $C(i)$ in $\mathcal{N}_C(n_1, m_1)$ is linear in the size of r_{nf};*
2. *numbers n_1 and m_1 are exponential in the size of r_{nf};*
3. *the size of the frame $\mathcal{N}_C(n_1, m_1)$ is exponential in the size of r_{nf}.*

Proof. (Sketch, see Fig. 7) We start with the model $\langle \mathcal{N}_C, V \rangle$ that refutes r (see Fig. 7 (a)). This model in general might have R-clusters of arbitrary cardinality. As the first step we filtrate them individually by disjuncts of the rule. After filtration (see Fig. 7 (b)) we obtain a model that have only finitely many non-isomorphic R-clusters, all of them of finite cardinality. Moreover, this cardinality is bounded as in item 1. of Lemma 2. The countably-indexed sequence of such clusters can be "rolled up" starting with some $C(i)$, $i \geq 3$. Not all of the obtained R-clusters are essential, and we can restrict our attention to R-clusters that are, in a sense, *maximal* with respect to the set of disjuncts of r (see Fig. 7 (c)). The resulting model has the form $\mathcal{N}_C(n_1, m_1)$ as required.

For the other direction we only need to show that given a model of the form $\mathcal{N}_C(n_1, m_1)$ refuting r, we can find for some $\mathcal{N}_C$ a valuation refuting r, using $\mathcal{N}_C(n_1, m_1)$ as a pattern.

Combining Theorem 1, Proposition 2 and Lemma 2 we obtain

Theorem 2. *The problem of satisfiability in LPD is decidable.*

The algorithm of verification for a formula φ to be satisfiable in LPD consists of the following steps:

(1) reduce the formula φ to the rule $x \to x/\neg\varphi$;

(2) transform the rule $r := x \to x/\neg\varphi$ to its reduced normal form r_{nf};

(3) check if the rule r_{nf} may be refuted in frames of the kind $\mathcal{N}_C(n_1, m_2)$ of sizes exponential in r_{nf}.

Overall complexity of the algorithm also includes the cost of reduction of rules $x \to x/\neg\varphi$ to normal reduced forms, but this complexity is single exponential, the same as for reducing Boolean formulas to disjunctive normal forms.

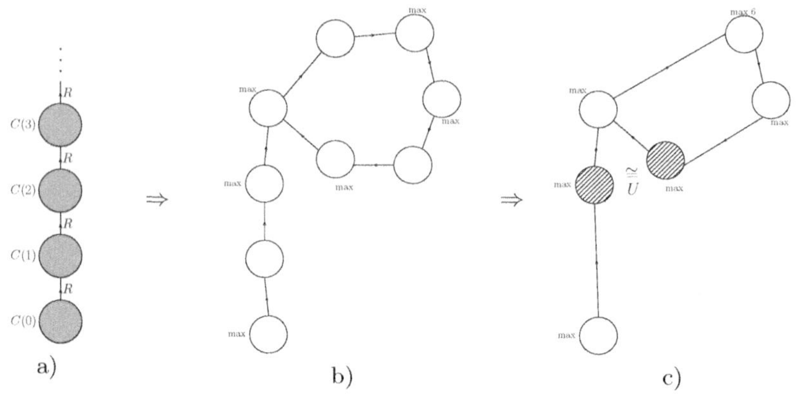

Fig. 7. Proof of Lemma 2 (schematically): a) original model $\mathcal{N}_C$ with (probably) infinite R-clusters; b) after filtration, all clusters are finite and there are only finitely many non-isomorphic ones; c) only clusters that are "maximal" with respect to our formula are needed.

6 Conclusion

The ongoing emergence of the global informational network is most prominently embodied by the development of the Internet. Its defining characteristics, such as low latency and low overhead cost of informational transactions combined with the problem of inherently unreliable and even outright malevolent participating agents, set a whole new set of practical and theoretical challenges for the industry and researchers alike.

One of the theoretical challenges is finding an adequate language for description of, at least, localized processes in such networks. Such language, if found, could be used for specification, verification and monitoring. This task faces a usual challenge of striking a balance between expressiveness, craved by practitioners and need of applications, and good logical properties, that are required by goals of scalability, maintenance and control. In this respect, a metaphor of multi-agent systems provides a convenient framework for further formal investigations of localized phenomena in a global network.

In this chapter we presented some logical tools that are suitable and can be useful for the purposes of behavioral description of multi-agent systems. These tools are based on apparatus of multi-modal propositional logic augmented with additional operators, the latter are specifically designed for monitoring and control. The particular method of frame refinements, that was suggested by one of the authors [29,30], allows us in many cases to preserve the good logical properties of the combined language, that is witnessed by the decidability of the logic LPD, provided as an example. Decidability is an important (but not necessary) condition which makes a logic and underlying language useful, at least theoretically, for verification and specification purposes.

There are certain drawbacks though inherent to multi-modal approach. They are witnessed by the example of LPD. The first-choice naïve decision algorithm has the double-exponential complexity bound, and there is a little hope for significant improvement. Even if we use the more refined tableaux-based methods, which seems can be applicable in this case, the theoretical complexity bound will stay the same (but there might be significant improvements for some formulas of special kind). This is the usual price to pay for using modal propositional logic. This double-exponential complexity bound makes the presented algorithm hardly practical for describing real life multi-agent systems, with dozens of interacting agents, and kilobytes of output data (we are talking here not about the total transmitted data, but rather the information pertinent to the behavior of the system). Nevertheless, it leaves the opportunity to investigate the basic underlying principles and structural properties of the multi-agent systems.

References

1. Arisha, K., Ozcan, F., Ross, R., Subrahmanian, V.S., Eiter, T., Kraus, S.: Impact: A platform for collaborating agents. IEEE Intelligent Systems 14(3), 64–72 (1999)
2. Avouris, N.M.: Co-operation knowledge-based systems for environmental decision-support. Knowledge-Based Systems 8(1), 39–53 (1995)
3. Blackburn, P., de Rijke, M.: Zooming in, zooming out. Journal of Logic, Language and Information 6, 5–31 (1997)
4. Blackburn, P., de Rijke, M., Venema, Y. (eds.): Modal Logic. Cambridge University Press, Cambridge (2001)
5. Bugarin, A.J., Barro, S.: Fuzzy reasoning supported by petri nets. IEEE Transactions on Fuzzy Systems 2(2), 135–150 (1994)
6. Caleiro, C., Carnielli, W.A., Rasga, J., Sernadas, C.: Fibring of logics as a univrersal construction. In: Gabbay, D.M., Guenther, F. (eds.) Combination of Logics. Handbook of Philosopical Logic, vol. 13. Kluwer, Dordrecht (2005)
7. Chagrov, A., Zakharyaschev, M.: Modal Logic. Oxford Logic Guides, vol. 35. Clarendon Press, Oxford (1997)
8. Cortés, U., Sánchez-Marré, M., Ceccaroni, L.: Artificial intelligence and environmental decision support systems. Applied Intelligence 13(1), 77–91 (2000)
9. Dubois, D., Prade, H.: The three semantics of fuzzy sets. Fuzzy Sets and Systems 90(2), 141–150 (1997)
10. Fagin, R., Halpern, J., Moses, Y., Vardi, M.: Reasoning About Knowledge. MIT Press, Cambridge (1995)
11. Fine, K.: Logics containing k4, part i. Journal of Symbolic Logic 39(1), 31–42 (1974)
12. Gabbay, D.M.: Fibred semantics and the weaving of logics, part 1: Modal and intuitionistic logics. The Journal of Symbolic Logic 61(4), 1057–1120 (1996)
13. Gabbay, D.M.: An Overview of Fibred Semantics and the Combinations of Logics. In: Frontiers of Combining Systems. Applied Logic Series, vol. 3, pp. 1–55. Kluwer Academic Publishers, Dordrecht (1996)
14. Gabbay, D.M.: Fibring Logics. Oxford University Press, Oxford (1999)
15. Gabbay, D.M., Kurucz, A., Wolter, F., Zakharyaschev, M.: Many-dimensional modal logics: theory and applications. Studies in Logic, vol. 148. Elsevier Science, Amsterdam (2003)
16. Goldblatt, R.: Logics of Time and Computation. CSLI Lecture Notes, vol. 7. CSLI (1992)

17. Goldblatt, R.: Mathematical modal logic: A view of its evolution. Journal of Applied Logic 1(5-6), 309–392 (2003)
18. Harmelem, F., Horrocks, I.: The semantic web and its languages-faqs on oil: The ontology inference layer. IEEE Intelligent Systems 15(6), 69–72 (2000)
19. Hendler, J.: Agents and the semantic web. IEEE Intelligent Systems 16(2), 30–37 (2001)
20. Hintikka, J., Vandamme, F.: Logic of Discovery and Logic of Discourse. Springer, Heidelberg (1986)
21. Kurz, A.: Coalgebras and Modal Logic. CWI, `http://www.cwi.nl/~kurz`
22. Libkin, L.: Elements of Finite Model Theory. EATCS. Springer, Heidelberg (2004)
23. Ohsawa, Y., McBurney, P. (eds.): Chance Discovery. Advanced Information Processing. Springer, Heidelberg (2003)
24. Rybakov, V.: Logic of discovery in uncertain situations—deciding algorithms. In: Apolloni, B., Howlett, R.J., Jain, L. (eds.) KES 2007, Part II. LNCS (LNAI), vol. 4693, pp. 950–958. Springer, Heidelberg (2007)
25. Rybakov, V.: Since-until temporal logic based on parallel time with common past. In: Artemov, S., Nerode, A. (eds.) LFCS 2007. LNCS, vol. 4514, pp. 486–497. Springer, Heidelberg (2007)
26. Rybakov, V.V.: A criterion for admissibility of rules in the modal system S4 and the intuitionistic logic. Algebra and Logica 23(5), 369–384 (1984)
27. Rybakov, V.V.: Admissible Logical Inference Rules. Studies in Logic and the Foundations of Mathematics, vol. 136. Elsevier Sci. Publ., North-Holland (1997)
28. Rybakov, V.V.: Construction of an explicit basis for rules admissible in modal system s4. Mathematical Logic Quarterly 47(4), 441–451 (2001)
29. Rybakov, V.V.: Logical consecutions in discrete linear temporal logic. Journal of Symbolic Logic 70(4), 1137–1149 (2005)
30. Rybakov, V.V.: Logical consecutions in intransitive temporal linear logic of finite intervals. Journal of Logic Computation 15(5), 633–657 (2005)
31. Rybakov, V.V.: Linear Temporal Logic with Until and Before on Integer Numbers, Deciding Algorithms. In: Grigoriev, D., Harrison, J., Hirsch, E.A. (eds.) CSR 2006. LNCS, vol. 3967, pp. 322–334. Springer, Heidelberg (2006)
32. Thomason, S.K.: Semantic analysis of tense logic. Journal of Symbolic Logic 37(1) (1972)
33. van Benthem, J.: Modal Correspondence Theory. PhD thesis, University of Amsterdam (1976)
34. van Benthem, J.: The Logic of Time. Synthese Library, vol. 156. Reidel, Dordrecht (1983)
35. van Benthem, J.: Modal Logic and Classical Logic. Bibliopolis, Naples (1983)

Multi-Agent Systems and Large State Spaces

Yann-Michaël De Hauwere, Peter Vrancx*, and Ann Nowé

Computational Modeling Lab - Vrije Universiteit Brussel
{ydehauwe,pvrancx,anowe}@vub.ac.be

Abstract. A major challenge in multi-agent reinforcement learning remains dealing with the large state spaces typically associated with realistic multi-agent systems. As the state space grows, agent policies become more and more complex and learning slows down. The presence of possibly redundant information is one of the causes of this issue. Current single-agent techniques are already very capable of learning optimal policies in large unknown environments. When multiple agents are present however, we are challenged by an increase of the state space, which is exponential in the number of agents. A solution to this problem lies in the use of Generalized Learning Automata (GLA). In this chapter we will first demonstrate how GLA can help take the correct actions in large unknown environments. Secondly, we introduce a general framework for multi-agent learning, where learning happens on two separate layers and agents learn when to observe each other. Within this framework we introduce a new algorithm, called 2observe, which uses a GLA-approach to distinguish between high risk states where the agents have to take each others presence into account and low risk states where they can act independently. Finally we apply this algorithm to a gridworld problem because of the similarities to some real-world problems, such as autonomous robot control.

Keywords: Multi-agent learning, reinforcement learning, large state spaces.

1 Introduction

Reinforcement learning (RL) has been shown to be a powerful tool for solving single agent Markov Decision Processes (MDPs). It allows a single agent to learn a policy that maximises a possibly delayed reward signal in an initially unknown stochastic stationary environment. While basic RL techniques are not suited for problems with very large state spaces, since they rely on a tabular representation for policies and enumerate all possible state-action pairs, several extensions have been proposed to reduce the complexity of learning. The use of temporally extended actions has recently been introduced as a possible solution [1,2]. Other

* Yann-Michaël De Hauwere and Peter Vrancx are both funded by a Ph.D grant of the Institute for the Promotion of Innovation through Science and Technology in Flanders (IWT Vlaanderen).

A. Håkansson et al. (Eds.): Agent and Multi-agent Technology, SCI 289, pp. 181–205.
springerlink.com

methods rely on function approximators, representing the agent's policy such as neural networks, decision trees and other regression techniques.

The non-stationary environment, agents experience and the uncertainty about the other agents' goal make the problem of large state spaces even more pertinent in multi-agent systems (MAS). Relatively little work has been done, however, on extending these RL techniques for large state spaces to MAS. One possible approach is to simply let each agent independently apply a single agent technique such as Q-learning, ignoring the other agents. This approach has limited applicability however, since in a multi-agent setting the typical convergence requirements of single-agent learning no longer hold. The other extreme is to let agents learn in a joint state-action space. In the cooperative case the problem can then be cast into a single agent MDP over the joint agent actions. This method assumes that centralised control is feasible and does not scale well since the joint actions are exponential in the number of agents. In the non-cooperative case typically one tries to learn an equilibrium between agent policies [3,4]. These systems need each agent to calculate equilibria between possible joint actions in every state and as such need each agent to retain estimates over all joint actions in all states. Other equilibrium learners, which do not learn in the joint action space exist [5,6], but even these systems still suffer from the large state spaces typical in realistic agent systems.

It is clear that some form of generalisation over the state space is necessary, to solve this problem. With this generalisation, a certain amount of accuracy must be traded in for acceptable learning times and memory considerations. In 1996, Boutilier already pointed out the interest and difficulties for using structured problem representations and generalisation techniques in multiagent environments [7] after having shown the usefulness of using Bayesian Networks (BN) to compactly represent the state transition function [8]. More recently, Guestrin et al. [9] introduced an algorithm for planning in cooperative MAS, using factored value functions and a simple message passing scheme among the agents to coordinate their actions.

However, all these approaches often assume that agent actively collaborate with each other or have full knowledge of the underlying transition and reward functions. Most research done so far towards learning to aggregate states in the RL problem has focused on learning the structure of the underlying problem [10,11,12]. Using this learned model, conventional techniques such as Dynamic Programming (DP) can be used to solve the problem.

Generalized Learning Automata (GLA) are capable of learning a generalisation over a large number of states, without needing to store large amounts of samples, perform computational intensive statistical tests or build a model of the environment. We will demonstrate the use of these GLA in large unknown environments and show that multiple GLAs can aggregate over states when more than one agent is present in the system. Furthermore we will introduce an alternative solution to the problem of large state spaces. Consider the problem of a robot trying to learn a route to a goal location. In a stationary environment the robot can simply rely on basic sensor input to explore the environment and learn a path. Suppose now that other mobile robots are added to this system. The robot must now take care to reach its goal without colliding with the other robots. However, even if the robot is provided with the exact locations of the other robots at all times, it does not make sense to always condition its actions on the locations of the others. Always accounting for the other robots means that both the state space as well as the action space are exponential in the number robots present in the system. For most tasks this joint state-action space representation includes a lot of redundant information, which slows learning without adding any benefits. For example, when learning how to reach a goal without colliding, robots need to take into account each other's actions and locations, only when it is possible to collide. When there is no risk of collision there's no point in differentiating between states based solely on the locations of other agents. Therefore we introduce a framework where agents will learn when to take the other agents into account and when it is safe to ignore their presence. This means that the robot relies on single agent techniques combined with its original sensor inputs to learn an optimal path, and coordinates with other agents only when collisions are imminent. We will use the GLA to let the agents learn when to opt for which technique. In the experiments section we demonstrate how our approach delivers significant savings, in both memory requirements as well as convergence time, while still learning a good solution.

Throughout this chapter we will begin by giving the necessary background information in Reinforcement Learning (RL) in Section 2 and multi-agent learning in Section 3 before presenting our work with GLAs in Sections 4 and 5. We conclude the chapter with a short discussion and guidelines for future work.

2 Reinforcement Learning

In this section we will explain some of the basic concepts of Reinforcement Learning (RL) and will elaborate in detail on one of the most well-known algorithms in this field: Q-learning.

Reinforcement learning solves the problem an agent encounters when he has to find a solution to a given situation. *Without some feedback about what is good and what is bad, the agent will have no grounds for deciding which move to make* [13]. Because of the absence of a teacher, telling which actions the agent must take, he must discover on its own, which actions yield the highest reward or reinforcement.

By repeatedly chosing random actions in the different game situations, the agent will eventually be able to build a predictive model of the environment and thus forsee the outcome of its actions. Note that this model is not necessarily an explicit one.

Furthermore, the rewards can either be immediate, i.e. the reward (or penalty) is granted immediately after an agent performed his action, for instance in the

n-armed bandit problem[1], or the reward can be delayed for a number of actions like in Chess for example, where the reward is given at the end of the game. So it is possible that an agent must sacrifice short-term gains for larger long-term gains [14,15].

The main advantage of reinforcement learning is its independence from an external supervisor. This means that it is able to learn in non-stationary environments using random actions to explore the environment. The main goal of reinforcement learning is to maximise the received reward. To do so, a trade-off must be made between exploring new actions and exploiting the actions found in the past with positive outcome. This trade-off is commonly known as the exploration-exploitation dilemma. We will not elaborate any further on this, but refer to the book by Sutton and Barto [14].

2.1 Markov Decision Processes

In a reinforcement learning system four elements can be distinguished next to the agent and the environment:

- a *policy* π_t. This is a function that maps states to actions, i.e., $\pi_t : S \rightarrow A$, where S is the set of states of the environment, and A is the set of actions available to the agent. $\pi_t(\sigma_t, a_t)$ denotes the probability that action a_t is chosen in state σ_t at time step t. This is the most important element of the reinforcement learning system because a policy alone is sufficient to determine the behaviour of an agent.
- a *reward function* R gives the quality of the actions chosen by the agent. It maps a state and an action to a real value (reward or penalty), i.e., $R : S \times A \rightarrow \mathbb{R}$. The reward signals are used by the agent for altering its policy.
- a *value function*, denoted $V(\sigma)$, is a mapping from a state to the reward an agent can expect to accumulate, when starting from that state σ. So this results in a long-term expected reward, whereas the reward function returns the immediate reward of a state. In reinforcement learning this is the function that agents will try to learn, as it is this function that will be used to obtain the highest reward.
- a *predictive model* of the environment. This will predict the behaviour of the environment (the next state and reward) when performing some action.

In the standard reinforcement learning model, depicted in Figure 1 proposed in [16], an agent interacts with its environment. This model consists of

- a discrete set of environment states S,
- a discrete set of actions from the agent A,
- a set of reinforcement signals R.

---

[1] In the n-armed bandit problem the agent is faced repeatedly with a choice among n different options, or actions. After each choice the agent receives a numerical reward chosen from a stationary probability distribution dependent on the action it selected. The objective is to maximize the expected total reward over some time period.

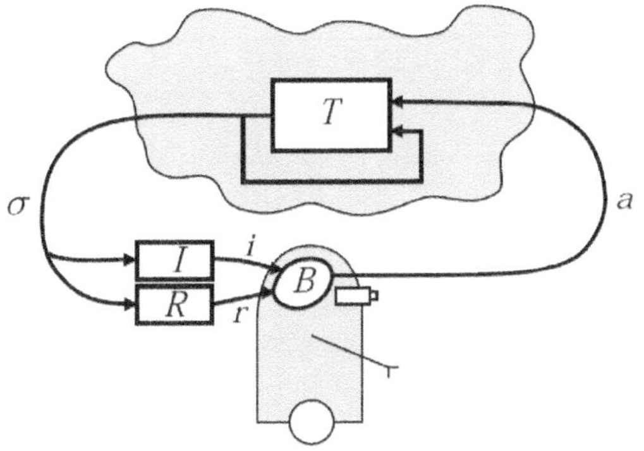

Fig. 1. The standard reinforcement learning model

On each time step the agent receives some information i about the current state s of the environment. Based on this information the agent chooses some action a. This action may change the state of the environment and the next state is communicated to the agent through a scalar reinforcement signal r. The agent will adapt his behaviour B in order to maximise the long-run measure of reinforcement.

A model for problems that are characterised by delayed rewards can be given as a *Markov Decision Process* (MDP). This brief overview is based on [14] and [16].

Definition 1. *A* ***Markov Decision Process*** *is a 4-tuple* (S, A, T, R)*, where:*

- S *is the state space,*
- A *is the action space,*
- T *is the state transition function,* $T : S \times A \rightarrow \Pi(S)$*, mapping states and actions on probabilities of making a transition to a next state. We will use* $T(\sigma, a, \sigma')$ *to represent the probability of reaching state* σ' *from* σ *after performing action* a*.*
- R *is the reward function,* $R : S \times A \rightarrow \mathbb{R}$*, mapping states and actions to a real numerical value.*

A Markov Decision Process has the following Markov property:

Definition 2. *A system is said to possess the* ***Markov property*** *if the system's state transitions are independent of any previous environment states or actions taken, more formally:*

$$T\left(\sigma_{t+1}|\sigma_t, a_t, \sigma_{t-1}, a_{t-1}, ..., \sigma_1, a_1\right) = T\left(\sigma_{t+1}|\sigma_t, a_t\right).$$

This property ensures that an agent can behave optimally by only observing its current state. It doesn't matter which actions were taken before, so no history needs to be stored. An agent behaving optimal, is often achieved through optimal value functions, which are defined in the following section.

2.2 Learning in MDPs

The optimal value of a state is defined as follows:

Definition 3. *The* ***optimal value of a state*** σ_t *is the expected infinite discounted sum of rewards the agent will receive when it starts in that state and follows the optimal policy from there, i.e.* $V^*(\sigma_t) = max_\pi E\left(\sum_{k=0}^{\infty} \gamma^k r_{t+1+k}\right)$.

An optimal policy is a solution to MDP's. It specifies which action should be chosen for every state. A policy is denoted as π, and the recommended action for state σ is $\pi(\sigma)$. An optimal policy is a policy that yields the highest expected reward. We use π^* to denote an optimal policy.

Values can be seen as predictions for future rewards. Often, actions are chosen based on these values, so one of the most important aspects of finding a solution to a reinforcement learning problem, is finding a method for estimating future rewards. In the following paragraphs two algorithms will be presented for calculating optimal policies. Both algorithms find their origin in Dynamic Programming (DP) techniques[2] as these are well known solutions for solving MDPs.

The idea of Dynamic Programming is to convert the famous Bellman equations (1957), given below in Equation 1, into update rules for improving approximations of the desired value function.

$$V^\pi(\sigma) = E\left[\sum_{k=0}^{\infty} \gamma^k r_{t+1+k} | \pi, \sigma_t = \sigma\right] \tag{1}$$

The optimal value function V^*, with π^* the optimal policy, are the solutions of the Bellman Optimality Equation (2) and are given in Equations 3 and 4.

$$V^*(\sigma) = \max_\pi E\left(\sum_{k=0}^{\infty} \gamma^k r_{t+1+k}\right) \tag{2}$$

$$= \max_a E\left(r_{t+1} + \gamma V^*(\sigma_{t+1}) | \sigma_t = \sigma, a_t = a\right)$$

$$= \max_a \left[R(\sigma, a) + \gamma \sum_{\sigma' \in \mathcal{S}} T(\sigma, a, \sigma') V^*(\sigma')\right], \forall \sigma \in \mathcal{S} \tag{3}$$

$$\pi^* = arg \max_a \left(R(\sigma, a) + \gamma \sum_{\sigma' \in \mathcal{S}} T(\sigma, a, \sigma') V^*(\sigma')\right) \tag{4}$$

For an arbitrary policy π, the state-value function V_π can be defined as:

$$V_\pi(\sigma) = E\left\{\sum_{k=0}^{\infty} \gamma^k r_{t+1+k}\right\}$$

[2] Dynamic Programming is the brainchild of the American Mathematician Richard Ernest Bellman, who described the way of solving problems where you need to find the best decisions one after another.

$$= E\left\{r_{t+1} + \gamma V_\pi(\sigma_{t+1}) | \sigma_t = \sigma\right\}$$
$$= \sum_a \pi(\sigma, a) \left[R(\sigma, a) + \gamma \sum_{\sigma' \in \mathcal{S}} T(\sigma, a, \sigma') V_\pi(\sigma') \right]$$

By improving V_π iteratively, V^* can be approximated. $V_\pi(\sigma)$-function itself can be obtained using successive approximations called iterative policy evaluation:

$$V_{k+1}(\sigma) = \sum_a \pi(\sigma, a) \left[R(\sigma, a) + \gamma \sum_{\sigma' \in (S)} T(\sigma, a, \sigma') V_k(\sigma') \right] \tag{5}$$

The ***policy iteration*** algorithm will manipulate a policy directly in order to improve it, based on the state-values that correspond to the current policy π. The algorithm is given below:

Algorithm 1. Policy Iteration

choose an arbitrary policy π
repeat
  $\pi \leftarrow \pi'$
  compute the value function V_π using iterative policy evaluation (Equation 5):
  improve the policy π at each state and store it in: π'
    $\pi'(\sigma) \leftarrow \text{argmax}_a(R(\sigma, a) + \gamma \sum_{\sigma' in \mathcal{S}} T(\sigma, a, \sigma') V_\pi(\sigma'))$
until $\pi = \pi'$

To improve the policy, the best action a in a state σ is identified, based on the current state values. So the policy π is improved in state σ by updating $\pi(\sigma)$ into the action that maximises the right hand side of Equation 5, resulting in a better policy π'. We thus obtain a sequence of monotonically improving policies and value functions:

$$\pi_0 \rightarrow V_{\pi 0} \rightarrow \pi_1 \rightarrow V_{\pi 1} \rightarrow \ldots \rightarrow \pi^* \rightarrow V^*$$

A finite MDP has only a finite number of distinct policies, at most $|A|^{|S|}$, and the sequence improves at each step, so this algorithm terminates in at most an exponential number of iterations. This means, that this process is bound to result in an optimal policy and an optimal value function, when no further improvements are possible.

The basic idea behind the ***value-iteration*** algorithm is to find the optimal value function $V^*(\sigma)$. The difference with policy iteration is that in this algorithm, no explicit policy evaluation will take place at each time step. This policy evaluation process is truncated, without losing convergence guarantees. This one-step truncated policy evaluation is given in Equation 6. The value iteration algorithm is given in Algorithm 2.

$$
\begin{aligned}
V_{k+1}(\sigma) &= \max_a E\{r_{t+1} + \gamma V_k(\sigma_{t+1}|\sigma_t = \sigma, a_t = a\} \\
&= \max_a \left[R(\sigma, a) + \gamma \sum_{\sigma' \in S} T(\sigma, a, \sigma') V_k(\sigma') \right]
\end{aligned} \tag{6}
$$

Algorithm 2. Value Iteration

initialise $V(\sigma)$ arbitrarily
repeat
  for all $\sigma \in S$ **do**
    for all $a \in A$ **do**
      $Q(\sigma, a) \leftarrow R(\sigma, a) + \gamma \sum_{\sigma' \in S} T(\sigma, a, \sigma') V_k(\sigma')$
      $V(\sigma) \leftarrow \max_a Q(\sigma, a)$
    end for
  end for
until Policy is good enough

As can be seen from the pseudocode, it is not strictly determined when to stop the outer loop. However this can be handled by the following condition: *If the maximum difference between two successive value functions is less than* ϵ, *then the value of the policy differs from the value function of the optimal policy by no more than* $2\epsilon\gamma/(1-\gamma)$. This provides an effective stopping criterion for the algorithm.

It is very common to reinforcement learning problems that the transition function or the reward function are unknown to the reinforcement learner. As such, it is hard to learn the optimal policy. So the problem here is, which evaluation function should be used. Let us define the evaluation function Q so that its value is the maximum discounted cumulative reward that can be achieved starting from state σ and applying action a as the first action.

$$Q(\sigma, a) \equiv r(\sigma, a) + \gamma V^*(\delta(\sigma, a)) \tag{7}$$

Note that with this equation, we can define the optimal policy as

$$\pi^*(\sigma) = \operatorname*{argmax}_a Q(\sigma, a) \tag{8}$$

and

$$V^*(\sigma) = \max_{a'} Q(\sigma, a') \tag{9}$$

and thus we have the following recursive definition for Q:

$$Q(\sigma, a) = r(\sigma, a) + \gamma \max_{a'} Q(\delta(\sigma, a), a') \tag{10}$$

So in order to find the optimal policy, one has to find this Q-function. Watkins described an algorithm to iteratively approximate it. In the ***Q-learning*** algorithm [17] a large table consisting of state-action pairs is stored. Each entry contains the value for $\hat{Q}(\sigma, a)$ which is the learner's current hypothesis about the actual value of $Q(\sigma, a)$. The $\hat{Q}$-values are updated accordingly to following update rule:

$$\hat{Q}(\sigma, a) \leftarrow (1 - \alpha_t)\hat{Q}(\sigma, a) + \alpha_t[r + \gamma \max_{a'} \hat{Q}(\sigma', a')] \quad (11)$$

where α_t is the learning rate at time step t. Which gives us more formally the following algorithm:

Algorithm 3. Q-Learning

For each $\langle \sigma, a \rangle$ pair, initialise the table entry for $\hat{Q}(\sigma, a)$ to zero.
Observe the current state σ
loop
- Select an action a and execute it
- Receive immediate reward r
- Observe the new state σ'
- Update the table entry accordingly to equation 11
- $\sigma \leftarrow \sigma'$

end loop

This algorithm was an inspiration for a whole range of new RL-algorithms and extensions. In the following section we will discuss some of the current work around RL for large state spaces before moving on to a multi-agent setting.

2.3 Reinforcement Learning in Large State Spaces

In the previous section we explained the basic Q-learning algorithm, as proposed by Watkins in 1992. The major drawback of this algorithm is that it relies on a tabular representation of all state-action pairs in the environment. It is clear that, as the environment grows due to increasing demands in detail and complexity, simply enumerating all these states is no longer feasible.

Several techniques have been proposed to deal with these ever increasing demands. These techniques typically rely on some type of generalisation method, which transfers knowledge between similar states (or state-action pairs). One way of doing this is by representing Q-values or policies by function approximators [18].

Other techniques use adaptive resolution methods. Here the learning agent uses statistic tests to determine when a greater granularity in state space representation is needed. One example of this kind of system is the G-learning algorithm [19], which uses decision trees to learn partitions of the state space. This technique assumes that a state is represented by a number of discrete valued variables. Starting out with a single state the technique keeps statistics on Q-values corresponding to the values of each variable. When it finds a significant difference in Q-values for different values of a variable, the decision tree is split based on this variable. The process is then repeated for the new leaves of the tree.

State information is often not presented in an optimal way and may contain a lot of redundant and/or useless information which slows the learning process down. A more efficient way is to represent the state information as a set of random variables $X = \{X_1, \ldots, X_n\}$, where every state variable X_i can take values in a finite domain $Dom(X_i)$ and every possible state in the system corresponds to a value assignment $x_i \in Dom(X_i)$ for every state variable, X_i. Such a system is called a Factored Markov Decision Process (FMDP) [8,7,20].

The transition function in such a system is described by a *Dynamic Bayesian Network (DBN)*. This is a two-layer directed acyclic graph where the nodes are $\{X_1, \ldots, X_n, X'_1, \ldots, X'_1\}$. In this graph, the parents of X'_i are denoted by $Parents_a(X'_i)$. With every node $X'_i \in G_a$, a Conditional Probability Distribution $CPD^a_{X_i}(X'_i|Parents_a(X'_i))$ is associated quantifying the DBN. This method benefits from the dependencies that exist (or don't exist) between the variables of the network.

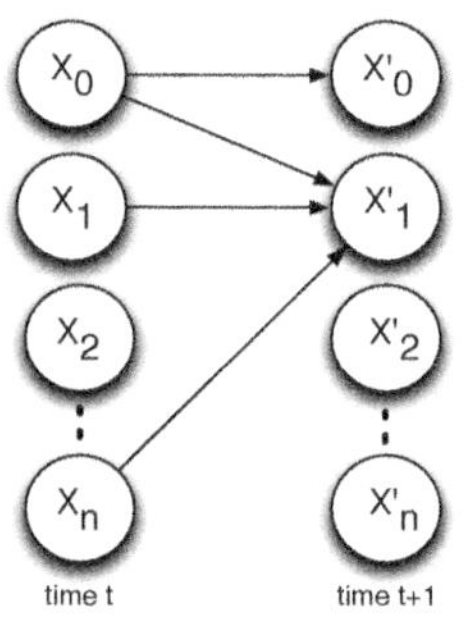

Fig. 2. DBN for action a_i

Figure 2 shows the dependencies between the variables at time t and time $t+1$ when action a_i is executed. According to this network, the Parents of X'_1 are X_0, X_1 and X_n.

When the algorithm is offered state information as such it can try to learn the state transition function and construct a model of the system, before applying planning algorithms on this model. This technique is commonly known as the DYNA-architecture [21] and has found its way to the FMDPs in [10]. Even without explicitely learning the model of the problem at hand, such a structured representation offers advantages to aggregate states and exploit the underlying structure of the system. One of the techniques that takes advantage of a structured representation are the GLA which we discuss in Section 4.

3 Multi-Agent Reinforcement Learning

In the previous two sections we explained some of the work done in RL and large state spaces. This section will focus on environments where multiple agents are present. Due to the presence of these agents, a large state space is inherently common to a MAS.

3.1 Markov Games

In the previous section we explained how a MDP is one of the key elements to solving RL problems. In MAS we can not use this model anymore. The Markov property fails, because there are different agents, changing the dynamics of the world. An extension of the single agent Markov decision problems (MDP's) to the multi-agent case can be defined by Markov Games [22]. In a Markov Game, actions are the joint result of multiple agents choosing an action seperately. Formally the Markov Game can be represented by the 5-tuple: $M = \langle \mathbb{A}, \{A\}_{\forall i \in \mathbb{A}}, \mathbb{S}, T, \mathbb{R} \rangle$, where:

- $\mathbb{A}$ is the set of agents participating in the game,
- $\{A_i\}_{\forall i \in \mathbb{A}}$ is the set of actions available to agent i,
- $\mathbb{S}$ is the set of states
- $T(s, \overrightarrow{a}, s)$ is the transition function stating the probability that a joint-action a will lead the agents from state s to state s',
- and $R_i : S \times A_1 \times \ldots A_{|\mathbb{A}|} \to \mathbb{R}$ is the reward function denoting the reward agent i gets for performing a joint action in the current state.

Note that the reward function R_i is now individual to each agent i. Different agents can receive different rewards for the same state transition. Since each agent i has its own individual reward function, defining a solution concept becomes more complex. Due to the existence of different reward functions, it is in general impossible to find an optimal policy for all agents. Instead, equilibrium points are sought. In an equilibrium, no agent can improve its reward by changing its policy if all other agents keep their policy fixed.

A special case of the general Markov game framework is given by the Multi-agent Markov Decision Process (MMDP) [23]. In this case, the Markov game is purely cooperative and all agents share the same reward function. This specialisation allows us to define the optimal policy as the joint agent policy, which maximises the payoff of all agents.

Because the agents share the same transition and reward function, one can think of the collection of agents being a single super agent with joint actions at its disposal and whose goal is to learn the optimal policy for the joint MDP. Since the agents' individual action choices may be jointly suboptimal, the added problem in MMDP's is for the agents to learn to coordinate their actions so that joint optimality is achieved. This model suffers from two major problems. Since all agents receive the same reward signal, it is not possible to apply this technique in situations where the agents have conflicting interest. The other drawback of this technique is that the states and actions of the MMDP increase exponentially in the number of agents, and thus it is not suitable for systems that already suffer from a large state space.

In Table 1 we present of a comparison of the state space complexity of some of the commonly used techniques in multi-agent learning.We compare a set of independent Q-learners to the MMDP-framework and the multi-agent extensions of Q-Learning, Nash and Correlated Q-Learning. The difference between these latter two and the MMDP lies in the fact that the agents still learn independently when using the Q-Learning variants. The major drawback of these two techniques is that they require a full observability about the actions and rewards of all the agents. On top of this, every agent stores not only his own Q table, but also a Q table for every other agent in the system. Even tough these techniques are a vast improvement over the MMDP, which also requires all the agents to select a joint-action, computationaly these techniques are not an improvement at all.

Table 1. Space complexity of different MAS techniques in function of the number of states, actions and agents

	Independent Q-learning	MMDPs	Nash Q-Learning	Correlated Q-Learning
#states	linear	linear	linear	linear
#actions	polynomial	polynomial	polynomial	polynomial
#agents	n.a.	exponential	exponential	exponential

In the following sections we will describe our solution in detail, which is computationally much less intensive than Nash and Correlated Q-Learning and whose space complexity is the same as independent Q-Learners without the need of communication at every time step.

4 GLAs for Multi-Agent Learning

In this section we introduce our approach for scalable multi-agent learning. The method is similar to the ideas used in single agent reinforcement learning for large state space, in that we are trying to generalise policies over similar states. In multi-agent systems additional problems arise, however. Since system control is distributed over multiple agents with possibly conflicting goals, coordination between agents is required. Furthermore, in addition to the large state space typically associated with realistic agent systems, the action space now also grows exponentially in the number of agents. This motivates the need for generalisation in the action space as well as the state space.

As a basic tool for our approach we use simple pattern matching units called Generalized learning automata. These automata have the advantage that they are computationally simple, can be combined into larger networks to offer greater flexibility and have theoretical foundations [24].

In the first instance we use GLAs to learn similar regions in the state space. A policy can then be represented by a mapping from each region to an action, rather than a mapping from each individual state.

In the next section we extend this approach to include adaptive state space resolution and action generalisations. Now the GLAs learn regions in the state space, but rather than directly selecting an action, they choose between different strategies. In state space regions were conflicts are possible the GLAs select multi-agent techniques in order to coordinate between agents, in other regions single agent strategies with limited state space visibility are used and agents effectively ignore each other.

4.1 Generalized Learning Automata

A GLA is an associative reinforcement learning unit. The purpose of a GLA is to learn a mapping from given inputs or contexts to actions. At each time step the GLA receives an input which describes the current system state. Based on this input and its own internal state the unit then selects an action. This action serves as input to the environment, which in turn produces a response for the GLA. Based on this response the GLA then updates its internal state.

Formally, a GLA can be represented by a tuple (X, A, β, u, g, U), where X is the set of inputs to the GLA and $A = \{a_1, \ldots, a_r\}$ is the set of outputs or actions the GLA can produce. $\beta \in [0, 1]$ again denotes the feedback the automaton receives for an action. The real vector u represents the internal state of the unit. It is used in conjunction with the probability g to determine the action probabilities, given an input $x \in X$:

$$P\{a(t) = a|u, x\} = g(x, a, u) \tag{12}$$

where g has to satisfy following conditions:

$$g(x, a, u) \geq 0 \; \forall x, a, u$$
$$\sum_a g(x, a, u) = 1 \quad \forall x, u$$

U is a learning algorithm which updates $\mathbf{u}$, based on the current value of $\mathbf{u}$, the given input, the selected action and response β. In this paper we use a modified version of the REINFORCE [25] update scheme. In vector notation this update scheme can be described as follows:

$$\begin{aligned} \mathbf{u}(t+1) = \mathbf{u}(t) + \lambda\beta(t)\tfrac{\delta ln\, g}{\delta \mathbf{u}}(\mathbf{x}(t), a(t), \mathbf{h}(\mathbf{u}(t))) \\ +\lambda\mathbf{K}(\mathbf{h}(\mathbf{u}(t)) - \mathbf{u}(t)) \end{aligned} \tag{13}$$

where $\mathbf{h}(\mathbf{u}) = [h_1(u_1), h_2(u_2), \ldots h_r(u_r)]$, with each h_i defined as:

$$h_i(\eta) = \begin{cases} L_i & \eta \geq L_i \\ 0 & |\eta| \leq L_i \\ -L_i & \eta \leq -Li \end{cases} \tag{14}$$

In this update scheme λ is the learning rate and $L_i, K_i > 0$ are constants. The update scheme can be explained as follows. The first term added to the parameters is a gradient following term, which allows the system to locally optimise the action probabilities. The next term uses the $h_i(u)$ functions to keep parameters u_i bounded within predetermined boundaries $[-L_i, L_i]$. This term is added since the original REINFORCE algorithm can give rise to unbounded behavior. In [26] it is shown, that the adapted algorithm described above, converges to local maxima of $f(\mathbf{u}) = E[\beta|\mathbf{u}]$, showing that the automata find a local maximum over the mappings that can be represented by the internal state in combination with the function g.

Originally these systems were proposed for classification problems, in which the context vectors represent features of objects to be classified and the GLA output represents class labels. We propose to use the same techniques in factored MMDPs. In such a system each agent internally uses a set of GLA to learn the different regions in the state space where different actions are optimal.

We use the following set-up for the GLA. With every action $a_i \in A$ the automaton can perform, it associates a vector $\mathbf{u_i}$. This results in an internal state vector $\mathbf{u} = [\mathbf{u_1}^\tau \ldots \mathbf{u_r}^\tau]$ (where τ denotes the transpose). With this state vector we use the Boltzmann distribution as probability generating function:

$$g(x, a_i, \mathbf{u}) = \frac{e^{\frac{x^\tau u_i(a_i)}{T}}}{\sum_j e^{\frac{x^\tau u_j(a_i)}{T}}} \tag{15}$$

with T a parameter that represents the temperature. Of course, since this function is fixed in advance and the environment in general is not known, we have no guarantee that the GLA can represent the optimal mapping. For instance, when using the function given in Equation 15 with a 2-action GLA, the internal state vector represents a hyperplane. This plane separates context vectors which give a higher probability to action 1 from those which action 2. If the sets of context vectors where different actions are optimal, are not linearly separable the GLA cannot learn an optimal mapping.

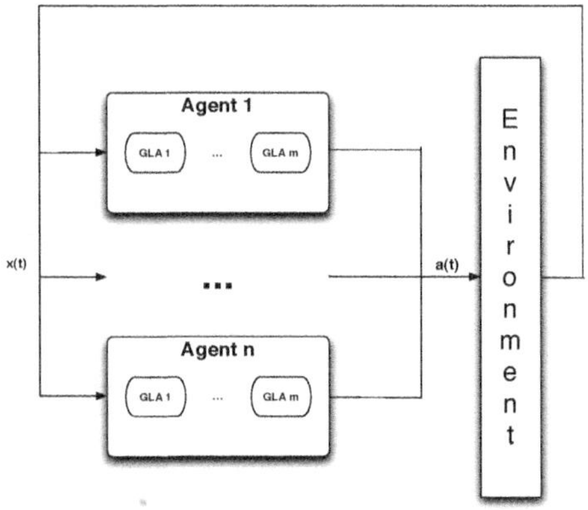

Fig. 3. Learning set-up. Each agent receives factored state representation as input. GLA decide action to be performed.

To allow a learner to better represent the desired mapping from context vectors to actions, we can utilise systems composed of multiple GLA units. For instance the output of multiple 2-action GLAs can be combined to allow learners to build a piecewise linear approximation of regions in the space of context vectors. In general, we can use systems which are composed of feedforward structured networks of GLA. In these networks, automata on one level use actions of the automata on the previous level as inputs. If the feedforward condition is satisfied, meaning that the input of a LA does not depend on its own output, convergence to local optima can still be established [24].

Figure 3 shows the general agent learning set-up. Each time step t a vector $\mathbf{x(t)}$ giving a factored representation of the current system state is generated. This vector is given to each individual agent as input. The agents internally use a set of GLA to select an action corresponding to the current state. The joint action $\mathbf{a(t)}$of all agents serves as input to the environment, which responds with a feedback $\beta(t)$ that agents use to update the GLA. One of the main advantages of this approach is that convergence guarantees exist for general feedforward GLA structures. In the common interest problems under study in this paper, a group of agents each internally using one or more GLA can be viewed as a single large network of GLA, thus ensuring convergence to a local optimum.

What follows are a demonstration of the capabilities of GLA in a number of relatively simple experiments. Our basic experimental set-up is shown in Figure 4. Two agents A and B move on a line between $[-1, 1]$. Each time step both agents select action *left (L)* or *right (R)*, move and then receive a reward based on their original joint location and the joint action they chose. Each agent then updates using only the reward signal and the joint location, without any knowledge of the action selected by the other agent.

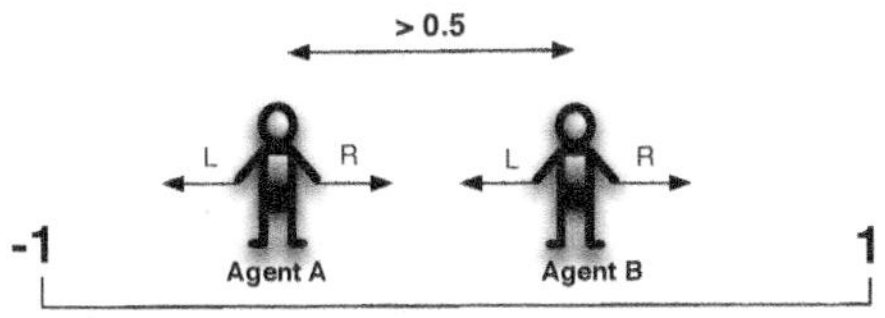

Fig. 4. Experimental set-up. Two agents move around on a line between positions -1 and 1. Each time step both agents take a step left or right.

4.2 Experimental Results

Agents using a single GLA. In this experiment the state space is divided in three regions, as shown in Figure 5(a). In region 1 Agent A is left of Agent B. In the second, Agent A is to the right of Agent B. The third region encapsulates all the states where the absolute value of the distance between the two agents is less than 0.5. Each agent has two possible actions, i.e. *Left* or *Right*. The reward scheme is as follows:

1. Region 1: A reward of +1 is given, when both agents choose action *Left*, 0 otherwise.
2. Region 2: A reward of +1 is given, when both agents choose action *Right*, 0 otherwise.
3. Region 3: A reward of +1 is given, when both agents move apart from each other, 0 otherwise.

For this experiment each agent uses a single GLA with 2 actions corresponding to the agent actions L and R. Each time step we give both agents an input vector $\mathbf{x} = [x1\ x2\ 1]$, where $x1$ is the position of agent A and $x2$ is the position of agent B. The GLA use a vector $\mathbf{u_i} = [u_{i1}\, u_{i2}\, u_{i3}]$ for each action i. The learning process of a GLA can then be seen as moving the line $(\mathbf{u_1} - \mathbf{u_2})^T\mathbf{x}$ which separates regions in the state space where the GLA prefers action 1(L) from those where it prefers action 2(R). Typical results obtained with this system can be seen in Figure 5(b). This result was obtained running the experiment for 100.000 iterations. Each iteration consists of a single action choice and update for both agents. After each move and subsequent learning update, the agents were reset to new random positions and the game was restarted. This was done to avoid the undersampling problem which occurs easily when dealing witch such large state spaces.

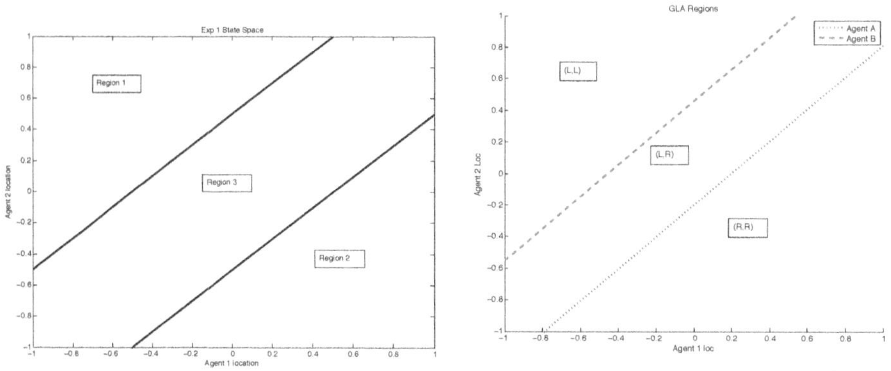

(a) (b)

Fig. 5. Experiment 1. (a) State space regions for experiment 1. (b) Typical result learnt by GLA. Lines separate regions where agents prefer different actions. Joint actions with highest probability are given in each region. Parameter settings where $\lambda = 0.01, K_i = L_i = 1$, $T = 0.5$.

Since GLA take context vectors as input, it is possible to present the state information in different forms to the agent. Figure 6 shows a comparison of the average reward obtained, with three distinct ways of information. We compared the use of the joint location described above, to an absolute distance metric $(AbsoluteValue(Pos(AgentA) - Pos(AgentB)))$ and a deictic distance metric $(Pos(AgentA) - Pos(AgentB))$. This experiment was run without tuning of the exploration of the Boltzmann action selection method, so these values are not necessarily measures for optimal performance of the GLA, but rather serve as a criterion to compare the influence of the information given in the context vectors.

The absolute distance metric clearly performs the worst due to the inability of making a distinction between different positions of the other agent. When presenting the agent with a deictic information to the position of the other agent, it outperforms agents using a joint location based state information. We

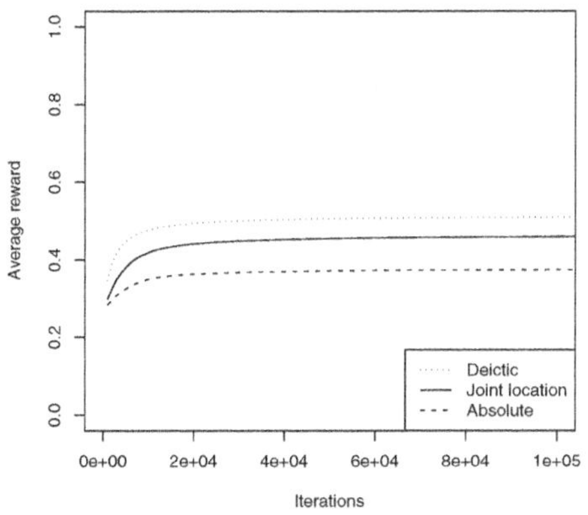

Fig. 6. Comparison of the influence of the state information given to the GLA

performed this experiment to show that, even though the same information is used, presenting it in different forms to the agent clearly benefits the learning results.

Agents Using Multiple GLA. In the second experiment we examine a situation where the different regions in the state space are not linearly separable. In such a case the agents cannot exactly represent the optimal mapping, but rather have to approximate it using hyperplanes. We use the same set-up as in the previous experiment, but now we consider two regions, as given in Figure 7(a). In region I, given by the inside of the parabola action (L, L) is optimal with a reward of 0.9. When the joint location of the agents falls outside the parabola, however, action (R, R) is optimal with reward 0.5. In both cases all other joint actions have a pay-off of 0.1.

Both agents use a system consisting of 2 GLA, connected by an AND operation. Both GLA have 2 actions: 0 and 1. If the automata both choose 1 the agents performs its first action L else it performs action R. Figure 7(a) shows 2 typical results for the boundaries that the agents learn to approximate the parabola. Figure 7(b) shows for both agents the evolution of probability of the optimal action L in region I. The probabilities in this plot where obtained by generating 100 points in the region with uniform probability and calculating the average probability over these points.

While it can be seen from the results in Figure 7 that the agents are able to approximate the desired regions, this experiment also demonstrated the limits of our approach. As was mentioned in the previous section the GLA are only guaranteed to converge to a local optimum. This means that the agent can get stuck in suboptimal solutions. Such a situation was observed when the reward for the optimal action in region II is increased. In this case it is possible for the agents to get stuck in a situation where they both always prefer the optimal action for region II and neither agent has a good approximation of the region inside the parabola. Since the rewards of both agents are based on their joint action, no agent can get out of this situation on its own. The agents can only improve their pay-off by switching actions inside region I together. In such a situation with multiple local optima, the final result obtained by the agents is depended on their initialisation.

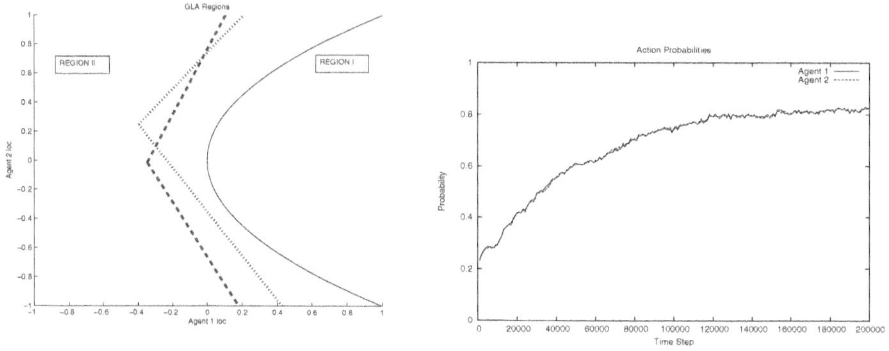

(a) (b)

Fig. 7. Experimental results for the second experiment. (a) Typical results for approximations for parabola learnt by agents. (b) Probabilities of optimal action in region I for both agents (average over 100 runs). Parameter settings where $\lambda = 0.005$,$K_i = L_i = 1$, $T = 0.5$.

5 Decoupling the Learning Processes

As demonstrated in the previous sections, learning in multi-agent systems is a cumbersome task. Choices have to be made wether to observe agents, or to communicate or even both. Most of these techniques make a very black and white decision for these choices, i.e. they either always observe all other agents or they never do. In many tasks however it is often enough to observe the other agents only in very specific cases. We propose to decouple the multi-agent learning process in two separate layers. One layer will learn when it is necessary to observe the other agent and select the appropriate technique. The other layer contains a single agent learning technique, to be used when their is no risk of influence of the other agent, and a multi-agent technique, when the agents will influence each other. Figure 8 shows a graphical representation of this framework.

5.1 First Level

The first layer of this framework is the important one, as this one decides the performance of the entire algorithm. A Q-learner could be installed on this level, taking the joint location as input, and choosing between the two techniques on the second level. This would however hardly be an improvement over a standard MMDP. Kok and Vlassis introduced a somewhat similar technique, where they would learn single agent when there is no need to coordinate, and learn using a

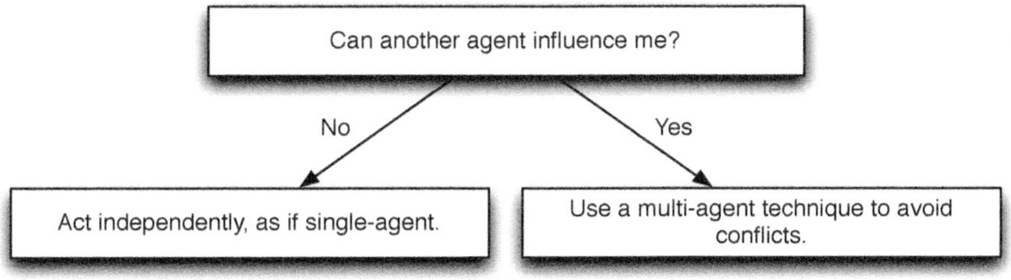

Fig. 8. Decoupling the learning process by learning when to take the other agent into account on one level, and acting on the second level

MMDP when they need to coordinate explicitely [27]. In this technique, called sparse tabular multi-agent Q-learning the list of states in which the agents had to be coordinated, needed to be defined explicitly beforehand. In our approach the main idea is that these states are learned as a function of the joint state. This gives us the benefit that we can generalise over states and just define danger zones as a function and not an explicit collection of individual states. It is for this reason that in our implementation of this framework we use *Generalized Learning Automata* (GLA).

Algorithm 4 contains the pseudo code of an algorithm that has been developed within this framework. It uses a GLA on the top level to decide if agents are within each others range of inference. If they are, a simple coordination mechanism is used for the next time step in order to avoid collisions. If the GLA judges that the agents will not interfere with each other in the next time step, both agents use Q-learning to select the action for the next time step.

Algorithm 4.

For each agent and every $\langle\sigma, a\rangle$ pair, initialise the table entry for $\hat{Q}(\sigma, a)$ to zero.
Observe the current state σ and the distance between the agents δ
loop
  if both agents decide to coordinate, based on δ **then**
    Agents coordinate on action to take to avoid a collision
  else
    Agents select action a, based on their action selection strategy
  end if
  • Execute the chosen actions
  • Receive immediate reward r_q and r_c, where r_q is the reward for the actions the agents chose and q_c is the reward for their coordination selection.
  • Observe the new state σ'
  • Update the Q-table entry accordingly to equation 11
  • $\sigma \leftarrow \sigma'$
  • Update the GLAs with r_c
end loop

The main advantage of using GLA is that the first layer of the algorithm can learn to determine which technique on the second level must be chosen, without explicitly storing estimates or samples of visited state-action pairs. All necessary information is encoded in the parameters of the GLA, which are typically much smaller in number than the states about which information needs to be kept. The possibility of combining GLA into a larger network which can learn more complex distinctions, also gives us the flexibility to adapt the complexity of the first layer to the needs of the problem at hand.

5.2 Second Level

At the second level of the learning algorithm two possible methods can be used, depending on the outcome of the first layer of the algorithm. If this first layer determines that other agents can safely be ignored a single agent technique is used, else agents use another technique that takes the other agents into account.

In this paper the single agent technique we use is an independent Q-learner. When it is deemed safe to do so agents act as if alone in the system, completely ignoring all other agents. Here we also assume that the states of our problem are represented by a set of state variables, some of which describe the local states of other agents (see Section 5.3 for an example). When learning in single agent mode, these superfluous variables are ignored, further reducing the size of the state-action space.

The multi-agent technique we adopt is a simple form of coordination through communication. If agents decide to coordinate their actions they first observe if a collision will occur if both agents would just choose their preferred action. If this is the case, one agent is selected randomly to perform its action, while the other selects a random other action. If no collision will occur, both agents can play their preferred action. Many other techniques are of course possible. Depending on the setting agents could try to bargain, learn an equilibrium policy or try to exploit each other.

In the following section we give some results of this algorithm and compare it to other RL algorithms.

5.3 Experimental Results

We test our approach by applying it in a gridworld game and test its performance against single agent Q-learning with independent agents and within the MMDP framework, using the joint location as the state. Even tough the problem seems an easy one, it contains all the difficulties of much harder problems and is widely used in the RL community [14]. Our gridworld represents a very simplistic version of a warehouse, where agents moving on tracks, have to get orders from the different aisles. Some of these aisles have only one track, so agents have to pass through them sequentially. Figure 9 shows a graphical representation of a gridworld where such problems are present. The two agents, in the upper and lower left corners, both have to reach the goal, G, which is surrounded by walls.

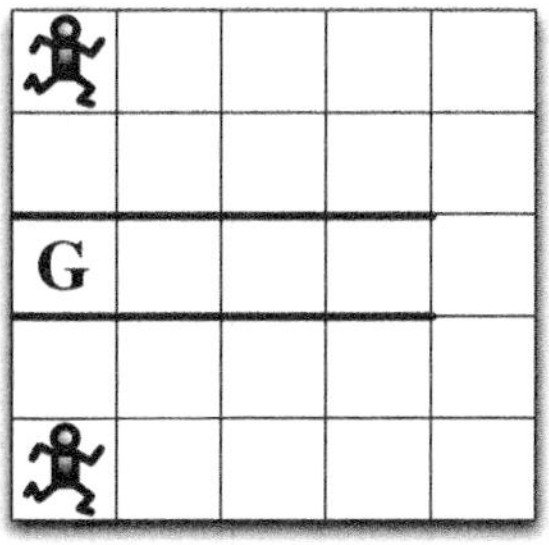

Fig. 9. Simple multi-agent gridworld problem

They have four actions at their disposal (N,E,S,W) for going up, right, down and left respectively. If both agents would use the shortest path to the goal, without considering the other agents, they would collide at the entrance of the passageway to the goal.

Before giving the results of the different techniques, we analyse the state-action spaces used by the different approaches. The independent Q-learners do not take into account any information about the other agents learn resulting in a state space consisting of only 25 states with 4 actions each. The joint state learners learn in a state space represented by the joint locations of the agents resulting in $(25)^2 = 625$ states, but select their actions independently, so they have 4 actions each. The MMDP learner also learns in the joint state space of the agents but with 16 actions (all possible combinations of the 4 individual actions). Our approach uses the same state action space as the independent Q-learners, unless the first level GLAs indicate, that a conflict is possible. In this case the agent communicate in order to select a safe action for next time step.

Since GLAs are able to learn regions in the state space, the actual size of their state space is not relevant.

All experiments were run with a learning rate of 0.05 for the Q-learners and Q-values were initialised to zero. An ϵ-greedy action selection strategy was used, where ϵ was set to 0.9. The GLA have a learning rate of 0.01, use a boltzmann action selection strategy and were initialised randomly. All experiments were run for 200.000 iterations, where an iteration is the time needed for both agents to reach the goal, starting from their initial positions. If an agent reached the goal, it receives a reward of $+100$. For the MMDP learner the reward of $+100$ was given when both agents reach the goal position, but once an agent is in the goal, its actions no longer matter, since the goal is an absorbing state. If an agent collided with another agent, it was penalised by -10. Bumping into a wall was also penalised by -1. For every other move, the reward was zero. For every collision, wether it was with a wall or with another agent, the agent is bounced back to its original position. The GLA were rewarded individually according to Figure 10.

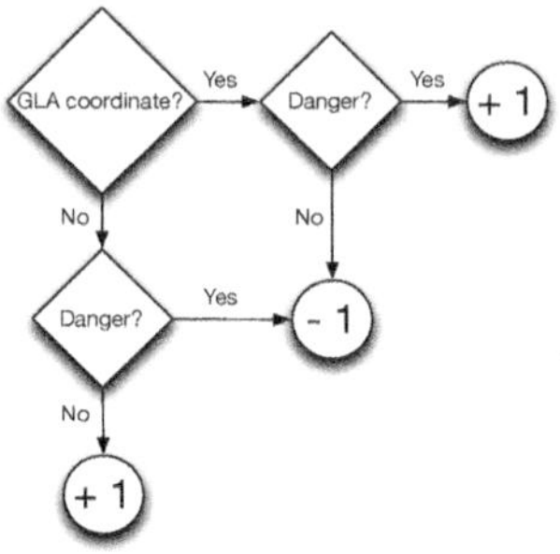

Fig. 10. Rewards for the GLA of the 2observe algorithm

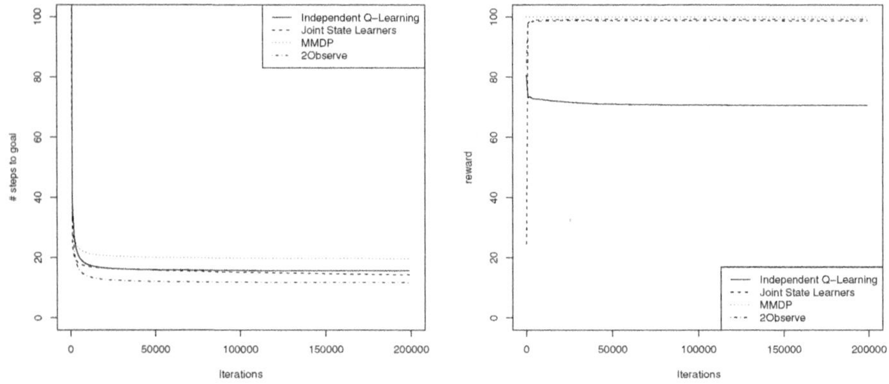

(*a*) (*b*)

Fig. 11. (a) Average number of steps from start position to goal position and (b) the average reward of both agents for the different techniques

If on top of this, both agents decided to coordinate when there was no need, they are penalised according to following rule: $0 - (\delta/10)$ where δ is the distance between both agents.

Figure 11(a) shows the average number of steps both agents needed to reach the goal, as well as the average reward they collected in Figure 11(b). Our technique converges not only to a better solution in terms of the number of steps to the goal, but also converges faster than both other techniques. When looking in terms of the reward, we see that our technique performs almost as good as the MMDP and much better than the independent Q-Learners. The reason for the quite poor performance of the latter can be seen in Figure 12(a).

Both multi-agent approaches find a solution with zero collisions relatively fast, whereas the independent learners never do. They converge to an average of three collisions per iteration. This means that they can only converge to a maximum reward of 70. In Figure 12(b) we illustrate that our approach converges to an average of approximately four coordination actions per iterations. This means that our agents only communicate explicitely four times per iteration, about which action to choose.

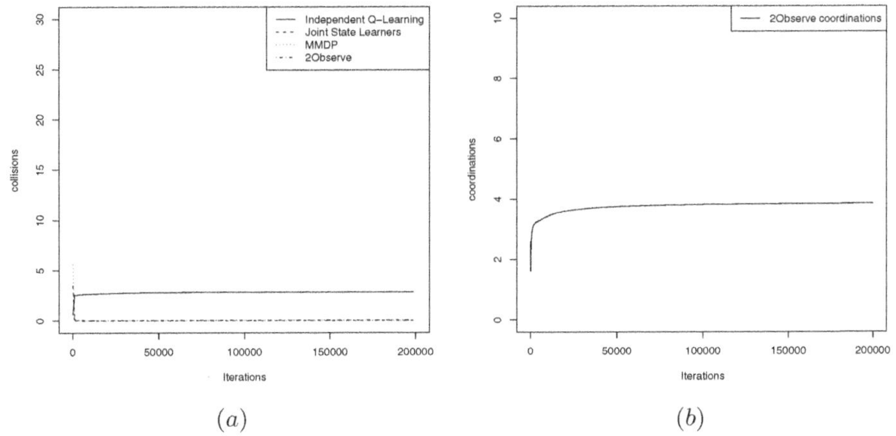

Fig. 12. (a) Average number of collisions and (b) the average number of coordinations per iteration

6 Discussion and Future Work

When dealing with large state spaces and the presence of multiple agents, the learning task becomes difficult. In this chapter we introduced two solutions for these problems. First we demonstrated how GLAs can be used to aggregate states in unknown environments and showed the importance of the way the state information was presented to the agent. Second, we introduced a general framework for learning in multi-agent environments based on a separation of the learning process in two levels. The top level will learn when agent's need to observe each others' presence and activate the appropriate technique on the second level. If no risk of interference is present, the agents can use a single agent technique, completely ignoring all the other agents in the environment. If the risk of interfering with each other is true, a multi-agent technique will be activated in order to deal with this increased complexity.

The main advantage of this framework is that it can be used as a foundation for using existing single-agent and multi-agent techniques, adapting the learning process wherever needed.

We implemented a concrete instantiation of this framework, called 2observe, which uses a generalized learning automaton on the top level, a Q-learner for the case where agents do not interfere with one another and a simple communication based coordination mechanism when the agents need to take each others' presence into account. We showed empirically that our technique was able to reach a solution, almost as good as MMDP learners but without the need of full observation and in a much smaller state space.

The possibilities for future work are wide. Many techniques exist to incorporate in our framework. On the second level, the entire range of single agent and multi-agent techniques can be used. On the first level also many alternatives exist. We chose to use GLA in this paper due to their simplicity and low computational costs without the need to store previously seen samples. However, appropriate statistical tests could be used on this level to measure the influence two agents have on each other. Another interesting research track is to use the rewards given on the second level as feedback for the first level. This would mean that the GLA could learn from delayed rewards using a monte-carlo updating scheme. In this way, a wider range of problems can be solved and state information can be used even more wisely.

References

1. Sutton, R.S., Precup, D., Singh, S.P.: Between MDPs and semi-MDPs: A framework for temporal abstraction in reinforcement learning. Artificial Intelligence 112(1-2), 181–211 (1999)
2. Stolle, M., Precup, D.: Learning options in reinforcement learning. In: Koenig, S., Holte, R.C. (eds.) SARA 2002. LNCS (LNAI), vol. 2371, pp. 212–223. Springer, Heidelberg (2002)
3. Hu, J., Wellman, M.P.: Nash q-learning for general-sum stochastic games. J. Mach. Learn. Res. 4, 1039–1069 (2003)
4. Greenwald, A., Hall, K.: Correlated-q learning. In: AAAI Spring Symposium, pp. 242–249. AAAI Press, Menlo Park (2003)
5. Vrancx, P., Verbeeck, K., Nowe, A.: Decentralized learning in markov games. IEEE Transactions on Systems, Man and Cybernetics (Part B: Cybernetics) 38(4), 976–981 (2008)
6. Vrancx, P., Verbeeck, K., Nowé, A.: Optimal convergence in multi-agent mdps. In: Apolloni, B., Howlett, R.J., Jain, L. (eds.) KES 2007, Part III. LNCS (LNAI), vol. 4694, pp. 107–114. Springer, Heidelberg (2007)
7. Boutilier, C.: Planning, learning and coordination in multiagent decision processes. In: Theoretical Aspects of Rationality and Knowledge, pp. 195–201 (1996)
8. Boutilier, C., Dearden, R., Goldszmidt, M.: Exploiting structure in policy construction. In: Mellish, C. (ed.) Proceedings of the 14th International Joint Conference on Artificial Intelligence, pp. 1104–1111. Morgan Kaufmann, San Francisco (1995)
9. Guestrin, C., Koller, D., Parr, R.: Multiagent planning with factored mdps. In: 14th Neural Information Processing Systems, NIPS-14 (2001)
10. Degris, T., Sigaud, O., Wuillemin, P.H.: Learning the structure of factored markov decision processes in reinforcement learning problems. In: Proceedings of the 23rd International Conference on Machine learning, New York, NY, USA, pp. 257–264 (2006)
11. Strehl, A.L., Diuk, C., Littman, M.L.: Efficient structure learning in factored-state mdps. In: AAAI, pp. 645–650. AAAI Press, Menlo Park (2007)
12. Abbeel, P., Koller, D., Ng, A.Y.: Learning factor graphs in polynomial time and sample complexity. Journal of Machine Learning Research 7, 1743–1788 (2006)

13. Russel, S., Norvig, P.: Artificial Intelligence, a Modern Approach. Prentice-Hall, Englewood Cliffs (1995)
14. Sutton, R.S., Barto, A.G.: Reinforcement Learning: An Introduction. MIT Press, Cambridge (1998)
15. Mitchell, T.: Machine Learning. McGraw-Hill Companies, New York (1997)
16. Kaelbling, L.P., Littman, M.L., Moore, A.P.: Reinforcement learning: A survey. Journal of Artificial Intelligence Research 4, 237–285 (1996)
17. Watkins, C.: Learning from Delayed Rewards. PhD thesis, University of Cambridge (1989)
18. Lin, L.: Programming robots using reinforcement learning and teaching. In: Proceedings of AAAI, vol. 91, pp. 781–786 (1991)
19. Chapman, D., Kaelbling, L.: Input generalization in delayed reinforcement learning: An algorithm and performance comparisons. In: Proceedings of the 12th International Joint Conference on Artificial Intelligence, pp. 726–731 (1991)
20. Guestrin, C., Hauskrecht, M., Kveton, B.: Solving factored mdps with continuous and discrete variables. In: Proceedings of the 20th conference on Uncertainty in artificial intelligence, pp. 235–242 (2004)
21. Sutton, R.S.: Reinforcement learning architectures. In: Proceedings ISKIT 1992 International Symposium on Neural Information Processing (1992)
22. Shapley, L.: Stochastic Games. Proceedings of the National Academy of Sciences 39, 1095–1100 (1953)
23. Claus, C., Boutilier, C.: The dynamics of reinforcement learning in cooperative multiagent systems. In: Proceedings of the Fifteenth National Conference on Artificial Intelligence, pp. 746–752. AAAI Press, Menlo Park (1998)
24. Phansalkar, V., Thathachar, M.: Local and global optimization algorithms for generalized learning automata. Neural Computation 7, 950–973 (1995)
25. Williams, R.: Simple Statistical Gradient-Following Algorithms for Connectionist Reinforcement Learning. Reinforcement Learning 8, 229–256 (1992)
26. Thathachar, M., Sastry, P.: Networks of Learning Automata: Techniques for Online Stochastic Optimization. Kluwer Academic Pub., Dordrecht (2004)
27. Kok, J.R., Vlassis, N.: Sparse tabular multiagent Q-learning. In: Nowé, A., Tom Lenaerts, K.S. (eds.) Proceedings of the Annual Machine Learning Conference of Belgium and the Netherlands, Brussels, Belgium, pp. 65–71 (January 2004)

Automatically Creating Hierarchies of Agents and Meta-agents Using Clustering

Anne Håkansson

Department of Computer and Systems Sciences,
Stockholms University,
Forum 100,
SE-164 40 Kista, Sweden
Anne.Hakansson@dsv.su.se

Abstract. This chapter describes an approach to automatically create hierarchies of intelligent agents and meta-agents in networks. The hierarchy contains intelligent agents at the bottom level, constituting leafs, and abstract meta-agents at the top levels. The commonality between the agents in the hierarchy is that higher-level agents comprise lower level agents with more distinct semantics, which together constitute information chains for solutions. The test environment is the web since it produces of vast amount of information that needs structuring in order to be manageable and accessible. The results of the search infrastructure are automatically cast into a more manageable and understandable hierarchy by using a clustering method.

1 Introduction

Searching for information on the web usually implies finding lots of web sites in which software agents search for information responding to a user request. Agent technology is an efficient method to search for requested information on the web. When agents are searching and collecting information, they often generate a vast amount of URLs to the web pages. Very often searching for web sites in a search engine, like Google, gives several thousands to millions of pages. As an example, we can look at searching for web sites containing the words "Agents in hierarchy", which results in 3 220 000 hits (June 2009). The result not only includes the words in the search string, it also includes similar words like "Agent", "agent's", "multi-agent systems" and "hierarchies". This is beneficial since the users may find other relevant pages, at the same time as it has drawbacks. The number of pages increases, tremendously.

Commonly, the hits are presented as links, in the browser, together with one sentence or two sentences, sometimes in fractions, containing the search words. The information is presented without mutual order of precedence or relevance and searching for relevant information among the links can be a difficult and tedious job. This work increases if finding information requires exploring many of the links. It hardly seems probable, or possible, to explore all 3 220 000 hits. A solution to handle all this information can be to sort the links and information into manageable pieces of

A. Håkansson et al. (Eds.): Agent and Multi-agent Technology, SCI 289, pp. 207–228.
springerlink.com

information and structure the result in hierarchies. Hierarchies support structuring information with subordinate entities, vertically and horizontally, which makes it easier to navigate and process.

Hierarchies are often a prior choice for structuring information since they utilise formal methods that are well established [1]. The hierarchies are good solutions for mapping larger data collections into abstract structures and, therefore, are used to structure complex software. For the web, hierarchies have been used to structure and present hypertexts, web sites and data collection applications. Moreover, files, documents, functions, options and computerised lists can be organized as hierarchies.

This chapter presents hierarchies of different levels of agents, searching for required information, and how these can be constructed, automatically. Automatically creating hierarchy is required to manage massive and complex environments with a vast amount of information, such as deoxyribonucleic acid (DNA), and the web. Typically, hierarchies have been built for IP addresses but for our research, we use paths resulting from searching for information on the web as test environment.

Automatically creating hierarchies of paths to the relevant web pages is a more complicated and complex task than creating hierarchies to structure URLs build on IP addresses as structured by Kumar et al., (1999) [2] and Eiron and McCurley, (2004) [3]. One of the problems is the environment that is constantly changing. Naturally, working with the IP addresses are rather static since they are attached to the computer connected to the Internet. The found web sites and the paths to the web sites, on the other hand, are highly dynamic and of varying nature since the result of the user request is often different as well as the contents of the web sites. Hence, a result from a search cannot be specified and organized in advance. Another problem is that the relevant web sites can be unreachable, while searching for the information or the messages can be erroneous. As a result, the hierarchy will have different content almost every time is created. Yet another problem, which is also the most intricate, is that it is not possible to classify the results before hand. IP addresses are often built in a logic structure often classified by countries so they can use the logic structure. This is not the case for the paths leading to the web sites and the content of the web sites. An additional problem, besides handling a vast number of web sites, is that it is not obvious how to classify the web pages. The classification needs to be made when the agents are returning with the result, preferably, at the browser's web site from which the user had sent the request. The classification will cluster similar information, which is partitioning the data set into subsets, i.e., clusters where the subsets, ideally, have some commonality.

In this research, we use a vast number of agents to find the paths to the relevant and related web sites. Each agent works at one particular part of a path leading to a web site. The intelligent agents are collected from searching solutions to a task, where each intelligent agent performs one sub-task and the result from the agents together constitutes the solution. Because of the sub-tasks, many intelligent agents will be used for finding a solution. To keep track of all intelligent agents, meta-agents are created from the intelligent agents that successfully complete their tasks. Thus, the meta-agent is created from the first launched intelligent agent and, then, continues expand for each new used agent. From these intelligent agents and meta-agents, hierarchies are automatically created where the intelligent agents are the leaves and meta-agents are the internal tree nodes. Since the meta-agents are built on the agents, they are superior of the agents. Therefore, the meta-agents become the parent nodes of the intelligent agents at the leaf level. The challenge is to categorize the information in the hierarchies.

The chapter is structured as follows: Section 2 gives a brief description of hierarchies together with structuring information in the hierarchies. Section 3 describes a multi-agent system for the web and Section 4 presents agent systems working with tasks on the web together with agents and meta-agents. Since hierarchies require some method of clustering, an algorithm for clustering agents is provided, in section 5.

2 Hierarchies

For the multi-agent system to be effective, well-ordered and context-sensitive hierarchies are required. In general, the hierarchies are either strict superior to the lower levels or mixed superior where the lower levels can be presented in the upper levels of the hierarchy [4]. The strict superior is easier to work with because it enables a direct control scheme to direct the overall problem solving [5]. This is good for problems like search that are divide-and-conquer algorithms, which is also used for our current multi-agent system. However, a mixed superior scheme may allow better solutions for some multi-agent system using hierarchies since it breaks the direct control and allows the agents to interact across the partitions of the hierarchy, resulting in a more opportunistic problem solving scheme. However, this requires special treatment of classification of the agents.

Some agents are built on top of each other and, therefore, should have relationships that correspond to a top down structure with more concrete information at the bottom level. Commonly, the relationships in hierarchies can be transitive, irreflexive or asymmetric [6]. The transitive relationship has superior entities throughout the hierarchy, while irreflexive is without superior entities. Asymmetric relationship has a strict order of superior or subordinate entities, but can also be without either superior or subordinate entities. Since the agents are built on each other, transitive hierarchy is the best solution.

Hierarchies can link entities directly or indirectly in a vertical or horizontal manner. The direct links are to the immediate superior or subordinate entities whereas the indirect links are extended vertically with multiple links in the same direction. The agent hierarchy will be extended both vertically and horizontally and will, therefore, contain a directly and indirectly entities. The leaves will have an indirect relation since the agents are a chain of actions to reach a goal. The agents are represented separated but they depend on each other in a way that the first agent, left in the hierarchy launches the second agent to the left in the hierarchy. Thus, these agents have a indirect relationship between each other.

Working with the web is to work in a distributed and non-hierarchical environment [7]. The environment is hierarchic since the IP addresses are organized in a hierarchy, which can be used to find information and handle the fast growing network [8]. Hence, the idea of using hierarchies on the web is old, but using hierarchies for keeping information on the web is newer. Only putting IP addresses into hierarchies is too trivial when working with finding and clustering information on the web. The IP addresses, themselves, do not add valuable information to the hierarchy. Nonetheless, the agents are searching for information on the web and will utilise the IP addresses to be able to return to the same web pages when requested. These addresses will be kept in the hierarchy together with the agents.

Commonly, representation of hierarchies is in trees [9]. Often, these trees are represented as connected directed acyclic graphs (DAGs) or undirected acyclic graphs [10] with a designated initial root at the top and the leaves at the bottom. In directed graph the already visited edges have to be different for further inspection and, hence, the edges cannot reuse again as in undirected graphs [11]. Thus, if it is an edge *(i, j)*, it is not an edge *(j, i)* in that graph. In undirected graphs, the same edge *(i, j)* and also be *(j, i)*. For creating hierarchies of agents in our work, the trees are represented as connected acyclic graphs but can be either directed or undirected. For the moment, we have not restricted the graph to be directed. If the user wants to categorise information in a specific manner, the graph may have to be directed.

The graph is completely connected if it has *an edge* between every pair of nodes (or vertices), $E = V * V$ where E is edge and V is set of nodes [11]. A graph is strongly connected if there exists *a path* from any node to any other node. Hence, it is possible to reach the entire graph from any node *i*. The completely connected and strong connected together with the simplicity of only having nodes and edges that one appears once, are basic characteristics for our automatically created hierarchies.

There is a unique simple path between the nodes in the tree [11]. In the path, there are a set of ancestors for a node *i*, except for root node *r*, and descendants (or subordinates) of all nodes to the node *i*, excluding the leafs. Thus, to each ancestor, there are one or more descendants. These ancestors and descendants are present in our hierarchy as agents: meta-agents (or meta-level agents) and software agents. The meta-agents are the ancestors and the software agents are the descendants. Also, the unique paths in the hierarchy are important since they support finding more specific information as moving further down in the hierarchy. This is especially important for the automatically created hierarchies since they must assure strict hierarchies so the users know what to expect.

Hierarchies play a crucial role in complex systems [12]. Large systems can contain hundreds of individual objects and the connections grow exponentially with the number of objects, thereby, producing complexity. The hierarchy can reduce complexity by applying modularity to simplify management of large systems.

Modularity, or clustering, reduces combinatorial complexity, which is useful for problem solving [13]. Related information can be grouped into the same cluster to make the information more compact and easier to scan. Moreover, clustering can support the efficient execution of tasks in a multi-agent system and help reduce communication costs between closely connected agents. Using clusters can also speed up the search for data in networks and graphs, as well as, support the task handling by the agents and reasoning at different levels in hierarchy.

The structure of the hierarchies can either be single or multiple. Commonly, hierarchies are built as single hierarchies where a child node only can have one parent. However, there are also other hierarchies, multiple hierarchies, where a child node can have more than one parent node. The multiple hierarchies tend to be more complex in structure, at the same time as they allow more flexibility. Moreover, multiple hierarchies can be managed in parallel. For our work, this supports building the hierarchies in parallel, saving a considerable amount of time. The problem is keeping track of the information overlap, for example nodes that have already been visited.

3 Multi-agent Systems on the Web

Multi-agent systems are a powerful solution for problems where the tasks can benefit from the work of several agents [14]. Using several agents is essential for tasks handling information on the web. These systems are powerful and can become even more efficient if the tasks can be performed simultaneously with parallel computation. The parallelism is especially interesting in distributed problem solving, since it supports efficiently finding solutions at the same time as it can also bring about problems with assembling the result correctly.

In agent-based systems, any complex system can be represented as graphs or networks [12]. Commonly, the nodes in the graphs or networks correspond to agents and the edges are the interactions between agents. There are also other alternatives for representation, where the edges correspond to connections between goals or sub-goals at which the agents move and the nodes are goals to be met [15; 16].

To find information in complex graphs, one solution is to disassemble the problem into smaller more manageable chunks and conquer one at the time. Then, each chunk or part of the problem is matched against the content in the graph and the results from the matching must be assembled into a useful representation. This works with direct corresponding matching. However, if the solutions are not obvious, the graph needs to be traversed to reveal what can be offered as solutions. In this case, the system must investigate the content of the graph using a structured planning. The planning can ensure that the agents perform and complete the whole task and return several solutions, which needs to be assembled. For keeping track of the agents and solutions, we propose applying a hierarchy of agents in the multi-agent system.

Some hierarchies arise from interactions that impose order on the agents [12]. In our multi-agent system, the hierarchy of the agents is built on agents in a dynamic environment and cannot be determined in beforehand. Thus, the arrangement of the hierarchy can differ each time the search is executed search, even for the same problem or request of information.

Automatically clustering the agents that are performing similar tasks, render possible automatic construction of agent hierarchies. The hierarchy is automatically built over all the agents that together solve a problem [17]. The hierarchy is updated by the agents performing tasks in a network while moving between nodes. The number of agents and interactions, i.e., used in the network, increase with the number of required and utilised nodes, which introduces complexity.

In the multi-agent system, the actual path, formed by the intelligent agents in the network, is the solution to a specific problem. The order of the agents is an important aspect of the solution and needs to be captured in the hierarchy. These intelligent agents and their performances become the leaves in the hierarchy.

However, only using intelligent agents for handling tasks in complex environment is not efficient. Therefore, we need to introduce several levels of abstraction. From the intelligent agents' performances, agents at a meta-level are automatically created. These meta-agents comprise the successful intelligent agents together with the performance costs. The agents at meta-levels, so called meta-agents, are then used for monitoring the intelligent agents and coordinating the result of the intelligent agents with the other intelligent agents. In the hierarchy, the meta-agents become parent nodes in the hierarchy, i.e., at a higher level over the intelligent agents in the network and up to root level.

3.1 The Agents in the Network

In the multi-agent system, the intelligent agents work to solve tasks in networks. The agents are capable of autonomous actions, situated in different environments [14] where some are intelligent and, thus, adaptable to the environment. The environment, i.e., the web, has some characteristics that the agents observe and act upon. Up to date, the web is a vast network with over 100 million web sites where 74% are commercial sites or other sites operating in the .com domain.

The environment on the web is complex but we consider that the agents, in our system, work in a rather uncomplicated task environment. The environment is partially observable, stochastic, episodic, continuous and semi-dynamic as described in Russell and Norvig (1995) [18], and Wooldridge (2002) [14].

In a partially observable environment only some of the information is known [18]. Networks, such as the web, are considered to be partially observable because of continuous growth. However, the agents must still be able to find the significant information needed for finding solutions. Finding the necessary information is important since the agents must be able to obtain complete, accurate and up-to-date information.

The environment, such as the web, can appear to be stochastic [18]. The agent cannot predict the behavior of the environment since the state, which will result from performing an action, will be unclear. Thus, there can be stochastic elements that randomly appear. The agents' tasks are searching for information on different web sites but the "hits", i.e., web sites, will vary with time. Moreover, the results of the agents do not necessary have to be different, i.e., they come back with the same result of a search but can the paths to find the result can differ for each search.

There might be uncountable many states arising from the continuous environment [14]. Continuousness can be a problem for the agents since the number of states and actions can be enormous, which may not be found within an acceptable time range. This requires special treatment of the agents. We use an execution suspension to control the agents' performance, which is an execution suspension for a short time interval followed by a resumption of the search.

The agents' choice of action in each episode depends on the episode [18]. The agents perform one task at the time while moving between commercial sites finding information. As mentioned above, the task itself can be more complex, but when divided into smaller, single tasks, the task can be applied to one agent. Each performed single task, complete the work of the agents and constitute the result.

Semi-dynamic, in this perspective, includes both static characteristics and dynamic characteristics in the environment. Dynamic characteristics change constantly whilst static characteristics change rarely. In a dynamic environment, the environment can change while agents are searching its contents [18; 14], which is beyond the agents' control. To handle these dynamic environments, the agents need to continuously interact with the environment and check the surroundings to act properly. However, the web environment might be static over short time intervals, thus, remaining unchanged while the agents consider their course of action between two states. Although the agents will find the information at some web sites, they still need to monitor the other web sites for each task. This is to get the best solution for a given problem.

The task determines which part of the network is exploited. The user denotes the starting position and the goal position and the agents search in the network to find a path between those two positions [15; 16]. From the starting position, the node can have one or several edges to adjacent vertices. At each edge, an agent works by moving between the connecting vertices. The agents have no priori knowledge of the other agents, as in pure contract nets [19]. However, the agents do not have to bargain to perform the task. Instead, all the agents connected to the starting position launch simultaneously following only one edge to the connected vertices. Thus, if there are five adjacent vertices, there will be five agents working.

Some edges might not interesting to explore due to some unwanted characteristics of the edge. If there are conditions that the users want the agents to meet, the users can denote these conditions. There can also be conditions that make the edge unwanted and, therefore, can be avoided. Nonetheless, there might be occasions when an unwanted edge will be used anyway, because it can contribute to find a requested solution. The entire paths that successfully found a solution must be used to decide the most optimal path through the network where the use of some edges in the path may be more inadequate than others.

Depending on the network, the vertices and edges contain certain information. This information is important to collect, as well as, act upon, correctly. The collected information is brought back to the user, which makes the agent only an information carrier. However, the conditions of the edges can require the agents to take some actions. An action can be matching the edges information to the user given information and deciding whether the edge is worth being explored in the first place.

The ground structure of the agents has the following construction:

$$\forall Ag(x)\ \exists x(Task(x) \wedge Environment\ Information\ (x))$$

For all agents, *Ag(x)*, in the system, every agent has an specific task to accomplish *Task(x)*. Each agent has a unique task assigned, which is moving along the edge assigned to that specific agent. *Environment information(x)* is information about the task environment. The environment information about the communication links can be hardware and software such as cables, cable performance, operating system, possibility to use the links, bit error ratio, speed, delay times and much more.

The *Task(x)* contains information about the agents' position in the network. For all agents, this includes the vertices, or nodes, the agent is moving between and identification of the arc, or edge, and commonality. The commonality identification connects several different edges together to constitute a commonality between the edges. Commonality can, for example, be used to assure the agents to move around in a local network instead of hopping between several different networks. The *Task (x)* is structured as:

agent(Node A, Commonality Id, Arc Id, Node B)

When the system is launched with the starting position, the agents work until they reach the goal, or end position. The agents act accordingly to the environmental information and are, therefore, somewhat intelligent. For example, if it is possible, they should avoid the bad conditions in the environment. When each agent reaches a node, another agent is launched. It will be a network of several agents working simultaneously.

Environment information(x) is information about the static parts of the environment and dynamic parts. The structure of the environment information(x) is:

$$\forall\, Environment\ Information(x)(\exists x(Static\ part(x)\ \wedge\ Dynamic\ part\ (x))$$

The static part is constant during the execution and do not change often at all. The static parts can make it easier, but also faster, to find some information, i.e., if the agents are using the communication links that are known to lead to the web sites, at which requested information has been found earlier. The dynamic parts change frequently and can change during the execution. The dynamic parts have to be explored to find information. The dynamic parts can convey new important information to a solution. The dynamic parts include new web sites as well as the unused communication links needed for finding unknown web sites containing requested information.

Contemplating the environment in broader perspective, it contains both static and dynamic parts, where static can become dynamic and vice versa. Therefore, it is a semi-dynamic environment.

Moving on the edges between the connected vertices provides information about execution time. A benefit of knowing the execution time is the possibility to calculate the time it takes for all the agents to move between the vertices to find the solution.

If the users have specified conditions that the agents have to meet, these must be matched against the environment. The agents that are not meeting the conditions, or are close to the given conditions or intervals of the conditions, should be set aside but not deleted immediately. Those agents might actually be needed later on.

Keeping track of all agents requires a supervising facility that control and manoeuvre the agents' performances, so-called meta-level agents. These meta-agents can support finding the agents that have been successfully performed a task but also speed up searching for data in the network. They can also keep track of the agents that do not meet the user-given conditions but are still interesting for later execution.

3.2 Meta-agents

In the multi-agent system, the meta-agents are superior to the intelligent agents. The meta-agents are built from the intelligent agents and then used to monitor the intelligent agents and their environment. From the information comprised by the intelligent agents, the meta-agents are able to schedule actions, solve conflicts and plan future interactions [15; 16; 20; 21].

Meta-agents become the strategic level as they capture the overall strategy of the solutions being tried. The meta-level agents adapt to the ground-level intelligent agents by learning from their performances and strategies. The adaptation occurs when agents learn from each other or change strategies as they gain experience [22]. The meta-agents build up experience about the task environment and the result of the agents' search. The experience can be used to guide the agents' future actions and build up reliable paths through the network, which may be somewhat faster.

We use meta-agents as logic programming uses meta-rules. The meta-rules are meta-knowledge about how problem-specific rules are used and reflect the body of knowledge gained from these problem-specific rules [23]. Meta-rules are used for meta-reasoning. Meta-reasoning is a technique that supports a system to reason about its own operation. In a meta-agent system, meta-reasoning implies implementing the agents' capability to reason about other agents. This reasoning can be used to reconstruct behaviour [24] and help with the interaction among different parts of the system and in the implementation strategies or plans for responding to requests [25]. By using the meta-rule technique for the meta-agents can introspectively monitor reasoning to evaluate and explain computation and recover from erroneous agents.

The meta-agents have the following structure:

meta_agent(Initial_state, End_state, RouteList, Costs).

When the user has stated the request, the starting position becomes the initial state and the end position becomes the end state. The agents start moving along the vertices from the user's starting position. When an intelligent agent reaches its node, the meta-agent is built on that agent. Typically, the starting position has several connecting vertices and nodes and for each launched and successful performed agent, a meta-agent is built. Thus, if there are five vertices of which four is successful, four meta-agents will be built. At the reached nodes, the agent's success becomes the launching events for the next agents to move. Each built meta-agent will be expanded with the next moving agent that has successfully performed its task. All the agents included in the meta-agent, compose a route-list, which is the path through the network.

Besides including the agents, the performance cost is noticed and stored. For each added agent, the cost for the agent is included in the meta-agent. Also the total costs will be accumulated.

One of the strengths of the meta-agents is that each meta-agent is unique. Thus, every meta-agent that has reached the end position has a unique path through the network. This benefits in comparing different paths through the network, taking into account the environment information together with costs and security.

A problem with the meta-agents is that many agents are produced. In a highly connected network, the intelligent agents can almost take any path and still reach the goal. These paths can be more or less useful. To speed up finding solutions and searching for particular information, we need to introduce a structure on the agents, clustering information and locating the most general agent at the top.

A benefit of dividing the task into small separate pieces is that the agents can work with the task independently using parallel computing. The agents are not dependent on the other agents and perform their tasks autonomously. The only dependence is the agent that launches the other agents. The results of the computing can efficient be reassemble by the hierarchy of agents.

4 Clustering Agents

Clustering can be defined as grouping sets of elements that share some similarity [26]. A simple way of cluster elements is obtaining maximum intra-cluster similarity and inter-cluster dissimilarity. Clustering needs to deal with membership and amount of clusters.

Clustering agents in multi-agent systems can group the agents that are working in localized domains. The clustering supports parallelism, by grouping closely related agents into the same node in the multi-processing system. To automatically create clusters of agents, we need some patterns of activities, i.e., ways that clusters of agents are created. There are two different ways of clustering: hierarchies and partitioning [27]. Hierarchical algorithms work by splitting existing groupings according to some criteria. Partitioning algorithms work by using a definition of a group and then work to divide the agents in to proper groupings. In our work, the formation of the hierarchy can provide a basis to perform the clustering. While either approach could work, the hierarchical algorithms are a more natural fit.

Hierarchical algorithms can either be agglomerative or divisive [28]. Agglomerative algorithms start with the single elements as separate groups or clusters and then merge these single elements into successively larger clusters. Agglomerative builds hierarchies of clusters and is a bottom-up algorithm. Divisive algorithm starts with the complete set and divides this set into successively smaller clusters. Divisive breaks up hierarchies of clusters and is a top-down algorithm. In our system, we are using an agglomerative algorithm since the activated intelligent agents are known and, therefore, possible to put into a hierarchy.

The easiest method of building up the clusters is clustering the closest neighbour agent. When an agent has reached its node (Node B) successfully, it is added as a separate cluster. This agent launches all the agents that are connected to that node (B) and when these have reached their goals, they will also be added as separate clusters. In a more complex case, the size of the cluster can be used as a criterion to divide the cluster. That is the cluster is given a maximum size, which triggers the division.

This clustering will proceed until the goal or end position has been met and the number of solutions is acceptable. The first solution found may not necessary be a good solution and the system needs to find more paths. If there are conditions that the users want the agents to meet, these conditions can be stated in the beginning of the execution, i.e., at the same time as the starting and goal positions are denoted. Hence, the clustering is created from the closest agents but can also be clustered taking the conditions into account.

Next step is to incorporate the building of meta-agents into the clusters. These meta-agents include the intelligent agents and are, therefore, at a higher level in hierarchy. The meta-agents either contain one or merge several agents together. Moreover, a cluster can be a part of several other clusters, i.e., if the cluster is close to several clusters simultaneously.

5 Creating Hierarchies of Agents

Hierarchies arise from the organisation of systems on interactions ordering the agents on different scales [12]. With hierarchies, the connectivity in multi-agent is ensured but the interactions linking up and down the hierarchy is limited.

In scale-free networks the number of edges differ. Most of the nodes have few edges linking them to others but some have large number of connections. In our system, there are no constraints on the number of connections a node can have. Therefore, a node that has n edges departing from it and, consequently, the node has the same number, n, of intelligent agents.

Hierarchies are created in response to the problem being solved. The hierarchies give two advantages. First they can be used to cluster the meta-agents into sets that represent groups of meta-agents that communicate strongly, yielding more efficient processing. Secondly, the hierarchies represent the problem being solved and provide a control structure to support reasoning over the agents to locate the best solutions.

The meta-reasoning used to create the hierarchy is a natural part of the processing logic. One approach is to generalize the current solution space. As branch points in the algorithm cause the creation of new meta-agents, the new agents become the children of a higher-level meta-agent.

As example we illustrate a network with several agents, see Figure 1. In the network, the agents do not have an internal order but they are working between the nodes. For example, between node A and node B, the agent A-B is placed and between the nodes B and C, the agent B-C works.

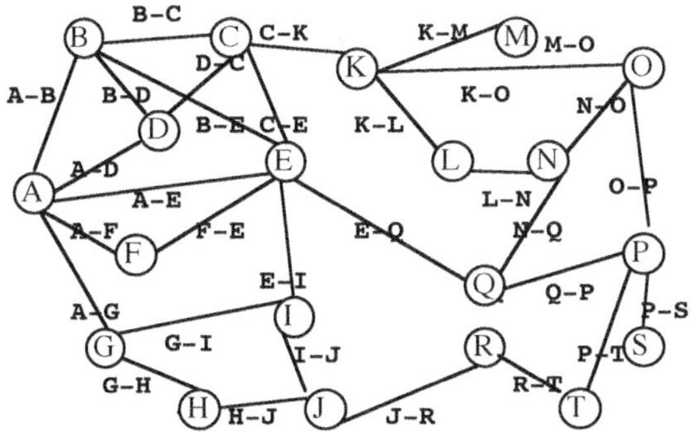

Fig. 1. Network with several intelligent agents

Assume a user is at the node A and wants to get information from node T, the agents must find at least one path between A and T in the network. However, from this request to the network in Figure 1, several different paths are to be found and several clusters in a hierarchy are developed. Examples of a paths found between A and T are, Figure 2 a-f:

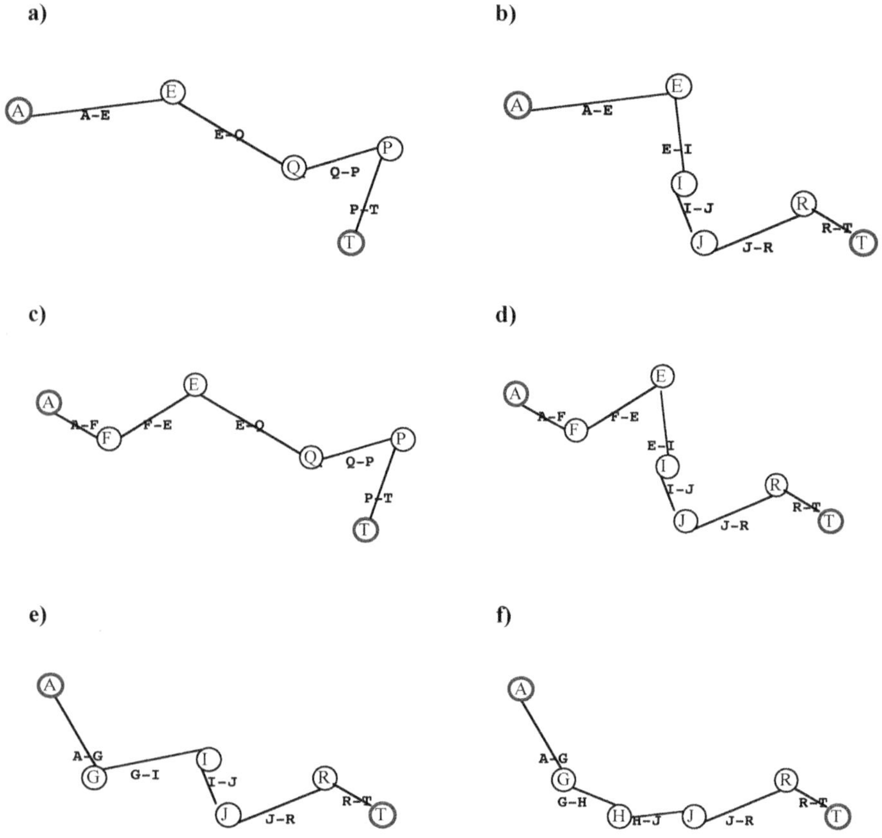

Fig. 2. Different paths between A-T in the network

The number of possible combinations gives many more possible paths through the network. However, we only show some examples that are close related and share some of the intelligent agents in the network.

For each of these paths, a merged cluster is created in the hierarchy. The intelligent agents, involved in the searching process, are added as single clusters in the hierarchy of agents, at lowest level, so called leaves, see Figure 3.

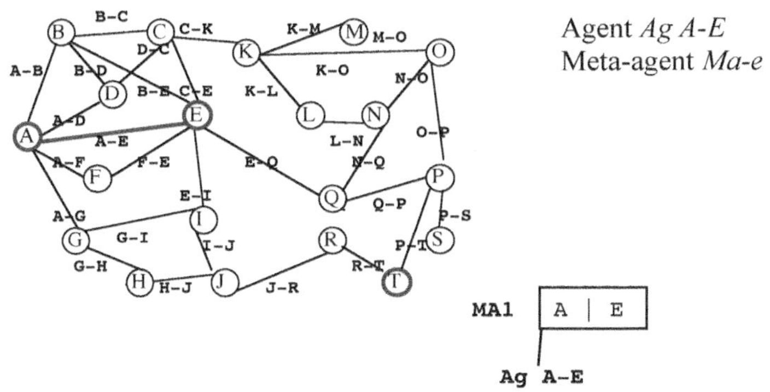

Fig. 3. Intelligent agents and meta-agents for the path A-E

When the intelligent agent Ag A-E has moved from node A to node E, a meta-agent MA1 is automatically created from the path between node A to node E, as illustrated in the path in Figure 2a. Besides containing the information about the intelligent agent, the meta-agent also has information about the performance time. When the meta-agents are developed from the agents, they are added as merged clusters above the intelligent agents in the hierarchy. The simplest form of created meta-agent is the meta-agent MA1, which also becomes a merged cluster is A | E.

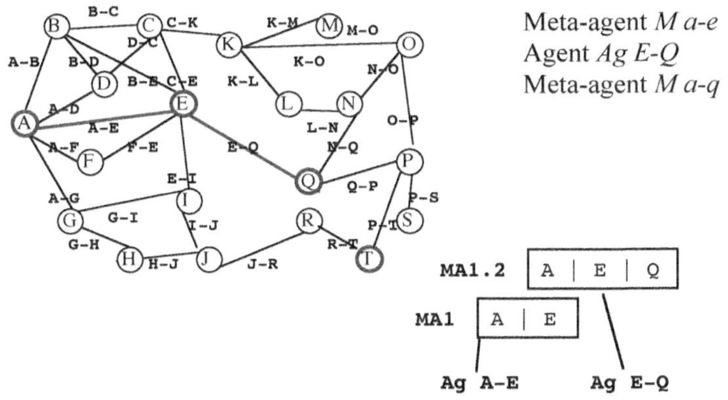

Fig. 4. Intelligent agents and meta-agents for the path A-Q

From the node E, the intelligent agent Ag E-Q moves from node E to node Q, which creates the meta-agent MA1.2. This meta-agent MA1.2 is an expansion of the meta-agent MA1, created earlier. Thus, the meta-agent MA1 has been used to expand the meta-agent, MA1.2 with information about the path between the nodes E and Q. The meta-agent, MA1.2 becomes the cluster containing A |E |Q, which is added on top of earlier meta-agent and, thereby, become the parent of MA1, since MA1.2 contains parts of MA1.

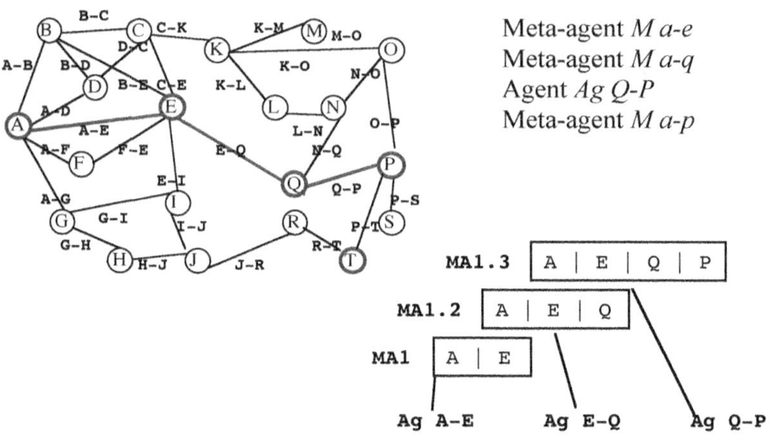

Fig. 5. Intelligent agents and meta-agents for the path A-P

The next created meta-agent, MA1.3, which also become merged clusters for A |E |Q |P and the meta-agent MA1.4 become merged cluster for A |E |Q |P |T.

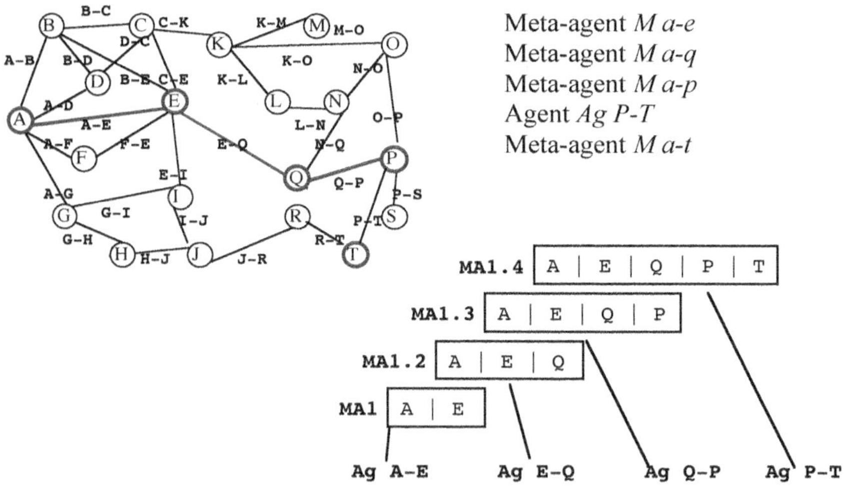

Fig. 6. Intelligent agents and meta-agents for the path A-T

Another example of a hierarchy from a path A-T is presented in Figure 4. This hierarchy corresponds to the second path above.

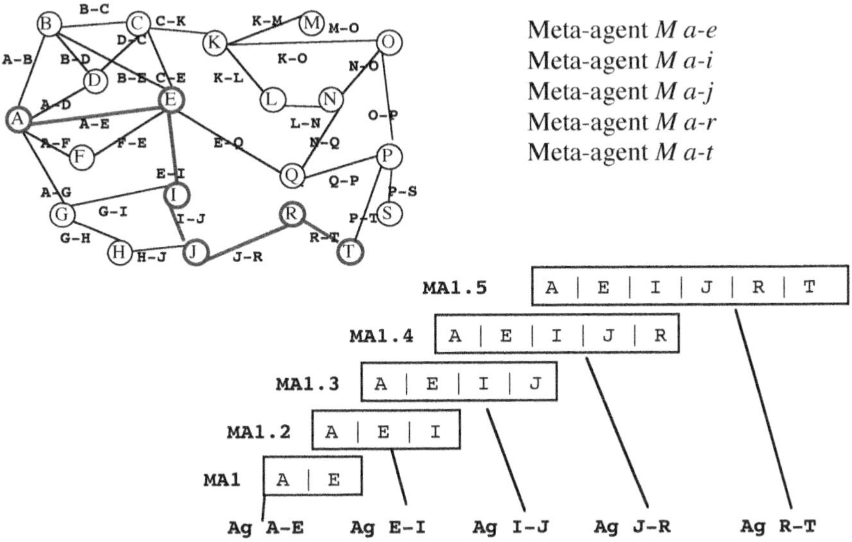

Fig. 7. Alternative hierarchy for the path A-T

As in the hierarchy in Figure 3, the first agent (Ag A-E) becomes a leaf in the hierarchy presented in Figure 4. The agent Ag A-E is moving from node A to node E and the meta-agent MA1 is created. Next, agent Ag E-I is moving from E to I and the meta-agent, MA1.2, is created. Again, the first meta-agent MA1 is included in MA1.2, which has the merged cluster A |E |I. Then, the agents Ag I-J, Ag J-R and Ag R-T are moving to their respective node J, R and T. The meta-agents MA1.3, MA1.4 and MA1.5 is created and added to the hierarchy. The top-level meta-agent MA 1.5 becomes the cluster containing A |E |I |J |R |T.

As the reader may have noticed, we have different contents in the meta-agents MA1.2, MA1.3 and MA1.4 presented in Figure 3d and Figure 4. In Figure 3c, the meta-agent MA1.2 cluster is A |E |Q, meta-agent MA1.3 cluster is A |E |Q |P and MA1.4 cluster is A |E |Q |P |T. In Figure 4, the MA1.2 cluster is A |E |I, MA1.3 cluster is A |E |I |J and MA1.4 cluster is A |E |I |J |R. This is, of course, not acceptable because it generates inconsistency in the hierarchy and, therefore, the created meta-agents with the same name but different contents have to be renamed. The easiest solution was to include an additional number, MA1.21, MA1.22 before include it in the hierarchy. But to avoid the problem with inconsistency, all created hierarchies begin with a unique number to be sure that they use unique names in the meta-agents.

The hierarchies presented above have two different clusters of the same solution but, as mentioned above, there are many more paths in the network. The hierarchies are built up in parallel but when a solution is found, the process of building hierarchies halts. Nonetheless, it is possible to get additional hierarchies. Beside the two presented above, the other hierarchies that correspond to the paths in Figure 2, c-f are presented in Figure 8.

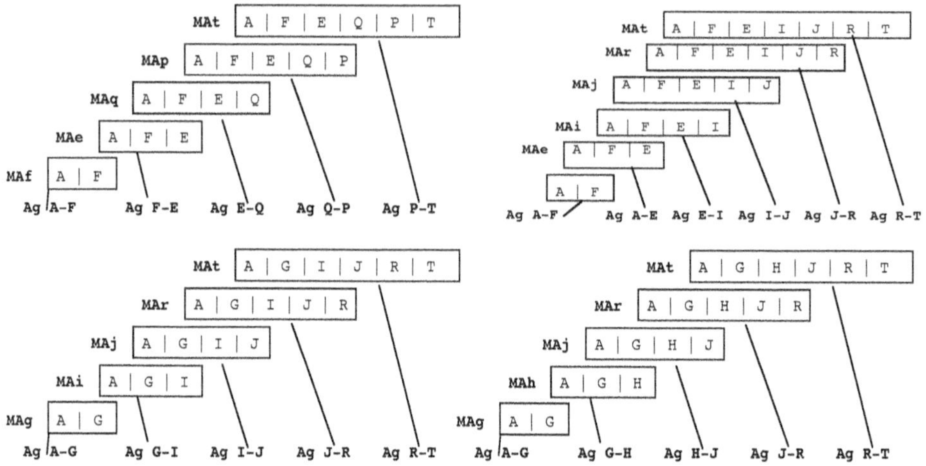

Fig. 8. Other hierarchies that correspond to the paths in Figure 2, c-f

As long as the user wants to try more paths, several paths will be generated and added to the hierarchies. The process continues as long as there are more available paths in the network.

As the opposite of the expansion, the number of paths can be limited, i.e., if the user denotes some conditions the agents must meet. For example, if the denoted conditions force some of the agents, in the beginning or in the middle of the path, to be discharged, the chain of agents will break down, which cause the total path to be unused. For example, if the agent Ag R-T, illustrated at the top in Figure 4 cannot fulfil the conditions, and it is the only connection between those two nodes, the complete path will be discharged.

5.1 Multiple Hierarchies of Agents

In some cases, one hierarchy for the path through the network is not enough. In these cases we need multiple hierarchies. Multiple hierarchies will be used in the cases where there is a set of top-level meta-agents. Thus, the multiple hierarchies are built when a node has more than one connection. When building the hierarchy, the starting node becomes the first part of the hierarchy. Then, the first connection that is visited by an intelligent agent expands the hierarchy. Next, for each connection, a separate hierarchy is initiated and built up and placed at the same level as the first hierarchy. The expansion of the hierarchies continues as long as we need to find the solution. As long as the connected agents concerns the hierarchies, each of these hierarchies will be expanded with their connected agents.

To illustrate the multiple hierarchies with top-level meta-agents, we present one hierarchy using several meta-agents. The example is a part of the network, as presented in Figure 1. The starting position is node A from which the agent A-B has move to its destination and the meta-agent MA1 has been created, see Figure 9.

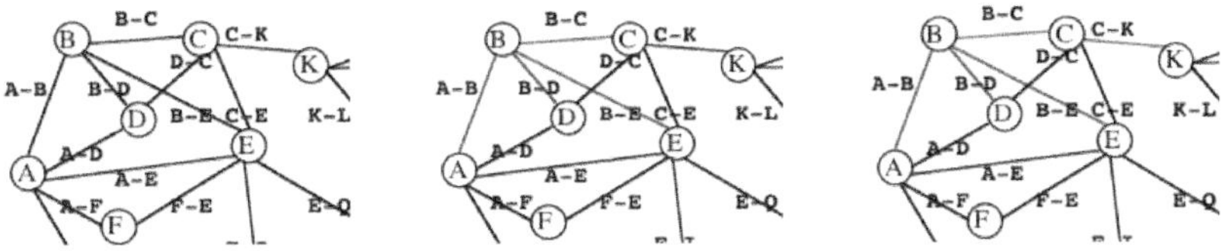

Fig. 9. Part of the network for a multiple hierarchies example

Assume that we are at node B, at the end of path A-B. Suppose that the node B has connections to node C, one to node D, and one to node E, as presented in the network in Figure 9. A is consider to be the starting point from with the Ag A-B moves and becomes the leaf in the hierarchy. From the node B, multiple hierarchies are created as presented in Figure 10.

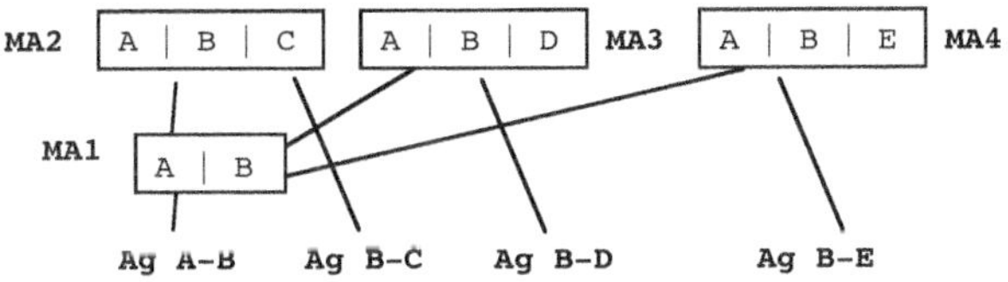

Fig. 10. Multiple hierarchies for several meta-agents

At this point, the system creates a tree of meta-agents that lead to node B, a so-called meta-agent MA1. The meta-agent MA1 is that first abstract level and has a child A on the paths back from A to the start point. As we move to next connected node, node C, we create meta-agent MA2, (upper left in the Figure), which has the child MA1. As we move to next connect node to B, i.e., node D, we create meta-agent MA3, which has the child MA1. Finally, as we move to the last connect node, node E, we create MA4, again with the child MA1. This means there has been a set of top-level meta-agents created automatically, which have merged into multiple hierarchies.

The next step in this work is to proceed with MA2. MA2 has three different connections: K, D, and E, see Figure 11.

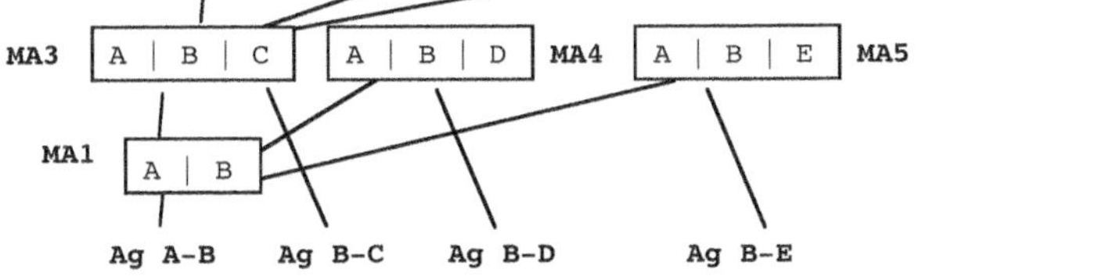

Fig. 11. Multiple hierarchies for several levels of meta-agents

Since the MA2 already include the node B, it is unnecessary to consider that node again. Also node D has been visited so either on of the paths, A->B->D or A->B->C->D (MA7 in the Figure) should be discharged since they lead to the same solution but with slightly different paths. Which path to be discharged can be decided by time limits or user given constraints. The one to discharge should be the one that is the most expensive one or includes any of the user-given conditions. Multiple paths to same node also the case for node E, which has the paths A->B->E and A->B->C->E, (MA8 in the Figure). Hence, only the node K is interesting to consider and the hierarchy A | B | C is be expanded with K to become A | B | C | K, see Figure 11.

After MA3 is finished, the work is to continue with MA4, MA5 and MA6. MA4 has connections to the nodes A and C, which have been visited already by other meta-agents, whereas MA5 has connections to the nodes A, F, C and Q. In this case, the most interesting are node F and node Q, even though F leads back to node A but the system is not aware of the case in beforehand. The procedure of visiting nodes and expanding the hierarchies continues until the goal is found and the alternative paths through the network.

Meta-agents can be discharged on some conditions and made inactive on others:

- We discharge the meta-agents that produce cycles.
- We can make meta-agents inactive if some preferred condition is not satisfied. An inactive agent can become active if conditions change.

When the meta-agent reaches a goal or an end node, the expansion of the hierarchy is completed. Then, the meta-agents announce the successful completeness together with the total cost for the path. A central control program receives these messages and keeps them in a sorted list. The control program also has a list of inactive meta-agents from which the program can invoke the inactive agents and/or the inactive meta-agents. A meta-agent will become active, i.e., if its children have a change of conditions.

5.2 Presenting the Hierarchy

Once the hierarchies are built, they are presented at the users' web site. By presenting closely related information in hierarchies, the users' grasp of the contents can be supported. The presentation must be more than just texts; it needs to have pictures of the hierarchy as well. In fact, it should be a combination of pictures and describing text to the pictures, as well as, information found at the sites. So far, we have worked with UML's diagrams that have been reformed to suit multi-agent systems but they do not really handle the hierarchy shape.

The problem with the presentation is the vast number of intelligent agents and meta-agents that will be present. UML has useful facilities for handling large amount of information but information sets will still be massive. Therefore, the presentation of the hierarchies can only be fragmented. The presented fragments may not give the required information and for this reason we need an excellent search algorithm.

6 Searching in the Hierarchy

To find agents in the hierarchy, we need a searching algorithm. The construction of the hierarchy is performed by starting to build up the lower-level and then using the result of the lower level to build up the higher-level [29]. The search and retrieval process works in opposite direction, the process proceeds from the higher-level hierarchy to lower-level. At the higher-level, the searching is mapping the request of path with the content of the meta-agents. At some point, the starting point of the path matches the meta-agent's starting node, from which the searching algorithm commences. The search must check that each agent corresponds to the request, which is an exact determination method.

Fetching information from a lower part than first specified can be required if more detailed information is searched for, which cannot be supplied at higher levels. Retrieving the information from the agents can be at three different possible levels compared to the specification of the level where the agents will be found. Either the information is found at the specified level of hierarchy, i.e., according to the specification, or at a more general level than current level. But the information can also be fetched from a more specific level of the current specification. That level can be at or beneath the level of meta-agents but not lower than the level of intelligent agents level.

In addition to searching and retrieving information, the hierarchy can support reasoning at different levels in the hierarchy by dividing the overall task for the agents in the hierarchy. As in hierarchical behaviour-space [30], the search algorithm can determine which level of abstraction to use for the problem step at hand. The higher-level node can inform the lower-level node, which activities they are to pursue and determine how redundancies are avoided. The relationships can aid the coordination process and specify which actions the agents are going to undertake.

The search algorithm can run from the central control structure, described above. The control structure has the top agents of the hierarchy, including those placed on an active status. From this top agent, the search algorithm can traverse the hierarchy down to the leaf agents. All agents, down through the hierarchy, following the branches that lead to the depth in the hierarchy, have relations to each other.

The search algorithm can either search either sequentially or in parallel. In many cases the number of top-level meta-agents is small and simple sequential search is sufficient. The reasoning referenced above can be included in the search steps. This allows the search to do an analysis for conditions that is more complex and cannot be evaluated during the meta-agent formation process, for example conditions that are complex functions over the edges of the search graph. In the case of parallel search, the clusters formed will divide the agents into the nodes of a multi-processor. This will give a natural implementation for the parallel search. The clusters will have closed related nodes, so the search will require minimal communication between the nodes.

The sequence of steps is important since the tasks break in to multiple steps that must be solved in sequence. These agents must be represented in the correct order to make use of later. Thus, the agents at the starting position are to the left in the hierarchy and the agents at the end position are to the right. In between the agents must be placed in the right order to perform tasks and find the correct path.

Moreover, a parallel set of steps will be present in the hierarchy. The search algorithm needs to find all these paths that can lead between two nodes. However, if there are alternative paths, only one needs to secede for success.

If we view agents as providing services, then part of the solution is locating the proper agents. Proper in this case means the agents that provide required services. The simplest form of providing agents is an extra set of constraints on the solution. However in a more complex case, the agents cannot be selected in isolation and require some reasoning on the path of agents in the solution.

7 Related Work

There are numerous examples in the literature of clustering and structuring hierarchical organizations in agent systems. For, e.g., clustering agents, Davis and Kraus (1999) [31] describe the uses of a separate cluster agent to collect and select the agents in a system into clusters. This clustering differs from our approach since in our work the clustering is performed in a distributed algorithm used by the meta-agents in the system.

For hierarchical organizations, there are examples of how to use of hierarchies to achieve coordination [32]. One example shows how coordination uses upper levels to send coordination information downward and status information to flow upward in the hierarchy. Similarly, there are examples of a hierarchical behaviour-space [30]. A hierarchical behaviour space uses a hierarchy of agents, which present different levels of abstraction. A search algorithm determines which level of abstraction to use for the problem step at hand. This is often used as the representation in a planning system.

One view of the ways hierarchies and clusters can be formed comes from Computational Organization Theory (COT) [33]. COT applies the lesions learned from the study of human organizations. The observation is that the organization enhances the abilities of the agents to solve problems beyond that of the collection of individual agents. It is the old adage "the whole is greater that the sum of the parts". And the reason is the organization has properties that support the individual activities of the agents. While this work is in the early stages and few general results are known, there are several basic tools, which support some basic organizations structures. These are embedded in experimental test-bed frameworks.

There are a couple of existing tools built on COT, e.g., MACE, SDML and Multi-agent SOAR. MACE [33] is a general domain independent multi-agent modelling framework. An interesting idea from MACE is recursive agent structure, which can compose a group of agents together and then viewed the agents, one at the time.

SDML allows team formation and structures of teams into a hierarchy. SDML is flexible enough to represent sophisticated agents at different levels, which can read and write rules of other agents. Multi-agent SOAR is a general architecture from general intelligence. A key aspect of this system is the social level to the agents. This can include expectations about the other agents goals and actions. Of these approaches, SDML and Multi-agent SOAR seem best suited for the problems considered in our work since agents at meta-level include the other agents' goals and actions.

8 Conclusions and Further Work

In this chapter we have presented an approach to create hierarchies of agents, automatically. The agents, both intelligent software agents and meta-agents, are represented in a hierarchy to cluster similar agents and optimize locality. Moreover, the cluster can support avoiding the unsuccessful paths and make the search efficient. In addition, the hierarchy supports reasoning at different levels in hierarchy by dividing the overall task for the agents in the hierarchy.

For the moment, the system works well for a small number of intelligent agents and meta-agents. Also the creation and expansion of hierarchies work well. However, the creation of multiple hierarchies can provide problems since we need to split up those in an acceptable manner. Moreover, we need to run the system with a large number of intelligent agents to check how fast it will handle all agents, meta-agents and building of hierarchies. From the test made on the current system, not surprisingly, it is found that it becomes slow with a small number of agents. The parallel computing must be implemented to make the system more efficient.

References

1. Waloszek, G.: Fighting (with) Hierarchies – Part I: Basics, `http://www.sapdesignguild.org/community/design/hierarchies.asp`
2. Kumar, R., Raghavan, P., Rajagopalan, S., Tomkins, A.: Extracting Large-Scale Knowledge Bases from the Web. In: VLDB, pp. 639–650 (1999)
3. Eiron, N., McCurley, K.S.: Link Structure of Hierarchical Information Networks. In: Leonardi, S. (ed.) WAW 2004. LNCS, vol. 3243, pp. 143–155. Springer, Heidelberg (2004)
4. Ahl, V., Allen, T.F.H.: Hierarchy Theory. Columbia University Press, New York (1996)
5. Adams, C., Lloyd, S.: Understanding PKI: concepts, standards, and deployment considerations. Addison-Wesley Professional, Reading (2002)
6. Izar, P., Ferreira, R.G., Sato, T.: Describing the organization of dominance relationships by dominance-directed tree method. American Journal of Primatology 67, 1–19 (2006)
7. December, J.: Web Develop, `http://www.december.com/web/develop/character.html`
8. Telecom, Lecture Notes, Chapter 16, IP: Internet Protocol Addresses, `http://www.cs.nyu.edu/artg/telecom/fall99/lecture_notes/lecture_notes.html`
9. Celko, J.: Trees and Hierarchies in SQL for Smarties, 1st edn. Morgan Kaufmann, San Francisco (2004)
10. Machanic, A., Kornelis, H., Rubbelke, L.: Expert SQL Server 2005 Development. Apress, Berkeley (2007)
11. Davis, T.A.: Fundamentals of Algorithms. Directed Methods for Sparse Linear Systems. Society for Industrial and Applied Mathematics, Philadelphia, PA, USA (2006)
12. Green, D.G.: Hierarchy, complexity and agent based models. In: Our Fragile World: Challenges and Opportunities for Sustainable Development. UNESCO, Encyclopedia of Life Support Systems (EOLSS), Paris, pp. 1273–1292 (2001)
13. Green, D.G., Bransden, T.G.: Complexity theory. In: McGraw-Hill Encyclopedia of Science and Technology, pp. 507–511. McGraw-Hill, New York (2006)
14. Wooldridge, M.: An Introduction to MultiAgent Systems. John Wiley & Sons Ltd., New York (2002)
15. Håkansson, A., Hartung, R.L.: Using Meta-Agents for Multi-Agents in Networks. In: The 2007 International Conference on Artificial Intelligence, The 2007 World Congress in Computer Science, Computer Engineering, & Applied Computing, vol. II, pp. 561–567. CSREA Press, U.S.A. (2007)

16. Håkansson, A., Hartung, R.L.: Calculating optimal decision using Meta-level agents for Multi-Agents in Networks. In: Apolloni, B., Howlett, R.J., Jain, L. (eds.) KES 2007, Part I. LNCS (LNAI), vol. 4692, pp. 180–188. Springer, Heidelberg (2007)
17. Håkansson, A., Hartung, R.L.: Autonomously creating a hierarchy of intelligent agents using clustering in a multi-agent system. In: The 2008 International Conference on Artificial Intelligence, The 2008 World Congress in Computer Science, Computer Engineering, & Applied Computing, vol. II, pp. 89–95. CSREA Press, U.S.A. (2008)
18. Russell, S., Norvig, P.: Artificial Intelligence: A Modern Approach. Prentice-Hall, Inc., Englewood Cliffs (1995)
19. Ferber, J.: Multi-Agent Systems. An introduction distributed artificial intelligence. Addison-Wesley Professional, Reading (1999)
20. Chelberg, D., Welch, L., Lakshmikumar, A., Gillen, M., Zhou, Q.: Meta-Reasoning For a Distributed Agent Architecture, `http://zen.ece.ohiou.edu/~robocup/papers/HTML/SSST/SSST.html`
21. Tozicka, J., Zelenzny, F., Pechoucek, M.: Modelling of agents' behaviour with semi-collaborative meta-agents. In: Pěchouček, M., Petta, P., Varga, L.Z. (eds.) CEEMAS 2005. LNCS (LNAI), vol. 3690, pp. 572–575. Springer, Heidelberg (2005)
22. Holland, J.H.: Adaptation in Natural and Artificial Systems: An Introductory Analysis with Applications to Biology, Control, and Artificial Intelligence. The MIT Press, Cambridge (1992)
23. Negnevitsky, M.: Artificial Intelligence – A guide to Intelligent Systems. Pearson Education, London (2002)
24. Pěchouček, M., Štěpánková, O., Mařík, V., Bárta, J.: Abstract Architecture for Meta-reasoning in Multi-agent Systems. In: Mařík, V., Müller, J.P., Pěchouček, M. (eds.) CEEMAS 2003. LNCS (LNAI), vol. 2691, pp. 84–99. Springer, Heidelberg (2003)
25. Costantini, S.: Meta-reasoning: a survey. In: Computational Logic: From Logic Programming into the Future: Special volume in honour of Bob Kowalski. Springer, Berlin (2002) (in print)
26. Restivo, A.O., Reis, L.P.: Clustering Agent Optimization Results in Dynamic Scenarios. In: 4th European Workshop on Multi-Agent Systems, pp. 289–297, CEUR-WS.org (2006)
27. Härdle, W., Simar, L.: Applied Multivariate Statistical Analysis, `http://www.quantlet.com/mdstat/scripts/mva/htmlbook/mvahtmlnode80.html`
28. Everitt, B.S., Landau, S., Leese, M.: Cluster Analysis, 4th edn. John Wiley & Sons, Inc., Chichester (2001)
29. Cheng, B.H.C., Jeng, J.J.: Formal methods applied to reuse. In: 5th Workshop in Software Reuse (October 1992)
30. Durfee, E.H., Montgomery, T.A.: Coordination as Distributed Search in a Hierarchical Behaviour Space. IEEE Transactions on Systems, Man, and Cybernetics, Special Issue on Distributed Artificial Intelligence SMC- 21 6, 1363–1378 (1991)
31. Davis, E., Kraus, S.: Agents for Information Broadcasting. In: Jennings, N.R. (ed.) ATAL 1999. LNCS, vol. 1757. Springer, Heidelberg (1999)
32. Hatvany, J.: Intelligence and cooperation in heterarchic manufacturing systems. Robotics and computer-integrated manufacturing 2(2):22, 101–104 (1985)
33. Carley, K.M., Gasser, L.: Computational Organization Theory. In: Weiss, G. (ed.) Multiagent Systems - A modern approach to distributed Artifical Intelligence. The MIT Press, Cambridge (1999)

A Role-Based Cognitive Architecture for Multi-Agent Teaming

Jinsong Leng[1], Chee Peng Lim[2], Jiuyong Li[3],
Dongguang Li[1], and Lakhmi Jain[4]

[1] School of Computer and Security Science
Edith Cowan University, WA 6050, Australia
J.Leng@ecu.edu.au,
D.Li@ecu.edu.au
[2] School of Electrical and Electronic Engineering
University of Science Malaysia, Penang, Malaysia
Cplim@eng.usm.my
[3] School of Computer and Information Science
University of South Australia, SA 5095, Australia
Jiuyong.Li@unisa.edu.au
[4] School of Electrical and Information Engineering
University of South Australia, SA 5095, Australia
Lakhmi.Jain@unisa.edu.au

Abstract. Agent teaming is a subfield of multi-agent systems that is mainly composed of artificial intelligence and distributed computing techniques. Autonomous agents are required to be able to adapt and learn in uncertain environments via communication and collaboration in both competitive and cooperative situations. The joint intension and sharedPlan are two most popular theories for the teamwork of multi-agent systems. However, there is no clear guideline for designing and implementing agents' teaming. As a popular cognitive architecture, the BDI (Belief, Desire, and Intension) architecture has been widely used to design multi-agent systems. In this aspect, flexible multi-agent decision making requires effective reactions and adaptation to dynamic environment under time pressure, especially in real-time and dynamic systems. Due to the inherent complexity of real-time, stochastic, and dynamic environments, it is often extremely complex and difficult to formally verify their properties *a priori*. For real-time, non-deterministic and dynamic systems, it is often difficult to generate enough episodes via real applications for training the goal-oriented agent's individual and cooperative learning abilities. In this article, a role-based BDI framework is presented to facilitate optimization problems at the team level such as competitive, cooperation, and coordination problems. This BDI framework is extended on the commercial agent software development environment known as JACK Teams. The layered architecture has been used to group the agents' competitive and cooperative behaviors. In addition, we present the use of reinforcement learning techniques to learn different behaviors through experience. These issues have been investigated and analyzed using a real-time 2D simulation environment known as SoccerBots.

A. Håkansson et al. (Eds.): Agent and Multi-agent Technology, SCI 289, pp. 229–255.
springerlink.com

Keywords: Agents, BDI Architecture, JACK Agent Software, Role, Multi-Agent Teaming.

1 Introduction

Artificial intelligence (AI) and agent systems are tightly combined in some application domains. AI is concerned with studying the components of intelligence, while agents focus on integrating the components with properties of AI. Distributed AI (DAI) brings out some other issues to be solved: complexity, communication, cooperation, and so on.

Multi-Agent Systems (MASs) are composed of several agents that can interact among agents and the environment. This means that agents not only interact with the environment, but also interact with each other. MASs usually operate in real-time, stochastic, and dynamic environments. An agent team is a set of agents having a shared objective and a shared mental state [12,13]. Many theoretical and applied techniques have been investigated in agent architecture with respect to cooperation, coordination, adaptation, and communication, in order to provide a framework for implementing intelligent computing techniques and systems [41,47]. However, issues of complexity, uncertainty, and incomplete knowledge still need much attention when designing agent-based systems.

MASs inherit many DAI motivations, goals and potential benefits. They also inherit those of AI technologies that may have the ability to deal with incomplete information, or the ability for each agent to exhibit distributed control with decentralized data and asynchronous computation [24]. Owing to the dynamic nature of MASs, it is difficult to develop a teamwork model of agents that is applicable to various domains in general. For real-time systems, flexible multi-agent decision making requires effective reaction and adaption to dynamic changes under time pressure. However, agent teaming theories such as joint intension and sharedPlan do not provide a guideline in general for implementing agents' teamwork. Therefore, it is of great importance for investigating and building a flexible BDI framework to address the issues of multi-agent decision making and adaption to dynamic changes under real-time constraints.

Agent-Oriented Software Engineering (AOSE) aims to simplify the construction of complex systems by using abstractions of nature [23]. For example, Open Agent Architecture (OAA) [10] is a team-oriented agent architecture for integrating a community of heterogeneous software agents in a distributed environment. The research on agents teaming aims to improve concept understanding, in order to develop some reusable algorithms, and to build high-performance teams in dynamic and possibly hostile environments.

JACK [4] is an agent development environment with agent-oriented concepts, including agents, capabilities, events, plans, knowledge bases (databases), resource and concurrency management. The JACK Teams [5] provides a hierarchical team-oriented framework using some unique concepts: teams, roles, and team plans.

In this article, a role-oriented BDI architecture is proposed for addressing the issues of real-time multi-agent teaming. The extension of real-time adaption and learning has been made to JACK Teams. The team tasks are specified by a set of roles to guide the joint behaviors coherently at the team level. This framework provides a guideline to build an effective multi-agent team capable of dealing with incomplete information, adaption and learning in real-time systems. To demonstrate the framework, a role-based soccer team is implemented in a real-time 2D simulation environment known as SoccerBots [2].

This article is organized as follows: Section 2 gives an overview of MASs and related work. Section 3 introduces the BDI architecture. The major properties of roles and team formation are discussed in Section 4. Section 5 describes some agent development platforms. Section 6 details the JACK and JACK Teams; and the extension to JACK Teams is also presented. A soccer agents' team that is implemented using the JACK Teams framework is detailed in Section 7. Section 8 introduces related work Finally, conclusions and suggestions for future work are presented.

2 Multiagent Systems

2.1 Properties of Agent

The definitions of agent proposed by various researchers may vary from different prospectives. Here are some definitions:

- Russell and Norvig [31] define an agent as an entity that can be viewed as perceiving its environment through sensors and acting upon its environment through effectors.
- Coen [11] views software agents as programs that engage in dialogs, and negotiate and coordinate the transfer of information.
- Wooldridge and Jennings [45] state that an agent is a hardware and/or software-based computer system displaying the properties of autonomy, social adeptness, reactivity, and proactivity.

The ability to act without human intervention or other systems (Autonomy) is a key feature of an agent. Despite being no universal definition of agents, the properties that an agent is supposed to have may help classify the types of agents in useful ways. Franklin [17] lists some properties of an agent, as shown in Table 1.

Table 1. Properties of an agent

Property	Meaning
reactive	responds in a timely fashion to changes in the environment
autonomous	exercises control over its own actions
goal-oriented	does not simply act in response to the environment
temporally continuous	is a continuously running process
communicative	communicates with other agents, perhaps including people
learning (adaptive)	changes its behavior based on its previous experience
mobile	able to transport itself from one machine to another
flexible	actions are not scripted
character	believable "personality" and emotional state.

In [19], intelligent agents are described in terms of a space with the three dimension of agency, intelligence, and mobility, as shown in Fig. 1.

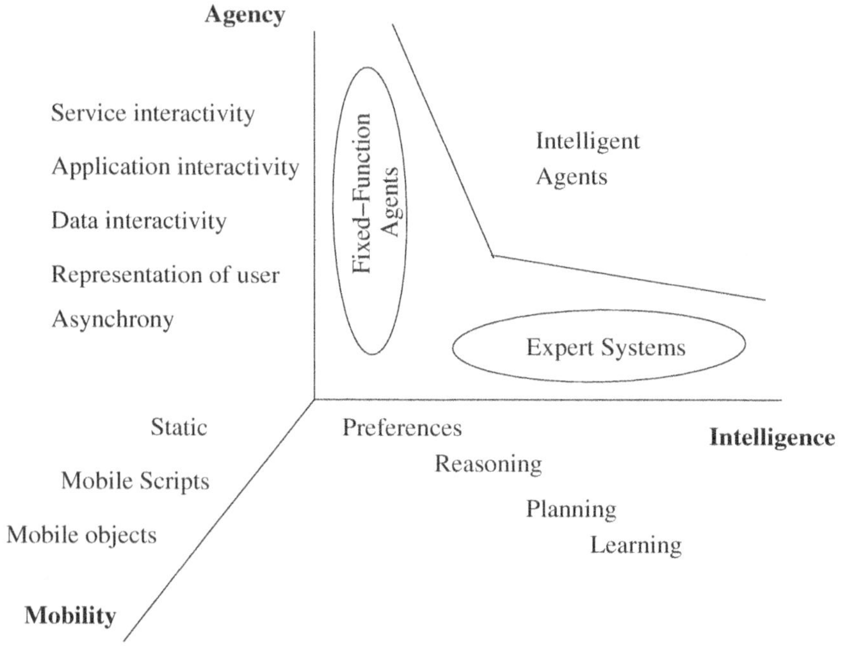

Fig. 1. The Scope of Intelligent Agent, taken from [19]

- Agency is the degree of autonomy and authority rested in the agent.
- Intelligence is the degree of reasoning and learned behavior.
- Mobility is the degree to which agents themselves travel through the network.

Agency may also be catalogized into:

- R*ational agency (logical)* [44] - Logical models using BDI, for example, knowledge-based inference systems;
- R*ational agency (economic)* - Economic model using preferences on the transition of state and action, the decision may depend on complete or partial uncertain knowledge;
- S*ocial agency* - Social commitments with other agents using cooperation, competition, and coordination;
- A*daptive agency* - Learning by interacting with other agents and environment.

2.2 Agent and Environment

A single-agent system includes the agent, the environment, and the interaction between them. Actually, the agent itself is part of the environment, but is considered as an independent entity with its own goals, actions, and knowledge (see Fig. 2).

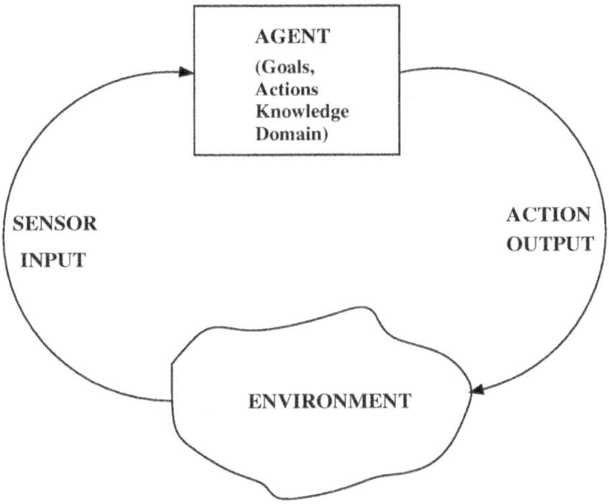

Fig. 2. A Single-Agent System

MASs differ from single-agent systems in that they are composed of several agents that can interact among agents and the environment. This means that agents not only interact with the environment, but also interact with each other. The environment of MASs can be changed dynamically by other agents. Heterogeneous agents need to communicate with each other and affect the environment in unpredictable ways, which significantly increases the complexity of the system, as indicated in Fig. 3.

Agent environments can be classified based on different properties that affect the complexity of the agent decision-making process [31].

- A*ccessible vs. inaccessible* – An accessible environment is one in which the agent can obtain complete, timely, and accurate information about the state of the environment;
- D*eterministic vs. non-deterministic* – In a non-deterministic environment, the state that will result from an action is not guaranteed even when the system is in a similar state. This uncertainty presents a greater challenge to the agent designer than deterministic systems;
- S*tatic vs. dynamic* – Static environments remain unchanged except for the results produced by the actions of the agent. Other processes operate in a dynamic environment, thereby changing the environment which is outside the control of the agent;
- D*iscrete vs. Continuous* – An environment is discrete if there are a fixed, finite number of actions and percepts in it.

2.3 Characters of Multi-Agent Systems

MASs take into account the ability to learn and solve problems, but also focus on social aspects - the ability to communicate, cooperate, and reach agreements

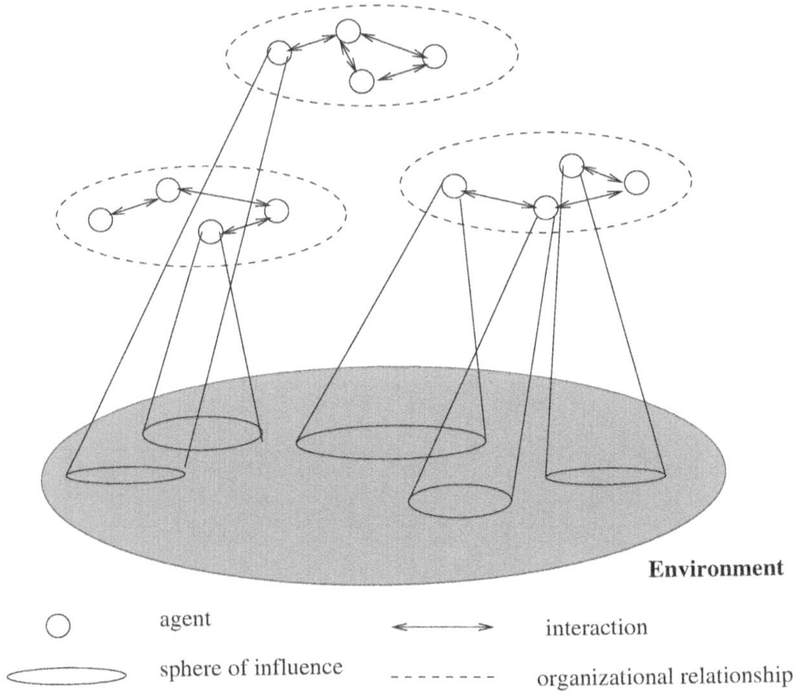

Fig. 3. A Multi-Agent System, taken from [23]

between peers. An MAS can be defined as a loosely coupled network of problem solvers that interact to solve problems that are beyond the individual capabilities or knowledge of each problem solver [14]. The characteristics of MASs are outlined as follows [39]:

- Each agent has incomplete information or capabilities for solving the problem and, thus, has a limited viewpoint;
- There is no global system control;
- Data are decentralized;
- Computation is asynchronous.

3 BDI Architecture

The BDI architecture has been widely used for designing MASs. The BDI model has the roots from cognitive architecture [9], which simulates the process of human cognition. A cognitive architecture is a computational paradigm that consists of some structures and cognitive processes: a set of primitive processes, memories, and control structures. The BDI model has been rigorously formalized on rational agency [30]. The mental state of an agent is explicitly defined in terms of beliefs, desires, and intentions: beliefs are the information of the environment;

desires are the goals that the agent is pursuing; intensions are the plans or actions that the agent is committing to perform.

At every time cycle, an agent controls its behaviors and mental states through two iterative steps at regular intervals: perceives the environment and updates its mental state, and then determines the commitments based on the current beliefs and executes the suitable plans, possibly resulting in further belief changes.

The BDI model is a goal-oriented framework for solving the control problems in MASs. The study of MASs may provide a guideline for designing complex and distributed systems. MASs inherit many DAI motivations, goals, and potential benefits. They also extend those of AI technologies that may have the ability to deal with incomplete information, or the capability for each agent to exhibit distributed control with decentralized data and asynchronous computation [24]. MASs emphasize the problem of optimization at the team level, in order to derive cooperative and coordinative joint behaviors. MASs bring some additional problems to be solved, e.g., complexity, communication, cooperation, and coordination. Ideally, the joint actions can be specified by using heuristic search [32]. One has to note that NP-complete (non-deterministic polynomial) problem is a fundamental question in complexity theory [18]. Consequently, it is impractical to specify a large state space, which may grow exponentially with the number of dimensions. One popular approach is to use approximation techniques to reduce the large state space, and to learn the optimal joint actions through experience [38].

The team-oriented agent architecture emphasizes team level optimization. Joint intension [12,13] and SharedPlan [20,21] are two well-defined formalisms for agents' teaming. The joint intention is based on the concept of joint persistent goal [12]. The core to form a coherent agent team is to find joint behaviors among agents. Cohen et al. defines the joint behaviors as the collective actions by agents who share a joint intension. Joint action of a team appears to involve more than just the union of simultaneous individual actions. The agents are jointly committed to its being the case that throughout the doing of the action, the agents mutually believe they are doing it. Typically, teams will be involved in joint activities that consist of many parts performed concurrently or in sequence [12,13]. The popular approach for teamwork focuses on finding the 'appropriate' joint actions of agents so as to pursue the common goals.

Agents need to share a common goal, and the roles performed in the team. Mutual understanding, mutual prediction, and conflict-resolution are important to construct an effective team. The requirements for effective cooperation and coordination are as follows [8]:

- Common ground: Common ground refers to the pertinent mutual knowledge, beliefs, and assumptions that support interdependent actions in the context of a given joint activity. This includes initial common ground prior to engaging in the joint activity as well as mutual knowledge of shared history and current state that is obtained while the activity is underway;
- Interpredictability: In highly interdependent activities, it becomes possible to plan ones own actions (including coordination actions) only when what

others will do can be accurately predicted. Skilled teams become interpredictable through shared knowledge and idiosyncratic coordination devices developed through extended experience in working together; bureaucracies with high turnover compensate for experience by substituting explicit predesigned structured procedures and expectations;
- Directability: Directability refers to the capacity for deliberately assessing and modifying the actions of other parties in a joint activity as conditions and priorities change.

Developing an effective agent architecture that facilitates some properties with respect to cooperation, coordination, and communication is demanded by various application domains. Basically, options to implement agents include using a dedicated open source platform, employing a commercial platform, or by directly implementing the agents from scratch by using some standard languages [42]. The well-known cognitive agent platforms include JADE [29], JACK [4], and SOAR [25]. Multi-agent teaming can be implemented in various team-based architectures such as STEAM (a Shell for TEAMwork) [40], JACK Teams [5], CAST [46], OAA (Open Agent Architecture) [10].

Normally, agents are designed as an open system, in which each agent has no explicit information or knowledge of other agents. The agents work as a team by negotiating with each other. This kind of approach forms a loosely-combined agent team, in which tasks and responsibilities are assigned and allocated dynamically on the fly. The cooperation process relies on Agent Communication Language (ACL) and Knowledge Query and Manipulation Language (KQML) as the message format and protocol to support run-time information and knowledge sharing among agents. An optional way is to construct a closed team in which agents and roles are clearly defined and fixed. Forming a coherent team depends on explicit models for mutual understanding, mutual prediction, and conflict-resolution [8].

4 Roles and Team Formation

4.1 Roles

There are several ways to form an agent team. One option is to construct a team by assigning tasks and responsibilities on the fly. Another option is to specify the tasks by a set of roles. If the team structure is known in advance, we are able to assign the responsibilities and capabilities to different roles that may be performed in the team. The role-based approach can make the team to perform more efficiently with the minimum communication and negotiation cost.

The concept of roles has been widely used for enterprise modeling. A role is defined as an abstraction of an entity that performs a specific set of operations and includes temporal constrains on the order in which the operations may be performed [48]. From the agent teaming perceptive, a role is an extension of the agents current knowledge, and plays in a (joint) process like problem solving, planning, or learning [28]. The role is a set of responsibilities to be taken by team members. Lind [28] describes a role as a logical grouping of atomic activities according to the physical constraints of the operational environment of the target system. In [28], the major characteristics of a role are outlined as follows:

- A role is a collection of expectations towards the behavior of the inhibitor of a particular position;
- There exist mutual dependencies between roles, some roles can only exist if other roles do exist as well;
- A member of a society can play several roles even at the same time. This property is called role multiplicity and can lead to so-called role conflicts.

4.2 Roles Assignment and Team Formation

For a specific agent-based system, one needs to specify the tasks and responsibilities, and then group them into roles. A set of roles is defined to perform the joint behaviors coherently at the team level. Each agent is capable of playing one or more roles in a team.

Each agent is also capable of playing several roles at the same time, and each role may be performed by different agent. Team formation is the process to map roles onto agents using either static assignment or dynamic assignment. Static assignment implies that the relationship between roles and agents is fixed when a team is formed. Dynamic role assignment reallocates the roles to agents during run time. Static role assignment does not consider the role changes when the agents' internal state changes, while dynamic role assignment needs some interaction protocol to guide the dynamic assignment process.

5 Platforms for Multi-Agent Systems

The BDI architecture provides a very popular paradigm for knowledge representation. There are various agent software platforms for constructing agent-based systems.

JADE is developed by Telecom Italia Lab, and stands for Java Agent Development Framework. JADE [29] is a software framework fully implemented in Java language. The agent platform can be distributed across machines (which do not even need to share the same operating system), and the configuration can be controlled via a remote graphical user interface. The communication architecture uses messaging, and agents can access their queue via a combination of several modes: blocking, polling, timeout, and pattern matching based. The communication model components have been fully integrated: interaction protocols, envelope, ACL, content languages, encoding schemes, ontologies, and, finally, transport protocols. JADE has also been integrated with JESS, a Java shell of CLIPS, in order to exploit its reasoning capabilities.

JACK is commercial tool developed by Agent Oriented Software Pty. Ltd, and stands for JACK Intelligent Agents^{TM}. JACK is an environment for building, running, and integrating MASs using a component-based approach. The JACK

Agent Language is a programming language that extends Java with agent-oriented concepts, including: Agents, Capabilities, Events, Plans, Agent Knowledge Bases (Databases), Resource and Concurrency Management. JACK provides a suite of graphical tools for constructing agent in Java language with agent-specific concepts. JACK allows access to all present and future Java capabilities.

AgentBuilder is a commercial integrated suite for building intelligent agents. AgentBuilder consists of two major components - the Toolkit and the Run-Time System, all implemented in Java. The AgentBuilder Toolkit includes tools for managing the agent-based software development process, analyzing the domain of agent operations, designing and developing networks of communicating agents, defining behaviors of individual agents, and debugging and testing agent software. The Run-Time System includes an agent engine that provides an environment for execution of agent software. Agents constructed using AgentBuilder communicate using the Knowledge Query and Manipulation Language (KQML) and support the formats defined for KQML. In addition, AgentBuilder allows the developer to define new inter-agent communications commands that suit his/her particular needs.

SOAR is a general cognitive architecture for developing systems that exhibit intelligent behaviors. Researchers, both from the fields of artificial intelligence and cognitive science, are using SOAR for a variety of tasks. The intention of SOAR is to support all the capabilities required by a general intelligent agent that uses an approximation of complete rationality. All decisions are made through the combination of relevant knowledge at run-time. In Soar, every decision is based on the current interpretation of sensory data, the contents of working memory created by prior problem solving, and any relevant knowledge retrieved from long-term memory. Through SOAR, there has been a single framework for all tasks and subtasks, a single representation of permanent knowledge (productions), a single representation of temporary knowledge, a single mechanism for generating goals (automatic subgoaling), and a single learning mechanism.

From the technical point of view, AgentBuilder is more like "BDI editors" that allows programming in high level abstraction, but is committed to single agent architecture. JADE is an agent programming infrastructure that is more versatile, but at the expense of coding and development. SOAR uses a cognitive architecture that defines a fixed set of mechanisms and structures, which are based on cognitive-based reasoning. JACK provides a good developing graphical user interface with some useful embedded functions. JACK is suitable for experimentation with new agent architectures to model the BDI knowledge and procedural logic reasoning processes.

6 The JACK Teams Based BDI Framework

We summarize the major properties of the BDI framework for constructing a coherent agents' team, as follows:

- A coordinated cooperation by explicitly designing team plan;
- A hierarchical team structure for constructing team;

- Roles for defining the relationship and dependency between agents;
- Team formation by assigning roles statically or dynamically;
- Beliefs transformation for sharing knowledge.

6.1 JACK Teams Development Environment

JACK [4] is commercial software tool with some embedded functions and building blocks to help construct BDI agents. Within this platform, each agent is triggered by an event, resulting in updating its beliefs about the world and desires to satisfy, driving it to form intentions to act.

An agent is an event-driven deliberation entity. An intention is a commitment to perform a plan. A plan is an explanation of how to achieve a particular goal, and it is possible to have several plans for the same goal. Normally, a JACK application can be constructed in terms of agents, plans, events, and beliefs. The details of functionalities are outlined as follows [15]:

- Agent – Agent has beliefs, desires and intensions;
- Beliefset – Beliefsets represent the agent's declarative beliefs in a first order, tuple-based relational form;
- Plan – Plans are procedure that define how to respond to events.
- Event – Events are the central motivating factor in agents. Both internal processing of an agent and external exchanges among agents are triggered by normal events and BDI events;
- Team – Team is extension of the BDI paradigm that facilitate the modeling of social structures and coordinated behaviors. A team is formed in a hierarchical structure that is composed of a team entity, a team belief, a team plan, and roles in a team. Each role in a team is performed by a sub-team entity.

BDI agents exhibit reasoning behaviors under both pro-active (goal directed) and reactive (event driven) stimuli. Adaptive execution is introduced by flexible plan choice, in which the current situation in the environment is taken into account. A BDI agent has the ability to react flexibly to failure in plan execution. Agents cooperate by forming teams in order to achieve a common goal.

Some reasoning and deliberation functions are provided in JACK, for instance, alternative approaches can be applied in case a particular strategy (plan) fails. For example, JACK agents can be employed to control the players' actions in a simulation environment known as Unreal Tournament (UT) [33]. The execution process is illustrated in Fig. 4.

The JACK Teams [5] platform provides a hierarchical team-oriented framework using some unique concepts: teams, roles, and team plans. The focus of this article is on the team programming model provided by the JACK Teams as a core technology. The JACK Teams model provides a natural way to specify team structure and coordinated behavior related to team structure in terms of BDI model entities. Roles are used to describe the requirements and capabilities.

The JACK Teams can provide the basic reactive behaviors, and is pro-active by using the integrated reasoning and deliberation functions. Joint behaviors are

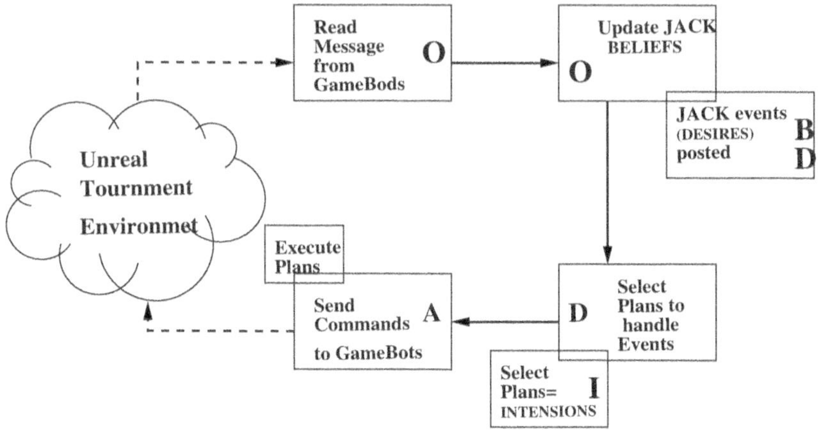

Fig. 4. JACK BDI Execution Model [33]

available by choosing the 'appropriate' plans. As discussed earlier, uncertainty is a common property and makes it infeasible to predict all possible situations a-*priori* in dynamic and non-deterministic systems. Automation and learning are two significant challenges, especially for human-agent interactions or intra-agent team.

6.2 Extension to JACK Teams

As described in Section 3, it is very difficult to specify the properties of complex and dynamic systems a-*priori*. The agents are required to adapt and learn through their own experience. The JACK Teams platform does not address the issues of adaption and learning. We construct a more flexible knowledge structure that can be improved through trial-and-error. The extension to JACK Teams with regard to learning algorithms over a large state space is useful for solving control problems in the complex and dynamic agent-based systems.

The team entity is able to coordinate the individual members' actions by finding the optimal joint behavior. In a real-time and dynamic environment, the team needs to make the individual agents to commit to the overall team goal under time pressure. The efficiency of multi-agent teaming depends on the performance and adaption of the techniques. The approximation techniques may improve the performance of the optimization techniques to solve complex and dynamic problems. Some extensions to JACK Teams have been made so as to enhance autonomy and adaption by learning through experience, as follows:

- The on-line self-learning or self-evolving property is particularly useful when designing complex, dynamic and real-time agent-based systems. An abstract class has been designed to enhance the on-line learning abilities. Decision theory can be applied to solve optimization control problems. In our previous work, we use temporal difference learning techniques to learn the control process in a soccer game [26,27]. In this BDI framework, the team entity

or sub-team entity can extend the abstract class LearningAgent.java, which includes some methods for on-line learning.
- The major difficulty is to deal with a large state space. An abstract class State.java is used to extend the Beliefset. Within the JACK Teams platform, the Beliefset is a rational database to save information and knowledge. The purpose of the abstract class State.java is to provide some function approximation techniques to deal with a large state space.
- JACK Teams provides the functions to synthesize the beliefs of sub-teams. In a noisy environment, individual members may have incomplete or inaccurate information. We need some other techniques to deal with the partial observation and information sharing problems.

7 Soccer Agents Team Implementation

7.1 Simulation System

Simulation in MASs plays an important role in MASs research, which can be used for developing AI techniques and investigating the behaviors of agent systems. The required scenarios can be simulated using testbeds to envision the real environment. Several aspects of the teamwork such as communication, knowledge-sharing, and effective coordination and cooperation will be tested.

Normally, computer games are dynamic, and use real-time environments that include complex 2D or 3D space, multiple interacting characters, and an uncertain environment. Unreal Tournament (UT) [1] and RoboCup soccer [3] are considered to be the best-known simulation systems for a variety of research in MASs.

SoccerBots is a part of the TeamBots(tm) software package [2], which provides an ideal testbed for MASs. The system can be modified by adding some soccer control and learning strategies. Each team can have 1 to 5 players, as shown in Fig. 5.

SoccerBots consists of three core components:

- The description file — the TBSim simulator reads in a description file to know what sorts of objects are in the environment, and their status at present.
- The simulation kernel — which runs the simulator at each step by drawing each object. Each object in the simulation includes two components: a drawing method, and a dynamic simulation method. The vision scope of soccer player and noise in the system can be defined.
- Robot control system — which controls the soccer players. This system can be modified by adding some soccer control and learning strategies.

The soccer game is a real-time, noisy, adversarial domain, which has been regarded as a suitable environment for MAS simulations [34]:

- Real-time: it acts in a dynamically changing environment.
- Noisy: which affects both sensors and effectors.

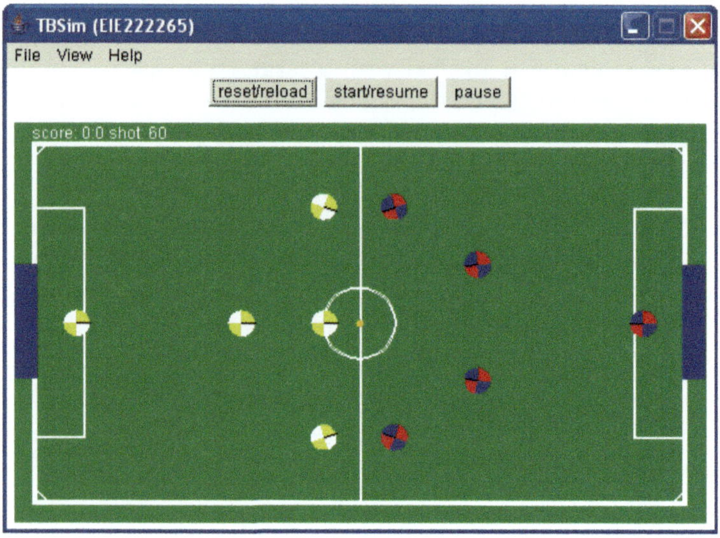

Fig. 5. The SoccerBots Simulation System

- Collaborative: the agents must cooperate (play as a team) to achieve the jointly desired goal.
- Adversarial: it involves competing with opponents.

7.2 Soccer Agent Architecture

In order to evaluate the realistic performance of the agents teamwork, we adopt a small-sized soccer league SoccerBots, which is one in the collection of application of TeamBots [2]. The soccer game players have to interact, coordinate, and cooperate with each other. The action that is sent back to each player is the deliberative outcome of the whole team. Each soccer agent has its own beliefs, desires, and intensions. The soccer team is illustrated in Fig 6.

Each player can observe the behaviors of other objects such as the ball, a teammate, an opponent, and their locations and motions via a sensor. The soccer game players have to interact, coordinate, and cooperate with each other. The action that is sent back to each player is the deliberative outcome of the whole team.

The SoccerBots promotes the following issues to be solved:

- Each player has a local and limited view of the world. Moreover, an agent team may have incomplete and imperfect knowledge of the world. The research question to be addressed is: How do we create techniques to filter raw data from the individual teammate, or to respond to deceptive agents in a hostile environment?
- In real-time situation, performance is a high priority to be considered to follow the changes of motions of the ball and the players. Owing to dynamic

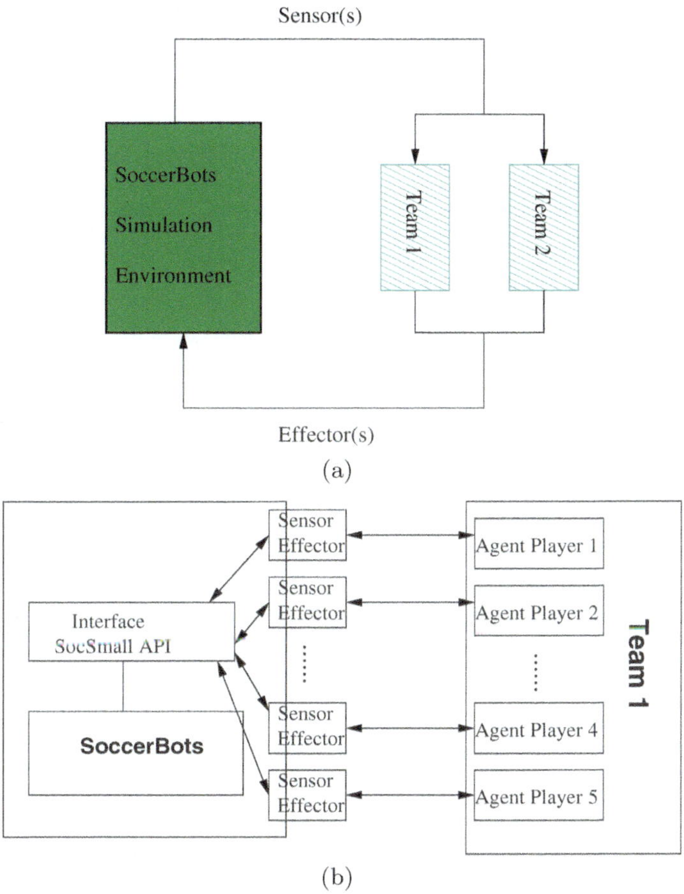

Fig. 6. The SoccerBots and Agents team. (a) SoccerBots. (b) Agent Team Architecture

changes of location and motion between the ball, the goal, and the players, we cannot completely predict or control events in order to take an action to deal with them. The question arises as to how the locations and motions can be mutually predicted by teammates let alone how they may be predicted by the opponent. How the optimal combination of behaviors can be achieved by adopting some approximate algorithms is another question to be addressed.

- Because of uncertainty in the dynamic environment, we investigate Bayesian learning within the cooperative context i.e. learning of optimal joint behaviors when the agents are on the same team. Moreover, we examine two sets of competing Bayesian learning teams when these teams are in opposition, but not taking deceptive strategies into consideration.

As such, we attempt to provide a role-based BDI framework with a soccer game to address the following issues:

1. Competitive skills: can the individual agents learn to intercept, shoot, dribble, clear.
2. Cooperative skills: when and how to pass; when and where to move to receive a pass.
3. Agent teaming: is there an interaction between the competitive skills and cooperative skills for the team to perform optimally.

7.3 Soccer Agents Teaming

The layered BDI agents are considered to operate as the soccer players. Each sub-team entity plays a role in the soccer team. The plans in a sub-team entity can be executed either reactively or proactively. For example, the forward plan can have reactive actions to shoot, intercept, and pass the ball. Meanwhile, it has some plans to perform the proactive actions, i.e., learning competitive skills through experience. The team entity provides a good coordination means to control the sub-team entities actions. The strategies of team can be specified in the team plans. The cooperative strategies can also be learned by using the learning algorithms.

The JACK team entity acts like a coach who is designed to control the actions of a soccer player performed in the SoccerBots simulation environment. The Soccer Team structure is illustrated in Fig. 7.

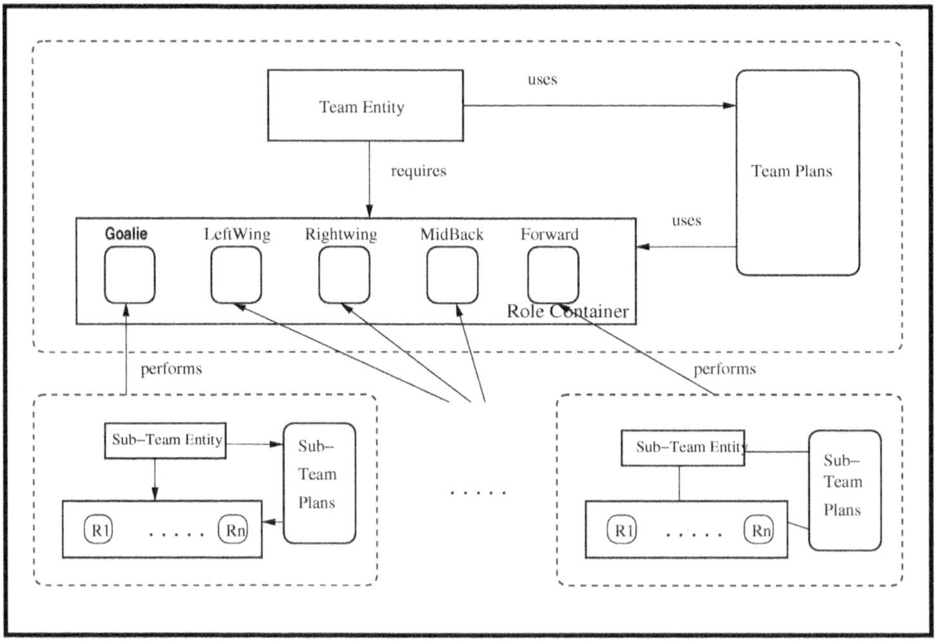

Fig. 7. An Overview of JACK Soccer Team Framework

```
1. Soccer Team
public team SoccerTeam extends Team {
#requires role Goalie ge(1,1);
#requires role LefwWing lw(1,1);
#requires role RighWing rw(1,1);
#requires role MidBack mb(1,1);
#requires role Forward fw(1,1);
#uses teamPlan TeamPlan;
.................
}

2. TeamPlan.plan
public teamPlan SoccerTeam extends TeamPlan {
#requires role Goalie ge(1,1);
#requires role LefwWing lw(1,1);
#requires role RighWing rw(1,1);
#requires role MidBack mb(1,1);
#requires role Forward fw(1,1);
.................
#reasoning method
estabish() { }
body()
{
@parallel( )
{
@team_achieve(ge, ge.DoGoalie.DefendGoal());
.................
@team_achieve(fw, fw.DoForward.Doaction());
}
}

3. GoalieInfo.td (Goalie's infomation)
teamdata GoalieInfo extends GoalieSensor{
#synthesis method ( ...)
{.................
#propagates changes GoaliePerception;
}
}
```

Fig. 8. Roles Definition in Soccer Team

A team reasoning entity (team agent) is a coordinative JACK agent with its own beliefs, desires, and intensions. Assume that a team consists of a set of roles $R = (R_1, R_2, \cdots, R_n)$, a role can be defined as:

$$\forall_{R_i \in R},\ R_i = (RN_i, ID_i, Plans_i, Bels_i) \tag{1}$$

whereas RN_i is the name of role, ID_i is the identification number, $Plans_i$ defines a set of plans that is being executed, and $Bels_i$ represents the beliefs or knowledge.

Every agent player can be regarded as a sub-team and performs a certain role that defines the relationship between team and sub-teams. The relation between soccer team and sub-team is shown in Fig. 8.

```
----------------------------------------------------------

1. Sub-team (Goalie)
public team SoccerAgent extends Team {
#performs role Goalie;
#performs role LefwWing;
#performs role RightWing;
#performs role MidBack;
#performs role Forward;
.................
}

2. Goalie.role
public role Goalie extends Role {
#handles event DoGoalie da;
}

3. DoGoalie.event
public event DoGoalie extends MessageEvent {
int numofPlayer;
#posted as
DefendGoal()
{ }
}

4. ProtectGoal.plan
public plan ProtectGoal() {
#handles event DoGoalie dg;
#reasoning method
body() { }
}

5. GoalieSensor.bel
public beliefset GoalieSensor extends Openworld
{
.................
}

----------------------------------------------------------
```

Fig. 9. Roles Definition in Soccer Sub-team

In Fig. 8, the declaration of #requires role Goalie ge(1,1) in class SoccerTeam to indicate the team requires the goalie with the minimum number of 1 and the maximum number of 1. The class SoccerAgent means that it has the ability to perform five different roles. However, we only assign one role to each agent by using a reasoning method called @establish(). The team beliefs can synthesize the sub-team entities' beliefs. For example, teamdata GoalieInfo can synthesize the belief of GoalieSensor in Fig. 9 , as shown in Fig. 8. The individual sub-team entity can have its own beliefs, plans, and intensions. An example of Goalie is given in Fig. 9.

Scaling plans onto a real-time and dynamic system is a challenge problem as well. The soccer team has a layered architecture that consists of some sub-systems. The agent can have individual skills and cooperative skills. The popular approach known as the divide-and-conquer strategy is to split a system into sub-systems.

The competitive skills are specified in the individual sub-team entities. We can use the hardcoded reactive behaviors or the learned behaviors. The reactive individual skills can be designed using a decision tree, and then coded as rules (if... then). A sample decision tree of the player (midback) is given in Fig. 10.

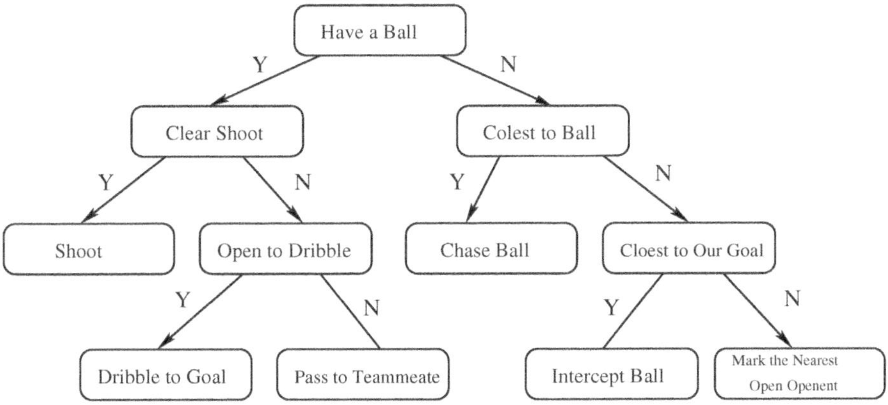

Fig. 10. An Soccer Agent Reactive Decision Tree

The cooperative behaviors are specified in the team entity. The team entity makes the team level decisions to assign the tasks to individual entities. The decisions are made according to the team beliefs: whether a player takes a cooperative actions or a player decide the action to be taken by itself. At each cycle, the beliefs of sub-team entities are updated, and the team entity synthesizes those beliefs to build the team entity beliefs. In a noisy and partial observation environment, the team entity can fuse the individual beliefs to obtain the entire information about the environment.

Based on the team level beliefs, the team entity coordinates the sub-team entities' actions. The team plan is used to control the actions of the soccer players performed in the SoccerBots simulation environment. The team controls the sub-teams behaviors in sequence or in parallel by using the method @parallel(). The execution sequence diagram is illustrated in Fig. 11.

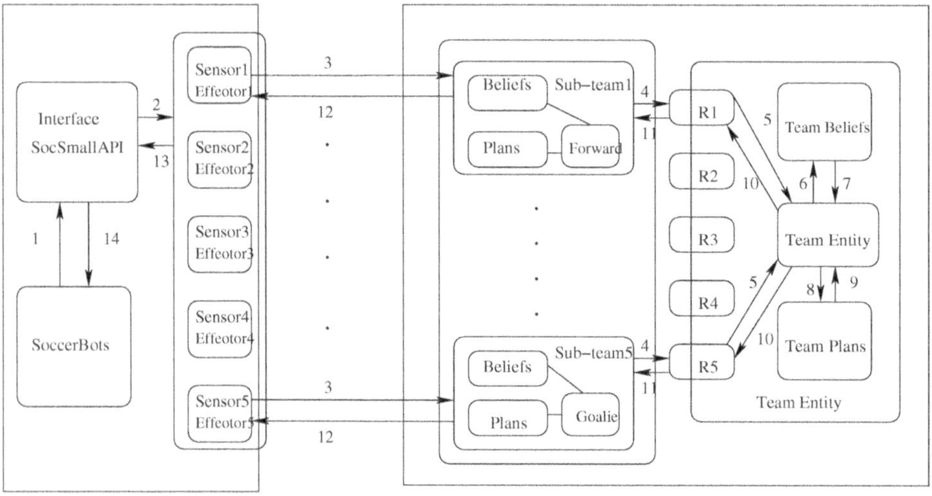

Fig. 11. An Execution Diagram of Soccer Agents Teaming

7.4 On-Line Learning

In the real-time situation environment, performance is a high priority to be considered to follow the changes of motion of the ball and the players. For every perception cycle, the team and sub-team entities initiate a reasoning process by populating their beliefs. In real-time and dynamic systems, specifying all possible state-action pairs is difficult, or infeasible, because of uncertainty and emergent behaviors. Sometimes, it is impractical to design and specify all possible plans. The ability to adapt to changes in the environment is highly demanded when designing the reasoning process in complex and dynamic environments. Reinforcement Learning (RL) [38] is an effective method to help agents learn the required skills through their own experience.

For most decision-making problems, the state changes randomly with time owing to stochastic processes in the environment. The state abstraction is used to choose different learning algorithms based on the current state or beliefs, as shown in Fig. 12. Normally, a stochastic and dynamic system involving decision-making can be formalized as an MDP [6]. An MDP can be defined as a tuple MDP = $\langle$ S, A, T, R $\rangle$, where S is a set of states; A is a set of actions; T is a transition function, T: S $\times$ A $\rightarrow \mathcal{P}$(S); R is a scalar reward function, R: S $\rightarrow \mathcal{R}$. The learning process in the team and sub-team entities can be modeled to learn an MDP, as shown in Fig. 12.

Learning algorithms suffer when dealing with large state spaces (the curse of dimensionality [7]). Function approximation techniques are effective methods to

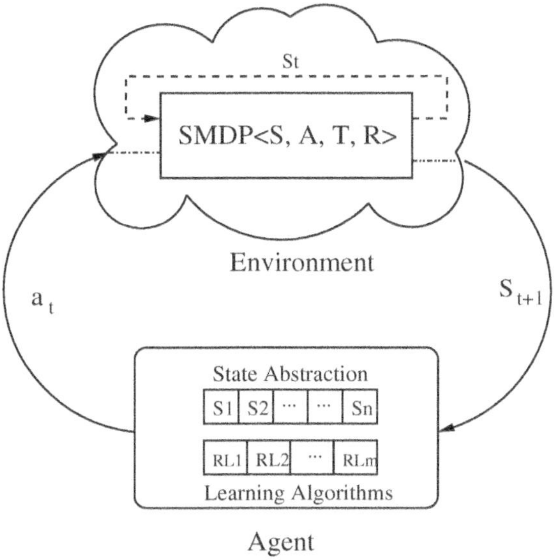

Fig. 12. Learning Process

StartEpisode (s_0):
1. Initialise $\boldsymbol{e} = 0$.
2. Get action $a_{LastAction}$ from s_0 using ϵ-greedy policy.
3. Calculate $Q_{LastAction}$ using s_0, $a_{LastAction}$.
4. For all i $\in \mathcal{F}(s_0, a_{LastAction})$
   $e_i \leftarrow 1$

ExecuteStep(s):
5. δ = reward - $Q_{LastAction}$.
6. Get action $a_{NewAction}$ from s using ϵ-greedy.
7. Calculate $Q_{NewAction}$ using s, $a_{NewAction}$.
8. Get $\mathcal{F}(s, a_{NewAction})$ using s and $a_{NewAction}$.
9. $\delta \leftarrow \delta + \gamma * Q_{NewAction}$.
10. Update all $\boldsymbol{\theta}$ and $\boldsymbol{e}$
    $\boldsymbol{\theta} \leftarrow \boldsymbol{\theta} + \alpha * \delta * \boldsymbol{e}$
    $\boldsymbol{e} \leftarrow \gamma * \lambda * \boldsymbol{e}$
11. For all i $\in \mathcal{F}(s, a_{NewAction})$
    $e_i \leftarrow 1$

StopEpisode (s'):
12. δ = reward - $Q_{LastAction}$.
13. $\boldsymbol{\theta} \leftarrow \boldsymbol{\theta} + \alpha * \delta * \boldsymbol{e}$.
14. End episode.

Fig. 13. Sarsa(λ) Control Algorithm using Approximation Function

deal with large, continuous state space, or partially observable environments. The purpose of function approximation is to define the state/action relationship by using fewer parameters. A detailed Sarsa (λ) control algorithm with replacing

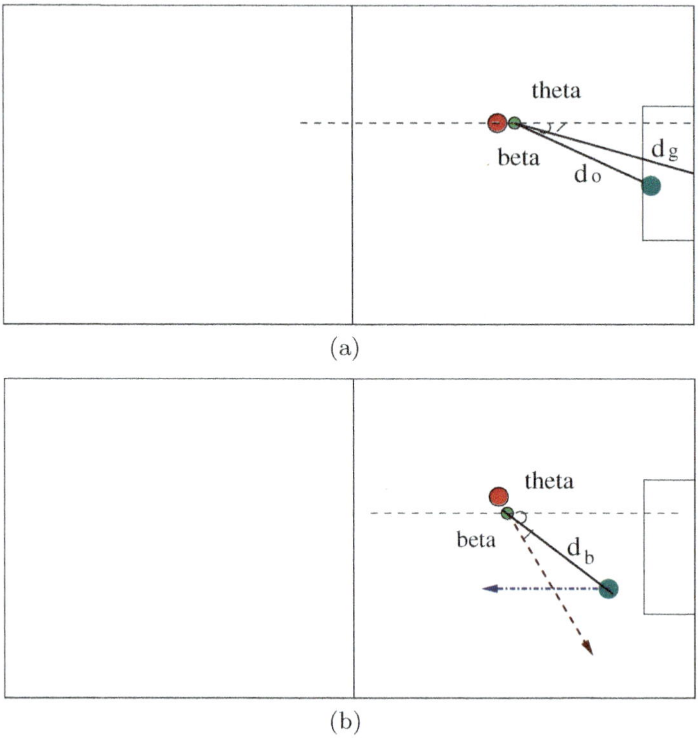

Fig. 14. (a) Shooting. (b) Intercepting.

traces is given in Fig. 13, where α is a learning rate, γ is a discount rate. The ϵ-greedy policy is the exploration strategy, i.e., the agent takes a random action with probability ϵ and takes the best action learned with probability (1 - ϵ).

Scoring goals and intercepting balls are the basic competitive skills for soccer agents. To map the reinforcement learning algorithm of Sarsa to SoccerBots, for scoring a goal, an attacker with a ball is placed in front of the goal, and a goalie is at the goalmouth moving north or south along the vertical axis. The player kicks the ball towards the goal ensuring the ball cannot be intercepted by the goalie. Noise influences the ball's velocity and direction at random. For intercepting a ball, it is assumed that the ball is kicked at certain angle and speed. The player is positioned at certain distance and angle from the ball. The player is trained to learn the optimal action at every step, in order to intercept the ball with the highest speed and with the fewest steps.

By running the simulation, the agent learns the required skills. The learned knowledge is represented in an approximation function. The performance of reinforcement learning algorithms with different parameter settings are shown in Fig.15.

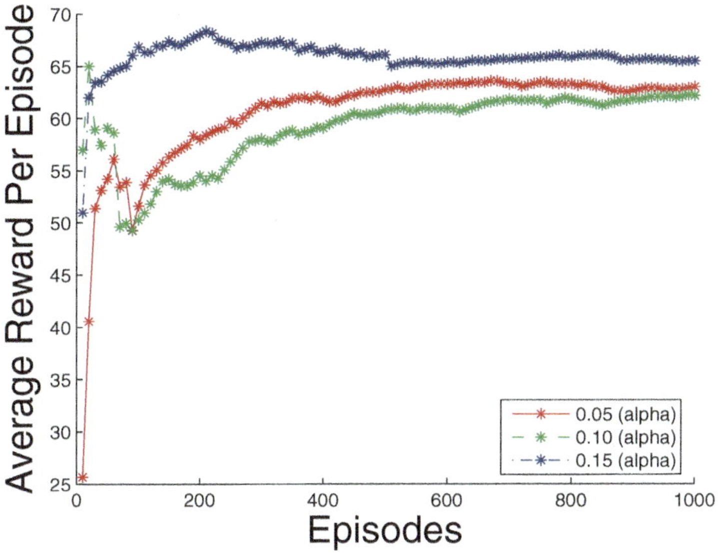

(a)

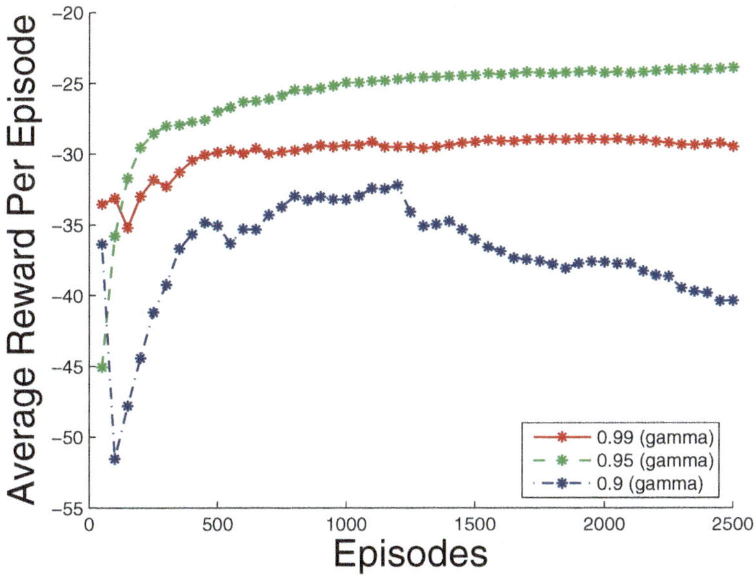

(b)

Fig. 15. The Diagram of Reward and Episodes (a) Scoring Ball. (b) Intercepting Ball.

As indicated above, convergence of the Sarsa (λ) algorithm with some specific parameter settings has been validated by the experimental results. It is demonstrated that the soccer agent can learn the optimal or near-optimal trajectory after running 500 episodes. Satisfactory experimental results are obtained with the adjusted values of those parameters.

8 Related Work

A real-time strategic teamwork is described in terms of the concepts of formations called the locker-room agreement in [37]. Homogeneous agents are able to switch roles within formations at a run-time simulator, known as the simulated robotic soccer. This teamwork structure includes [37]:

- Flexible agent roles with protocols for switching among them;
- Collections of roles built into team formations; and
- Multi-step, multi-agent plans for execution in specific situations: set-plays.

The agent keeps track of three different types of state: the world state, the locker-room agreement, and the internal state. The agent takes actions by choosing the predefined multi-agent plans according to team formations. This flexibility increases the performance of the overall team.

In [43], a layered learning paradigm is presented and allows different layers to be trained concurrently. The SoccerBots simulator has been used as a testbed, whereby the agents are trained by the neural networks. Stone et al. [35,36] present a simulation environment called Keepaway to help agents learn skills by using reinforcement learning algorithms.

STEAM (a Shell for TEAMwork) [40] is a hybrid teamwork model using the joint intension theory formalism, which is built on top of the Soar architecture [25]. STEAM starts with joint intentions, but then builds up hierarchical structures that are parallel with the SharedPlans theory, particularly, partial SharedPlans. STEAM's basis is to execute hierarchical reactive plans [40].

The MadKit platform [16,22] provides an Agent/Group/Role model in terms of organization (agent, group, role). An organization oriented MAS emphasizes on capabilities and constraints, on organizational concepts such as roles, groups, tasks, and interaction protocols. In [16], an organization consists of two aspects: a structural aspect and a dynamic aspect. The structural aspect is composed of a partitioning structure and a role structure. A partitioning structure defines how agents are assembled into groups and how groups are related to each other. A role structure is defined by a set of roles and their relationships. The dynamic aspect is related to the institutionalized patterns of interactions that are defined within roles. In [22], the AGR (Agent/Group/Role) model was proposed, which is based on three primitive concepts, i.e., Agent, Group, and Role:

- Agent: an agent is an active, communicating entity playing roles within groups. An agent may hold multiple roles, and may be member of several groups.

- Group: a group is a set of agents sharing some common characteristic. A group is used as a context for a pattern of activities, and is used for partitioning organizations.
- Role: the role is the abstract representation of a functional position of an agent in a group. An agent must play a role in a group, but an agent may play several roles.

9 Concluding Remarks

This article presents a role-based BDI framework to facilitate coordination and cooperation in MASs. Multi-agent teaming can be constructed in a layered architecture: individual agent behaviors (competitive skills) are specified at the low level, and team behaviors (cooperative skills) are addressed at the high level. This framework needs a team entity that has the entire knowledge of team members and the ability for coordinating with individual members. Such an assumption may not be suitable for some application domains, for example, negotiation in an ad-hoc manner.

While the described framework provides an abstract way to model multi-agent teaming, a challenge remains owing to the lack of effective techniques to solve the multi-agent behavior optimization problems. Another practical problem is that an explicit model for role assignment (either static or dynamic) is necessary to build an efficient team. Further work to be pursued includes to compare team level performance by using different approaches, i.e., hand-coded programs versus learning techniques.

Our ultimate goal is to develop an effective agent-teaming architecture with competitive and cooperative learning algorithms. Dynamic agents role assignment will be considered for agents' teaming. Owing to the stochastic nature of MASs, the use of some optimization techniques is required, in order to find the optimal team behaviors for building a robust multi-agent team.

Acknowledgment

The initial research findings have been published in KES 2008 proceedings. The authors gratefully acknowledge the valuable comments and suggestions of the reviewers, which have improved the quality and presentation of this article.

References

1. InfoGrames Epic Games and Digital Entertainment. Technical report, Unreal tournament manual (2000)
2. Teambots (2000), `http://www.cs.cmu.edu/~trb/Teambots/Domains/SoccerBots`
3. Humaniod Kid and Medium Size League, Rules and Setup for Osaka 2005. Technical report, Robocup (2005)
4. Agent Oriented Software Pty Ltd. JACK Intelligent Agents User Guide (2002)
5. Agent Oriented Software Pty Ltd. JACK Teams User Guide (2002)
6. Bellman, R.: A Markovian Decision Process. Journal of Mathematics and Mechanics 6 (1957)

7. Bellman, R.: Dynamic Programming. Princeton University Press, Princeton (1957)
8. Bradshaw, J.M., Feltovich, P., Jung, H., Kulkarni, S., Allen, J., Bunch, L., Chambers, N., Galescu, L., Jeffers, R., Johnson, M., Sierhuis, M., Taysom, W., Uszok, A., Van Hoof, R.: Policy-based Coordination in Joint Human-agent Activity. In: Proceedings of the IEEE International Conference on Systems, Man, and Cybernetics, The Hague, The Netherlands, pp. 2029–2036 (2004)
9. Bratman, M.E.: Intention, Plans, and Practical Reasoning. Harvard University Press, Cambridge (1987)
10. Cheyer, A., Martin, D.: The Open Agent Architecture. Autonomous Agents and Multi-Agent Systems 4(1-2), 143–148 (2001)
11. Coen, M.H.: Sodabot: A software agent construction system. Technical report. MIT AI Laboratory, Cambridge (1995)
12. Cohen, P., Levesque, H., Smith, I.: On Team Formation. Contemporary Action Theory (1998)
13. Cohen, P.R., Levesque, H.J.: Teamwork. Nous 25(4), 487–512 (1991)
14. Durfee, E.H., Lesser, V.: Negotiating Task Decomposition and Allocation Using Partial Global Planning. In: Gasser, L., Huhns, M. (eds.) Distributed Artificial Intelligence, pp. 229–244. Morgan Kaufmann, San Francisco
15. Evertsz, R., Fletcher, M., Jones, R., Jarvis, J., Brusey, J., Dance, S.: Implementing Industrial Multi-agent Systems Using JACK^{TM}. In: Dastani, M.M., Dix, J., El Fallah-Seghrouchni, A. (eds.) PROMAS 2003. LNCS (LNAI), vol. 3067, pp. 18–48. Springer, Heidelberg (2004)
16. Ferber, J., Gutknecht, O., Michel, F.: From Agents to Organizations: An Organizational View of Multi-agent Systems. In: Giorgini, P., Müller, J.P., Odell, J.J. (eds.) AOSE 2003. LNCS, vol. 2935, pp. 443–459. Springer, Heidelberg (2003)
17. Franklin, S., Graesser, A.: Is it an Agent, or Just a Program?: A Taxonomy for Autonomous Agents. In: Jennings, N.R., Wooldridge, M.J., Müller, J.P. (eds.) ECAI-WS 1996 and ATAL 1996. LNCS, vol. 1193, pp. 21–35. Springer, Heidelberg (1996)
18. Garey, M.R., Johnson, D.S.: Computers and Intractability: A Guide to the theory of NP-completeness. WH Freeman and Company, New York (1979)
19. Gilbert, D., Aparicio, M., Atkinson, B., Brady, S., Ciccarino, J., Grosof, B., O'Connor, P., Osisek, D., Pritko, S., Spagna, R., Wilson, L.: IBM Intelligent Agent Strategy. Technical report, IBM Corporation (1995)
20. Grosz, B., Kraus, S.: A Collaborative Plans for Complex Actions. Artificial Intelligence 86, 269–358 (1996)
21. Grosz, B., Sidner, C.: Plan for Discourse. In: Cohen, P., Morgan, J., Pollack, M. (eds.) Intensions in Communication, pp. 417–444. Morgan Kaufmann, San Francisco (1990)
22. Gutknecht, O., Michel, F., Ferber, J.: Integrating Tools and Infrastructure for Generic Multi-Agent Systems, Autonomous Agents 2001. ACM Press, New York (2001)
23. Jennings, N.R.: On Agent-based Software Engineering. Artificial Intelligence 117, 277–296 (2000)
24. Jennings, N.R., Sycara, K., Wooldridge, M.: A Roadmap of Agent Research and Development. Autonomous Agents and Multi-Agent Systems 1(1), 7–38 (1998)
25. Laird, J.E., Newell, A., Rosenbloom, P.S.: Soar: an architecture for general intelligence. Artificial Intelligence 33(1), 1–64 (1987)
26. Leng, J., Fyfe, C., Jain, L.: Reinforcement Learning of Competitive Skills with Soccer Agents. In: Apolloni, B., Howlett, R.J., Jain, L.C. (eds.) KES 2007, Part I. LNCS (LNAI), vol. 4692, pp. 572–579. Springer, Heidelberg (2007)

27. Leng, J., Jain, L., Fyfe, C.: Simulation and Reinforcement Learning with Soccer Agents. International Journal of Multiagent and Grid Systems 4(4), 415–436 (2008)
28. Lind, J.: Agents, Multiagent Systems and Software Engineering. In: Lind, J. (ed.) Iterative Software Engineering for Multiagent Systems. LNCS (LNAI), vol. 1994, pp. 9–33. Springer, Heidelberg (2001)
29. Pokahr, A., Braubach, L., Lamersdorf, W.: Jadex: Implementing a BDI-Infrastructure for JADE Agents. EXP – in search of innovation 3(3), 76–85 (2003)
30. Rao, A.S., Georgeff, M.P.: BDI-agents: from theory to practice. In: Proceedings of the First Intelligent Conference on Multiagent Systems (1995)
31. Russell, S., Norvig, P.: Artificial Intelligence: A Modern Approach. Prentice-Hall, Englewood Cliffs (1995)
32. Russell, S., Norvig, P.: Heuristic: Intelligent Search Strategies for Computer Problem Solving, A Modern Approach. Prentice-Hall, Englewood (1995)
33. Sioutis, C.: Reasoning and Learning for Intelligent Agents. PhD thesis, School of Electrical and Information Engineering, University of South Australia (2005)
34. Stone, P.: Layered Learning in Multiagent Systems: A Winning Approach to Robotic Soccer. MIT Press, Cambridge (2000)
35. Stone, P., Kuhlmann, G., Taylor, M.E., Liu, Y.: Keepaway Soccer: From Machine Learning Testbed to Benchmark. In: Bredenfeld, A., Jacoff, A., Noda, I., Takahashi, Y. (eds.) RoboCup 2005. LNCS (LNAI), vol. 4020, pp. 93–105. Springer, Heidelberg (2006)
36. Stone, P., Sutton, R.S., Kuhlmann, G.: Reinforcement Learning for RoboCup-Soccer Keepaway. Adaptive Behavior 13(3), 165–188 (2005)
37. Stone, P., Veloso, M.: Task Decomposition, Dynamic Role Assignment, and Low-Bandwidth Communication for Real-Time Strategic Teamwork. Artificial Intelligence 110(2), 241–273 (1999)
38. Sutton, R.S., Barto, A.G.: Reinforcement Learning: An Introduction. MIT Press, Cambridge (1998)
39. Sycara, K.: MultiAgent Systems. AI Magazine 19(2), 79–92 (1998)
40. Tambe, M.: Towards Flexible Teamwork. Journal of Artificial Intelligence Research 7, 83–124 (1997)
41. Tweedale, J., Ichalkaranje, N., Sioutis, C., Urlings, P., Jain, L.C.: Building a Decision Making Framework using Agent Teams. International Journal of Intelligent Decision Technologies 1(4), 175–181 (2007)
42. van Breemen, A.J.N.: Integrating Agents in Software Applications. In: Kowalczyk, R., Müller, J.P., Tianfield, H., Unland, R. (eds.) NODe-WS 2002. LNCS (LNAI), vol. 2592, pp. 343–354. Springer, Heidelberg (2003)
43. Whiteson, S., Stone, P.: Concurrent Layered Learning. In: Proceeding of the Second International Joint Conference on Autonomous Agents and Multi-Agent Systems (AAMAS 2003), Australia,, pp. 193–200 (2003)
44. Wooldridge, M.: Reasoning About Rational Agents. MIT Press, Cambridge (2000)
45. Wooldridge, M., Jennings, N.: Intelligent agents: Theory and practice. Knowledge Engineering Review 10 (1995)
46. Yen, J., Yin, J., Ioerger, T.R., Miller, M.S., Xu, D., Volz, R.A.: CAST: Collaborative Agents for Simulating Teamwork. In: IJCAI, pp. 1135–1144 (2001)
47. Zaki, Y., Pierre, S.: Mobile Agents in Distributed Meeting Scheduling: A Case Study for Distributed Applications. International Journal of Intelligent Decision Technologies 1(1-2), 71–82 (2007)
48. Zhang, Y.: Modelling role-based agent team. In: Kobti, Z., Wu, D. (eds.) Canadian AI 2007. LNCS (LNAI), vol. 4509, pp. 1–13. Springer, Heidelberg (2007)

Mobile Agents and Eventually Perfect Predicate Detection: An Intelligent General Approach to Monitoring Distributed Systems

Chunbo Chu[1] and Monica Brockmeyer[2]

[1] Franklin University, Columbus OH 43215, USA
chuc@franklin.edu
[2] Wayne State University, Detroit MI 48202, USA
mbrockmeyer@wayne.edu

Abstract. This chapter presents an application of mobile agents to solve distributed predicate detection by distributing and delegating monitoring tasks, result collection, and aggregation to overcome the scalability and flexibility limitations (as compared to most traditional centralized predicate detection approaches). Mobile agents are very useful for monitoring purposes because mobility gives the agents the autonomy needed to monitor the computation effectively. By restructuring the implementation of an eventually accurate failure detection sequencer into mobile agents, the functionalities of predicate detection in a failure-prone partially synchronous system are clearly separated from the monitored computation. The separation leads to additional benefits of enhanced flexibility, performance, and robustness. It provides a solution to create a general-purpose predicate detection infrastructure that provides the basic functionalities to achieve reasonable predicate detection semantics in a realistic distributed system.

1 Introduction

A distributed system [1] consists of several computer systems connected to each other by a communication network. Though much more complex and difficult to build, distributed systems have seen a rapid increase both in installation and research because of their manifold and significant advantages over traditional centralized approaches. Some of the advantages include: reduced price-performance ratio, increased reliability, availability, flexibility and computational performance. These features enable the success of Internet, a world-wide heterogeneous distributed system connecting large and small networks around the globe.

To achieve these advantages, it is essential to have the ability to identify the characteristics of distributed systems and to monitor their behavior. Many applications in distributed systems, such as automatic reconfiguration, deadlock detection and termination detection, raise the fundamental challenge of evaluating whether or not a pronouncement of a proposition (known as a *predicate*) of the status that a distributed computation may pass through has held in an

A. Håkansson et al. (Eds.): Agent and Multi-agent Technology, SCI 289, pp. 257–274.
springerlink.com

execution of the system. For example, a predicate could be "*At a moment in time, half of the processes have a CPU utilization of below fifty percent*". The challenge to evaluate a predicate in a distributed system has been known as *predicate detection.*

Over the past decades, a considerable amount of research has been carried out to address the problem of predicate detection [2,3,4,5,6]. However, the goal of an efficient mechanism to discover and understand the behavior of massive and highly dynamic distributed systems has not yet been realized. There are still great challenges to achieve desirable solutions to predicate detection, especially the following.

First, approaches based on asynchronous assumptions are unscalable and have limited semantics. Distributed systems are generally modeled as either synchronous or asynchronous based on the different roles time plays. In an asynchronous system, there is no bound on any system activity, such as message delivery delay, clock drifting rate or the time necessary to execute a step. In a nutshell, no assumptions about time are permitted. The constituent processes have neither common memory nor a common system clock. They can communicate only by message passing.

The asynchronous model was attractive because it has simple semantics; applications on the basis of this model are easier to port and there are several sources of asynchrony in reality such as variable or unexpected workload and network traffic. However, asynchrony implies that the timing relation on state transitions of a distributed system can only be established based on causality. Such state transitions cannot be totally ordered in this model since the causal relation imposed by message passing is only a partial ordering. Without perfect synchronization, processes cannot have consistent views of the global history of an execution, nor can they determine unambiguously the sequence of states through which the system passed. This leads to an obvious difficulty for detecting whether a predicate holds [7].

Unfortunately, in tradition and most research work, distributed systems have generally been modeled as fully asynchronous. Therefore it is impossible to precisely determine the actual state a distributed system, especially a wide-area one, was in at a given moment in time. For this reason, predicate detection approaches based on asynchronous assumptions usually have to build a lattice of *global states* defined as combinations of all concurrent local states permitted by the causal relation of events [4]. Evaluation of a given predicate has to be carried out on every global state in the lattice. The time complexity of these methods is exponential in the number of processes, which leads to the state explosion problem as the number grows. The asynchronous model introduces great inefficiency to predicate detection and prohibits the detection of certain predicates (e.g. those involving time constrains).

Second, existing approaches do not address failures in realistic systems. The challenge of efficient monitoring and predicate detection in distributed systems has not yet been met also because most detection approaches in the literature are not robust in faulty models involving message losses and/or process crashes. It is well known that crash failures are not easy to tolerate in asynchronous systems [8] since they cannot be detected with absolute certainty [9]. Moreover, since failures are usually observed in real computing environments, it is often desirable to detect predicates that involve states related to crashes. For example, "(process p_i holds a write lock) AND (p_i is crashed)". This kind of predicate is useful in operating systems and databases, but cannot be handled by approaches that are dependent on failure-free assumptions.

Finally, there has been a lack of a general-purpose infrastructure to solve predicate detection. Existing approaches for predicate detection are *ad hoc* because their design, implementation and validity are bound to the unique features that are particularly assumed in the underlying system model. For example, traditional approaches for the asynchronous systems mostly require the construction of a lattice of consistent global states. They are only good in small-scale and static systems. Approaches proposed to solve predicate detection with stronger modalities [10] take advantage of the real-time stamp from approximately synchronized clocks.

For distributed applications to date, there is no general-purpose framework that provides the basic functionality to support predicate detection. Building and maintaining an *ad hoc* solution from scratch is expensive, error-prone, and hard to transplant to a different context.

2 Contributions

In this chapter, I propose a general-purpose solution for predicate detection in a partially synchronous distributed system by using mobile agents [11]. This solution is derived from the approach to address predicate detection involving process failures in an Finite Average Response (FAR) time system [12].

The fully asynchronous assumption is overly restrictive for today's distributed systems. In fact, a number of high-performance clock synchronization protocols have appeared in the literature [13,14,15,16,17]. These protocols can keep the differences in process's clock readings at any instant within some known bound, which creates a rough global time base for distributed systems. Protocols like the Network Time Protocol (NTP) [18] can efficiently maintain synchronization of clocks to within a few milliseconds in most local-area distributed systems [15]. Even in extremely wide-area distributed systems like the Internet, clock synchronization can usually be maintained to within a few tens of milliseconds [18].

The improvement in clock synchronization justifies that a certain "amount" of synchrony does exist in many distributed systems. In recent years, *partial synchrony* has emerged in the research on some distributed system problems, such as the consensus [19]. Partial synchrony is a mixture of synchronous assumptions on some system activities and asynchronous assumptions on the other system activities. This concept faithfully reflects the fact that realistic systems are not always fully asynchronous.

The impact of partial synchrony on predicate detection has been studied [20] in three well defined partially synchronous models in the literature [21,22] for solving failure detection. The work proves the connections among partial synchrony, predicate modalities and detection semantics in a formal framework. It is demonstrated that predicate detection may provide stronger guarantees of detection semantics and modalities with partial synchrony in systems where otherwise only the weakest semantics and modalities can be achieved under fully asynchronous assumptions.

In Artificial Intelligence, agents are communicative, intelligent, rational and possibly intentional entities [23] which perceive their environment and act upon that environment. From the computing perspective, agents are autonomous, asynchronous, communicative, distributed, and possibly mobile [23]. Mobile agents are programs that can move autonomously from one computer to another [24]. Mobile agent technology has been considered as an enhancement of distributed computing due to the powerful and efficient mechanisms it provides for distributed and heterogeneous systems.

In this research, I apply the mobile agent technology to the approach that was designed to solve eventually perfect predicate detection in a weak partially synchronous system – the FAR system [12]. The outcome is a general-purpose framework to solve predicate detection in realistic applications. This work is not only a significant investigation of the behavior of predicate detection and innovative practical solutions to some of the most realistic monitoring contexts, but also pioneers a new way towards the ultimate goal of a scalable, high-performance, highly-available, and fault-tolerant distributed system.

Targeting cooperative distributed applications deployed in unreliable systems (*e.g.* the Internet), our approach has a multi-layer architecture. We make only minimal restrictions on detection algorithms. This means that applications have higher flexibility and independence from particular predicate modalities, compared to [25] that only works with *observer-independent predicates* [26]. Our work provides a desirable monitoring solution to many Internet applications because the stabilizing semantics of $\Diamond\Sigma$ leads to eventual detection of undesirable computation properties, while it is permitted by the limited amount of synchrony which is reasonably available in the Internet.

3 Applications

Many applications in distributed computing can benefit from my research as long as they can be modeled as executing some notification or reaction when the state of the system satisfies a particular condition. With the wide availability of low-cost computers and pervasive network connectivity, more and more applications require the ability of self-configuring and self-managing these computer resources in a large scale. Examples include distributed monitoring and debugging, detection of particular states, such as deadlock and termination, and dynamic adaptation of a program's configuration such as for load balancing.

Check-pointing and fault recovery improve resilience and reliability of distributed systems. They can be better supported by monitoring predicates of system state. When a fault occurs and is detected (maybe even anticipated by setting appropriate predicates), the system will be rolled back to a previous safe state. Distributed debugging tools could be built upon predicates detection mechanism to monitor the internal state of a system and to report any condition of interest. To solve these of problems accurately and efficiently is essential for the proper operation of distributed systems.

There are much more applications of predicate detection. In short, the ability to efficiently observe a distributed computation and evaluate a predicate constitutes the core of solutions to these applications.

4 Organization

This chapter is organized as follows: Section 5 is the related work. Section 6 introduces the problem to be solved. Section 7 describes the system model. The definition and properties of $\Diamond\Sigma$ in the FAR system are given in Section 8. In Section 9, we define the eventually perfect predicate detection semantics and reinvent the approach to achieve eventually perfect predicate in Section 10. Section 11 gives the conclusion.

5 Related Work

Cooper and Marzullo propose two modalities that apply to the entire distributed execution [4].

- *Possibly*: There exists a consistent observation of the execution such that the predicate holds in a global state of the observation.
- *Definitely*: For every consistent observation of the execution, there exists a global state of it in which the predicate holds.

Their algorithms for detecting predicates under the *Definitely* and *Possibly* modalities are based on a lattice of consistent global states in a distributed system. Such a lattice consists of n (the number of application processes) orthogonal axes, with one axis for each application process. A possible execution is a path in the lattice where the level of each subsequent node in the path increases by one. For every node in the lattice, there exists an execution path that passes through it. Hence, if any node in the lattice satisfies a predicate ϕ, then *Possibly* ϕ holds. The property of *Definitely* ϕ requires all possible executions to pass through a node that satisfies ϕ. Application processes send their local states to the monitor process. The monitor maintains sequences of these states, one sequence per process, and assembles local states into possible consistent global states by using vector time stamp.

Their algorithm for detecting the *Possibly* modality constructs all possible global states at a certain level which are reachable from the previous level and checks if there is a state that satisfies the predicate. The main drawback of this approach is the often excessive number of global states that may have be constructed, which is of $O(m^n)$ where m is the maximum number of events a monitored process has executed. The state explosion is due, among other causes, to the modeling of concurrency by interleaving, or, more accurately, to the exploration of all possible interleaving of concurrent events. This method is not scalable.

Gärtner and Pleisch have proved that it is impossible to solve predicate detection even with a perfect failure detector in crash-affected systems [27]. In their subsequent work [25], they attempt to remedy this impossibility by the proposal of a failure detection sequencer (Σ). Σ can report to applications both the crash of a process and the last state it was in before crashing. They prove that predicate detection can be solved using Σ based on its capability to provide necessary and sufficient information about failures by embedding crash events consistently into the causal order of events in a computation. However, they assume synchronous communications in the implementation of Σ, which is too strong to apply to many realistic distributed systems.

Fetzer *et al.* introduce the Finite Average Response time (FAR) model [28]. This model only makes minimal synchrony assumptions as the following: incrementing an integer takes some unknown time; communication channels have a finite buffer and use flow control, and the average response time converges and it is finite. This model has no assumptions about the relative speed of processes, no upper or lower bound on message delivery times, no assumption of eventual system stability, and no clocks. The FAR model is very suitable for realistic, large-scale distributed systems, such as those deployed as overlay networks on the Internet. This model is not purely asynchronous because Fetzer has demonstrated that the eventually perfect failure detector (also denoted as *EA-FD* in [28]) can be implemented in this model. Therefore, the consensus problem [29] can be solved in the FAR system as well. But it has not yet been established that a failure detection sequencer (or possibly in weaker forms) can be implemented in the FAR model, nor what predicate detection semantics can be satisfied.

An approach for continuously monitoring highly dynamic distributed environments using mobile agents has been presented in [30]. Their proposal considers a divide-and-conquer cooperation structure with agents that take into account the delays they experience and the environmental delays. Mobile agents are used to perform the monitoring tasks efficiently and effectively, thus reducing the network traffic and carrying the tasks to wherever the required data can be obtained with a good performance. There is a loose coupling among the agents (based on soft commitments), which increases the fault tolerance of the system, as it leads to a "graceful degrading" of performance when some of the agents fail to perform their tasks in time.

Motshegwa and Schroeder have suggested and presented a framework for studying determination of global states such as termination in a society of agents [23]. Agents are higher semantic level entities that execute protocols to engage in potentially complex interactions. Their proposed framework incorporates the use of protocols by agents. It incorporates a novel algorithm that allows a monitor to determine termination when presented additional information on partial behavior specification by agents. Possible criteria for evaluation are also suggested.

6 The Problem

Monitoring and predicate detection is a fundamental challenge of evaluating whether or not a predicate has held on the state of a distributed system in an execution. A predicate is a pronouncement of a proposition of the status a distributed computation may pass through. Ideally, a perfect predicate detection algorithm satisfies two properties: (1) safety: never erroneously detects a predicate, and (2) liveness: does not miss any occurrence of the predicate in the underlying computation.

Traditionally, *failure detectors* [31] are provided to determine which processes are up (operational) and which are down (crashed) in a computation. Failure detectors encapsulate the synchrony assumptions of an otherwise asynchronous system. Unfortunately, it is impossible to solve predicate detection even with a *perfect failure detector* ($\mathcal{P}$) [31] in a faulty environment [27]. The reason for this impossibility is that failure detectors are only functions of failures. They cannot causally order failure detection events of the system under observation.

In order to make predicate detection involving failures possible, a *failure detection sequencer* (Σ) has been introduced [25]. Σ enhances a failure detector by embedding crash events consistently into the causal order of events in a computation. It has been proved that perfect predicate detection (the liveness and safety properties are always satisfied) can be solved using Σ [25]. However, the implementation of Σ relies on a system model ensuring at least bounded and known communication delays. This assumption is too strong for modern communication systems, especially those in which routing is used for communications among non-contiguous processes, such as the Internet. Assuming that the communication delays in such systems are eventually bounded is questionable. In addition, very large systems might never stabilize completely. Ensuring and/or justifying such "global" assumptions in these systems is in fact intrinsically difficult.

It is claimed that [27] in asynchronous systems with crash failures, if the communication is reliable and FIFO, it is possible to achieve the *eventually perfect* predicate detection semantics (liveness is always and safety is eventually satisfied) with any plausible *eventually perfect failure detector* ($\Diamond\mathcal{P}$)[1]. However, neither a proof to this claim is given nor how to achieve this semantics is demonstrated.

The fundamental challenge to the detection of predicates involving failures in systems with weak communication assumptions is that predicate detection is strictly stronger than failure detection which is already tricky to solve.

We have studied eventually perfect predicate detection in a failure-prone system without assumptions about the bound on communication delays, specifically

[1] $\Diamond$ is a temporal logic operator meaning "eventually in the future".

the Finite Average Response time (FAR) model [12]. This model only makes the "local" assumption that the *average* response time for messages sent between a pair of correct processes is finite. This is much easier to justify because it is only necessary that failed links are eventually repaired.

We have proved that it is impossible to implement Σ in the FAR model. Then we defined and implemented a weaker form of Σ, called *eventually accurate failure detection sequencer* (denoted by $\Diamond\Sigma$) in the FAR model based on a $\Diamond\mathcal{P}$ which has been demonstrated to be possible in this model [28]. $\Diamond\Sigma$ is allowed to make a finite number of temporarily wrong decisions on the last state of a crashed process. Each wrong decision will eventually be remedied at the cost of one probing message. So eventually, $\Diamond\Sigma$ orders crashes causally in state transitions of the underlying computation.

The output of Σ is an n-tuple, where n is the number of processes under monitoring. The value at each index of this tuple indicates either a process is up or it has crashed in its final state at the query time. However, predicate detection requires a complete observation of a computation from the beginning to the querying time. Our $\Diamond\Sigma$ extends Σ by outputting a complete sequence of state transitions from the beginning of a computation, called a sequencer history, with failure events causally embedded. Therefore, unlike the approach in [25], no redundant observation of computation is necessary in the predicate detection module. Predicate detection based on $\Diamond\Sigma$ can be simpler and more efficient.

In the implementation of $\Diamond\Sigma$, a self-adapting timeout mechanism is applied to help $\Diamond\Sigma$ encapsulate the weak synchrony of the communications in FAR. With this timeout mechanism, the possibility of $\Diamond\Sigma$ making wrong decisions is well reduced. No failure detector in $\Diamond\mathcal{P}$ can either order failures casually in state transitions or is capable to reduce wrong decisions. We showed that perfect predicate detection is impossible in the FAR model. Instead, we proved that the eventually perfect predicate detection semantics [27] can be achieved using $\Diamond\Sigma$.

Despite all the nice features of the proposed $\Diamond\Sigma$ [12] to solve predicate detection involving failures, it requires every application process to have a special module to interact with the $\Diamond\mathcal{P}$ portion of the monitor. All these failure detection modules, as well as the one to implement $\Diamond\Sigma$ on the monitor have to be pre-installed and configured. They do not handle dynamic situations well. The single static monitor process may be the bottle neck of the detection. What makes this approach more costly is that when a new monitoring task is operated, all modules have to be reinstalled and configured.

7 Model

A distributed computation, $\mathcal{C}$, describes the execution of a distributed algorithm by a finite set of *application processes*: $\Pi = \{p_1, p_2, \ldots, p_n\}$. p_i can prematurely stop executing its actions if it crashes. Crashed processes do not recover.[2] p_i has a local state s_i and performs state transitions in sequence. Processes communicate

[2] In many real world applications, a process that crashes and recovers to the last state before the crash can be appropriately treated as a slow process.

solely by sending messages to each other over communication channels. Sending a message to one or more processes or receiving a message from a process also results in a state change. A state transition is known as an *event* that is assumed to be atomic. The *local history* of p_i is either an infinite sequence of local state transitions: $H_{p_i} = \{s_i^1, s_i^2, \ldots\}$; or a finite sequence of local state transitions ending in a crash event ($\top$): $H_{p_i} = \{s_i^1, s_i^2, \ldots, \top\}$. A *prefix* of a local history is a finite sequence $H_{p_i}^k = \{s_i^1, s_i^2, \ldots, s_i^k\}$. $H_\Pi = \bigcup_{i=1}^{n} H_{p_i}$ is the *global history* of $\mathcal{C}$. A *global state* is an n-tuple of local states $S = (s_1, s_2, \ldots, s_n)$, having one local state per process. S is *consistent* if and only if $\forall 1 \leq i, j \leq n, s_i \npreceq s_j$, where $\preceq$ is the causal order of local states [32].

To conform to the FAR model, we assume the following:

Each pair of processes can communicate through a fixed-sized, acknowledged *stubborn channel* [33] with a maximum message size ($MAXSZ$). The stubborn channel works as follows: If a correct process p_i sends a message m of a size $\leq MAXSZ$ to a correct process p_j via an acknowledged stubborn channel and p_i delays sending any other message to p_j until it receives an acknowledgment for m, then eventually m will be delivered to p_j and p_i will receive an acknowledgment that p_j has delivered m. The average time until the acknowledgment of a message sent between two correct processes arrives is finite. Also, incrementing an integer by 1 takes an unknown minimum amount of time GT ($GT > 0$). So a weak form of clock based on an integer counter can be implemented to establish some notion of time.

When an event occurred on a process, a *control message* containing the new local state is sent to a *monitor* process q. Control messages are also sent via stubborn channels with q.

The notion of time in our model is sufficient to provide a "subjective" notion of slow and fast messages: A slow message is one having an acknowledgment arriving after the timeout for that message, and a fast message is one having an acknowledgment arriving before the timeout.

8 Eventually Accurate Failure Detection Sequencer

A failure detection sequencer (Σ) was created to provide the necessary and sufficient information about process failures to solve predicate detection in faulty environments. Unfortunately, its design is dependent on synchronous communications, which is difficult to justify in realistic systems, such as the FAR system.

Since it is impossible to implement Σ in a FAR system [12], we proposed a weaker form of Σ, called eventually accurate failure detection sequencer (denoted by $\Diamond\Sigma$), and proved that it can be implemented in a FAR system to solve eventually perfect predicate detection involving process failures.

The following formal definitions are adopted from the work for failure detectors by Chandra and Toueg [31] and for failure detection sequencer by Gärtner and Pleisch [25]:

A failure pattern F is a mapping from the time domain $\mathcal{T}$ to the power set of Π. The value of $F(t)$ specifies the set of application processes that have crashed until time $t \in \mathcal{T}$. Without loss of generality, we assume that at any instant in time, at most one process takes a step. Two functions are introduced: a step function S_s from $\mathcal{T}$ to the set of all algorithm steps, and a process function S_p from $\mathcal{T}$ to Π. $S_p(t)$ denotes the process which takes a step at time t and $S_s(t)$ identifies the algorithm step which was taken [25].

8.1 Definition and Properties

Formally, an *eventually accurate failure detection sequencer*, or $\Diamond\Sigma$, maps a failure pattern F, a step function S_s and a process function S_p to a set of sequencer histories $H_{\Diamond\Sigma}$. If queried at time τ, $\Diamond\Sigma$ outputs an observation of H_Π through τ on monitor q, named *sequencer history*: $H_{\Diamond\Sigma}(\tau) = \bigcup_{i=1}^{n} H^i_{\Diamond\Sigma}(\tau)$, where $H^i_{\Diamond\Sigma}(\tau)$ is a sequence of events on p_i observed by $\Diamond\Sigma$ through τ : $H^i_{\Diamond\Sigma}(\tau) = \{(s^t_i, crashed^t_i)\}^{t \leq \tau}_{t \in \mathcal{T}}$. s^t_i is the *most recent state* of p_i at time t given S_s and S_p, denoted by $most\_recent\_step(p_i, t, S_s, S_p) : \exists t' \leq t, (S_s(t') = s^t_i) \wedge (S_p(t') = p_i) \wedge (\forall t' < t'' < t, S_p(t'') \neq p_i)$. $crashed^t_i = \top$ if $p_i \in F(t)$; otherwise, $\bot$.

$\Diamond\Sigma$ has the following three properties:

- **Consistency**: State transitions in the sequencer history are consistent with the local history of a process under monitoring. Formally,[3]
  $(\forall p_i, \forall k \in \mathbb{N}, \exists t \in \mathcal{T} \wedge \pi_{s_i}(H^i_{\Diamond\Sigma}(t)) = H^k_{p_i}) \wedge (\forall p_i, \forall t \in \mathcal{T}, \exists k \in \mathbb{N} \wedge H^k_{p_i} = \pi_{s_i}(H^i_{\Diamond\Sigma}(t)))$
- **Completeness**: If p_i crashes, then eventually $\Diamond\Sigma$ will permanently suspect p_i to be in some state (which is the final state p_i was in). As the Completeness property of Σ in [25], the Completeness property of $\Diamond\Sigma$ only concerns the crash of processes. So there is no need to include the state in the following formal description:
  $\forall p_i, \forall t \in \mathcal{T}, p_i \in F(t) \Rightarrow \exists t' \geq t, \forall t'' \geq t', crashed^{t''}_i = \top$
  The Completeness property is of the "eventual" sense: there exists an instant in time (t') after which every crashed process will be permanently (for any instant of time t'' after t') suspected. This formal definition of "eventually" will be used in the rest of this paper.
- **Eventual Accuracy**: Eventually no process is incorrectly suspected to be in a state. Formally,
  $\forall p_i, \forall t \in \mathcal{T}, \exists t' \geq t, \forall t'' \geq t', crashed^{t''}_i = \top \Rightarrow$
  $(p_i \in F(t')) \wedge (s^{t''}_i = most\_recent\_step(p_i, t', S_s, S_p))$

Our $\Diamond\Sigma$ extends Σ by outputting a sequencer history that embeds process failures causally in a sequence of local state transitions, compared to Σ which only outputs the most recent operational status of correct processes and the final states of crashed ones. The benefit of this enhanced sequencer history is that

[3] π_{s_i} denotes the projection operation onto local state s_i.

computation observation is separated from the predicate detection function. Redundant control message processing at the predicate detection module, as in the approach of [25], is eliminated. Note that $H_{\Diamond\Sigma}(\tau)$ relies on the control messages received on the monitor by time τ. Since the message delivery delay in a FAR system is unbounded, it is impossible to require a sequencer history to capture a computation in real time.

Gärtner and Pleisch implement Σ by using a perfect failure detector ($\mathcal{P}$) in a system with synchronous communications. Their algorithms assume FIFO messages sending and receiving, so the sequencer maintains the correct sequence of state transitions. Σ has a timer for each process. The timer for p_i is set when $\mathcal{P}$ suspects p_i. Upon the expiration of this timer, if $\mathcal{P}$ continues to suspect p_i and if there is no message received from p_i, Σ knows that p_i is crashed and the last state of p_i on record is its final state.

8.2 Timeout on Stubborn Channels

The weak synchrony of communications in the FAR model is characterized by the finite average response time assumption. Essentially, $\Diamond\Sigma$ needs to encapsulate this synchrony and to causally order failure events in state transitions. Unfortunately, it is impossible to assume any simple timeout value in the FAR model as in [25]. Since communication delays are not bounded, nothing can be assumed about the response time for an individual message. Therefore no matter what timeout value is chosen, $\Diamond\Sigma$ can make mistakes about the last state of a crashed process if the last control message from that process arrives after its timeout. However, we can reduce the chance of this error by improving the estimation of an appropriate timeout value in FAR. Fetzer *et al.* use a timeout, δ, per communication link [28] to implement $\Diamond\mathcal{P}$. δ is defined by the following equation:

$$\delta = \mathit{unacked}[p_i] + (1 + \mathit{numFast}[p_i]) \times (1 + \log(1 + \mathit{slowmsgs}[p_i]))$$

$\mathit{unacked}[p_i]$ is the sending time of the last unacknowledged message from the failure detector to p_i; $\mathit{numFast}[p_i]$ is the number of fast messages between two slow messages received from p_i; and $\mathit{slowmsgs}[p_i]$ is the total number of slow messages received from p_i.

The intuition behind this timeout scheme is that the *average* response time of a link increases slowly with the number of slow messages sent via this link. Meanwhile, the timeout is incremented quickly with the number of recent fast messages under the assumption that a potential slow message may be received from a process later. So the probability that the failure detector does not suspect a slow process is increased. Large timeout values will reduce the possibility of a false detection, at the expense of performance. But the value of timeout does not affect the eventual correctness because there cannot be an infinite number of slow messages, otherwise the average response time of the link would be infinite. The timeout value reflects tuning of the protocol to the underlying system.

This timeout mechanism is suitable for stubborn channels with only finite average response time. It has been applied to the implementation of $\Diamond\Sigma$ in the FAR model [12] as function "get-delta(p_i)" which returns the value of δ for the stubborn channel connecting the monitor q and process p_i.

9 Eventually Perfect Predicate Detection

In a faulty system, a predicate may involve the operational status of processes. To solve the detection of such predicates, we have actually extended the notion of local state s_i to be $(s_i, \mathit{crashed}_i)$ in Section 8.1, such that $\mathit{crashed}_i = \top$ if p_i crashes right after s_i; otherwise, $\bot$.

We model a predicate detection algorithm, $\mathbb{D}$, as an application level function layered on $\Diamond\Sigma$ (Figure 1). Orthogonal to the implementation of $\Diamond\Sigma$, $\mathbb{D}$ takes $\Diamond\Sigma$'s output ($H_{\Diamond\Sigma}$) and generates an event, detect_ϕ, when $\mathbb{D}$ detects that ϕ has held at time t in $H_{\Diamond\Sigma}(t)$. detect_ϕ holds valid until some time $t' > t$, if it exists, when $\mathbb{D}$ determines that ϕ has not held in $H_{\Diamond\Sigma}(t')$ and detect_ϕ will be cancelled. Our approach clearly separates the detection algorithm $\mathbb{D}$ from $\Diamond\Sigma$ by furnishing a complete sequencer history including processes' operational status.

Predicate detection semantics characterize the relationship of the detection of a predicate and the underlying computation. They can be specified based on two abstract properties: *safety* and *liveness*, given by Gärtner and Pleisch [27]:

- **Safety** (S): If a predicate ϕ is detected, then ϕ has held in the computation. Formally,
  S: $\mathit{detect}_\phi \Rightarrow \blacklozenge\phi$ ($\blacklozenge$ is a temporal logic operator denoting 'eventually in the past'.)
- **Liveness** (L): If ϕ has held in the computation, it will eventually be detected. Formally,
  L: $\phi \Rightarrow \Diamond \mathit{detect}_\phi$

Gärtner and Pleisch also prove that [27] in asynchronous systems with crash failures and any failure detector not strictly stronger than $\mathcal{P}$, it is impossible to achieve the perfect predicate detection semantics (defined by having $L + S$). This semantics is also unavailable in a FAR system because even $\mathcal{P}$ is impossible to implement in FAR according to Theorem 4.1 in [12].

It has been proved that the *eventually perfect predicate detection semantics* ($L + \Diamond S$) can be solved with $\Diamond\Sigma$ in a FAR system [12]. Also, $\Diamond\Sigma$ can extend any predicate detection algorithm that preserves safety and liveness properties to an FAR system, to achieve the eventually perfect semantics. Many monitoring applications can benefit from this semantics.

We want to point out that $\mathbb{D}$ detects whether ϕ *has held* in $H_{\Diamond\Sigma}$, not whether it currently *holds*. Even though the possibility of erroneous predicate evaluation cannot be eliminated entirely, the impact of such an occurrence is mitigated by the observation that whether or not ϕ is *stable*, the property that "ϕ has held" is stable. That is, once true, it remains true for the duration of the computation. The requirement that $\mathbb{D}$ preserves the safety and liveness properties with respect

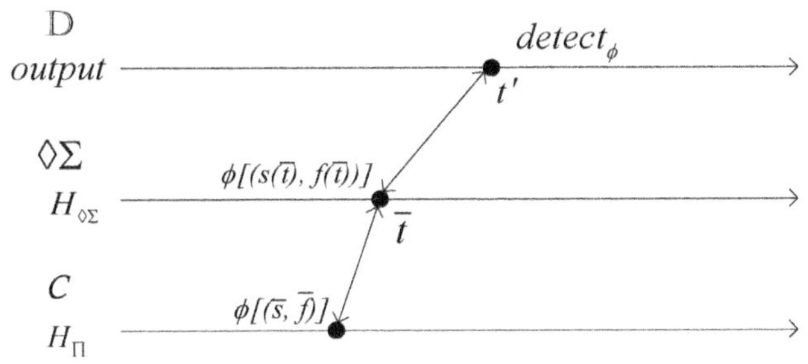

Fig. 1. Layered design for $\mathbb{D}$, $\Diamond\Sigma$, and $\mathcal{C}$

to $H_{\Diamond\Sigma}$ is practical because many algorithms in literature (*e.g.* [2,4]) qualify and can be adopted as $\mathbb{D}$. This means higher flexibility for applications and compatibility with new advances in research. Also, the detection is not restricted to certain predicate modalities, compared to [25] that only works with observer-independent predicates.

To solve predicate detection of at least as strong as the eventually perfect semantics [27], all events must be eventually observed by the monitor. This is why $\Diamond\Sigma$ must have the consistency property which is a non-issue for Σ because communications are implicitly assumed to be reliable in the implementation of Σ [25]. Reliable communication guarantees message delivery at a correct recipient process even if the sender crashes immediately after having sent a message [33]. Unfortunately, stubborn channels in the FAR model are weaker than reliable channels because they have no guarantee of message delivery when the sender crashes. To achieve the consistency property, we have to augment the system with the following assumption: the stubborn channels between application processes and the monitor are reliable.

10 Predicate Detection by Mobile Agents

The implementation of an eventually accurate failure detection sequencer is layered on $\Diamond\mathcal{P}$. It has been proved that Algorithm 1 in [12] implements an eventually accurate failure detection sequencer by satisfying the Consistency, Completeness, and Eventual Accuracy properties (Theorem 4.2 in [12]).

However, to achieve eventually perfect predicate detection, one has to add two layers of software on the monitor, each of which implements $\Diamond\mathcal{P}$ and $\Diamond\Sigma$ respectively. A module to interact with the monitor has to be added as well on every application process. Since these additional software modules are closely coupled to the distributed system in a static and *ad hoc* way, it can be considerably expensive to build a new predicate detection application from scratch for every monitoring task. The configuration of this approach is fixed per monitoring application, which means less flexibility and efficacy when the monitored computation gets unstable. For example, predicate detection may suffer temporary network disturbance or performance bottleneck on the monitor process.

Inspired by the work of using mobile agents to monitor distributed systems-[23,30,34], we strategically restructure the approach for solving eventually perfect predicate detection in [12] and reinvent it with the application of mobile agents. The mobile agent structure assumes that each process in the distributed computation has a set of agents residing and running on it as demonstrated in Fig. 2.

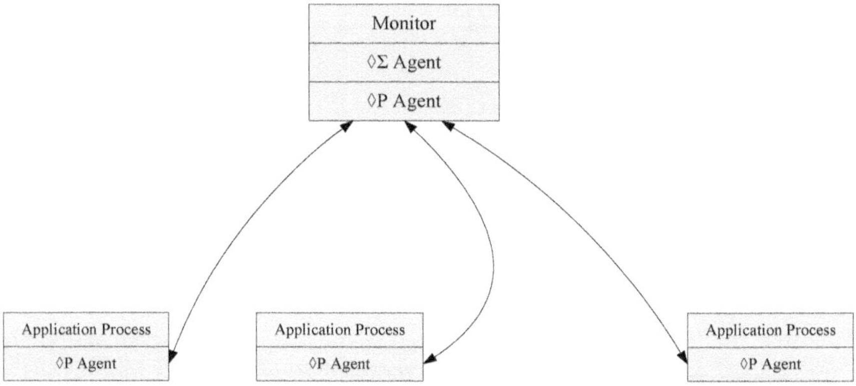

Fig. 2. The architecture of predicate detection by mobile agents

These agent types are:

- $\Diamond\mathcal{P}$ Agent (PA) executes the code that implements an eventually perfect failure detector. Depends on where they are running, there are two types of PA: monitor and application. Currently we only allow one monitor PA in a task. The monitor PA has two operation primitives for an $\Diamond\Sigma$ agent: **FD_Send** to send process liveness probing messages, and **FD_Receive** to receive and deliver control messages.

  An application PA simply replies to the probing messages from the monitor PA, as it does in the implementation of $\Diamond\mathcal{P}$. The current local state is piggy-backed in the reply.
- $\Diamond\Sigma$ Agent (SA) runs on the monitor process. It maintains a timer $r_q[i]$ for each application process p_i. SA starts the monitoring task by **FD_Send**ing a liveness probing message (through the PA residing on the monitor) to every process in Π. When SA **FD_Receive**s a reply from p_i before $r_q[i]$ expires, p_i is regarded as correct. So SA extracts the local state contained in the message, sets the operational status of p_i, $D_q[i]$, as $\bot$, and adds this update to the sequencer history. The monitoring continues with another "alive?" message sent to p_i.

  If the SA learns that p_i is suspected by the monitor PA, it resets timer $r_q[i]$ as the current value from get-delta(p_i). When $r_q[i]$ expires and p_i is still suspected, $D_q[i]$ is set to $\top$ and sequencer history is updated. The SA keeps observing the entire computation history including crash events via control messages. When queried at time t, the SA outputs an array $H_q[1..t][1..n]$, which is the sequencer history on monitor q through t: $H_{\Diamond\Sigma}(t)$.

These agents interact with each other according to Algorithm 1 in [12]. The PAs implement an eventually perfect failure detector, and the SA works as a $\Diamond\Sigma$. The new architecture for solving eventually prefect predicate detection by mobile agents is very flexible and useful because mobility gives agents the autonomy needed to monitor the environment effectively. All agents can be encoded at run-time. Thus they can make use of application specific information to permit more efficient detection. For example, to adjust the initial parameters of the timeout calculation to the measured network characteristics. In a rare but possible case in realistic computing, the SA and monitor PA can also move to another process to survive the degradation of current monitor process's performance or network traffic jam.

With this approach, there is no need to explicitly keep track of the processes/computers involved in the computation because this task is distributively carried out by the mobile agents. PAs will be created and destroyed as processes join, temporarily suspend, and leave the computation. This adaptive approach of mobile agents can eliminate unnecessary overhead associated with additional static software modules.

Deployed closely to the entity being monitored (the application processes), mobile agents offer the flexibility of distributing themselves and reduce the communication loads between processes and the monitor. Artificial intelligent based on mobile agents makes it possible to consider the amount of synchrony in the system and take a proactive behavior in monitoring. When needs arise, agents can also migrate across processes to accomplish more complicated detection tasks. Mobile agents provide suitable mechanisms to solve the predicate detection problem in a distributed, efficient, and convenient manner. Therefore, they can solve predicate detection in large-scale distributed systems with richer semantics.

More importantly, predicate detection approach that works on the basis of mobile agents is very flexible. Adding new monitoring functionalities is as easy as incorporating new types of mobile agents that will be move among the involved computers as needed [30]. This can be done dynamically without disturbing other monitoring tasks in operation in the same distributed system. Mobile agents can move across and interact with heterogeneous computing devices, which makes our approach even more favorable.

Since the FAR model does not make assumptions of the bound on relative speed of processes, and because it assumes flow control in the form of stubborn channel, communication of state changes from application processes to the monitor process may not keep up with the state changes as they occur in a run of the computation. In this research, we assume that application processes cannot take an unbounded number of state changes before receiving a probe from the monitor. While this would appear to be a limitation due to the FAR model and the requirement for stubborn channels, it is, in fact, a necessary condition for *online* monitoring. If the monitor can fall behind by an unbounded amount, then the detection is by definition offline.

A process may have multiple, but bounded number of local state transitions between two consecutive probing messages from the monitor. Since the stubborn channel requires a bounded and known upper limit on the size of messages, piggybacking multiple state change notifications in a single control message to the monitor may ameliorate the challenge to the monitor. The consistency property of $\Diamond\Sigma$ is preserved because events from a process are totally ordered.

11 Conclusions and Future Work

In this chapter, we extend the study of predicate detection in a failure-prone system assuming that only the average response time is finite. The extension includes applying the technique of mobile agent to implement an eventually accurate failure detection sequencer ($\Diamond\Sigma$) in the FAR model. $\Diamond\Sigma$ has an augmented sequencer history that embeds failure events causally in a complete sequence of state transitions. No redundant observation of computation is necessary in the predicate detection module. So the implementation can be simplified and has higher efficiency.

We demonstrate the power of the FAR model over the purely asynchronous model because $\Diamond\Sigma$ encapsulates the weak synchrony of communication and minimizes the possibility of making wrong decisions. No failure detector in $\Diamond\mathcal{P}$ can either order failures casually in state transitions or is capable to minimize wrong decisions. In our approach, $\Diamond\Sigma$ brings predicate detection and the underlying computation together in a layered model where applications have the flexibility to choose appropriate predicate detection algorithms with the possibility to solve different predicate modalities. It is more efficient and more reliable than solving the eventually perfect semantics using $\Diamond\mathcal{P}$ based on the claim in [27].

This research is an innovative contribution to predicate detection in large-scale distributed systems with crash failures and limited synchrony. By factoring in execution of protocols of agents, the reinvention of the original predicate detection approach [12] not only inherits all its great features, but also enhances its flexibility, scalability and robustness. It has good potential to be developed into a general-purpose predicate detection infrastructure that provides the basic functionalities to achieve reasonable predicate detection semantics in a realistic distributed system. Using our approach, many Internet applications could have eventual detection of some undesirable computation properties and identification of a potential "bug" in the system.

Our approach may also fit in distributed systems with stochastic communication networks where the components, links and nodes are not 100 percent reliable, as long as the probability and delay of exceptional slow messages do not violate the requirement of finite average response time. It is worth a future study to shred light on the impact of stochastic communications on $\Diamond\Sigma$, detection semantics and costs.

In separate research, we have articulated a taxonomy of predicate detection semantics that permits the establishment of a framework to illustrate the interplay between synchrony, predicate detection semantics and the properties of the detection function itself. The approach presented in this paper suggests that further study is warranted to explore these interactions.

References

1. Dollimore, J., Kindberg, T., Coulouris, G.: Distributed Systems: Concepts and Design, 4th edn. Addison-Wesley, Reading (2005)
2. Alagar, S., Venkatesan, S.: Techniques to tackle state explosion in global predicate detection. IEEE Transactions on Software Engineering 27(8), 704–714 (2001)
3. Chase, C.M., Garg, V.K.: Detection of global predicates: Techniques and their limitations. Distributed Computing 11(4), 191–201 (1998)
4. Cooper, R., Marzullo, K.: Consistent detection of global predicates. In: Proceedings of the ACM/ONR Workshop on Parallel and Distributed Debugging. ACM SIGPLAN Notices, vol. 26, pp. 167–174 (1991)
5. Garg, V.K., Mittal, N.: On slicing a distributed computation. In: Proceedings of the 21st IEEE International Conference on Distributed Computing Systems (ICDCS 2001), pp. 322–329 (2001)
6. Sen, A., Garg, V.K.: Detecting temporal logic predicates on the happened-before model. In: Proceedings of the 16th International Parallel and Distributed Processing Symposium (IPDPS 2002), vol. 116 (2002)
7. Stoller, S.D.: Detecting global predicates in distributed systems with clocks. In: Mavronicolas, M. (ed.) WDAG 1997. LNCS, vol. 1320, pp. 185–199. Springer, Heidelberg (1997)
8. Fischer, M.J., Lynch, N.A., Paterson, M.S.: Impossibility of distributed consensus with one faulty process. Journal of the ACM 32(2), 374–382 (1985)
9. Chandy, K.M., Misra, J.: How processes learn. Distributed Computing 1(1), 40–52 (1986)
10. Stoller, S.D.: Detecting global predicates in distributed systems with clocks. Distributed Computing 13(2), 85–98 (2000)
11. Waldo, J.: Mobile code, distributed computing, and agents. IEEE Intelligent Systems 16(2), 10–12 (2001)
12. Chu, C., Brockmeyer, M.: Eventually perfect predicate detection in crash-affected finite average response time systems. International Journal of Parallel, Emergent and Distributed Systems (2009) (to appear)
13. Corell, E., Saxholm, P., Veitch, D.: A user friendly TSC clock. In: Passive and Active Measurement Conference, Adelaide, Australia (March 2006)
14. IEEE: IEEE Standard for a Precision Clock Synchronization Protocol for Networked Measurement and Control Systems. ANSI/IEEE Std 1588-2002 (2002)
15. Mills, D.L.: Improved algorithms for synchronizing computer network clocks. IEEE/ACM Transactions on Networking 3(3), 245–254 (1995)
16. Veitch, D., Babu, S., Pàsztor, A.: Robust synchronization of software clocks across the Internet. In: Proceedings of the 4th ACM SIGCOMM Conference on Internet Measurement (IMC 2004), pp. 219–232 (2004)
17. Zhang, L., Liu, Z., Xia, C.H.: Clock synchronization algorithms for network measurements. In: Proceedings of the 21st Annual Joint Conference of the IEEE Computer and Communications Societies (IEEE INFOCOM 2002), vol. 1, pp. 160–169 (2002)

18. Mills, D.L.: Internet time synchronization: the network time protocol. IEEE Transactions on Communications 39(10), 1482–1493 (1991)
19. Dwork, C., Lynch, N., Stockmeyer, L.: Consensus in the presence of partial synchrony. Journal of the ACM 35(2), 288–323 (1988)
20. Chu, C., Brockmeyer, M.: Predicate detection modality and semantics in three partially synchronous models. In: Proceedings of the 7th IEEE/ACIS International Conference on Computer and Information Science, pp. 444–450 (2008)
21. Larrea, M., Arévalo, S., Fernández, A.: Efficient algorithms to implement unreliable failure detectors in partially synchronous systems. In: Jayanti, P. (ed.) DISC 1999. LNCS, vol. 1693, pp. 34–48. Springer, Heidelberg (1999)
22. Larrea, M., Fernández, A., Arévalo, S.: On the impossibility of implementing perpetual failure detectors in partially synchronous systems. In: Proceedings of the 10th Euromicro Workshop on Parallel, Distributed and Network-based Processing, pp. 99–105 (2002)
23. Motshegwa, T., Schroeder, M.: Interaction monitoring and termination detection for agent societies: Preliminary results. In: Omicini, A., Petta, P., Pitt, J. (eds.) ESAW 2003. LNCS (LNAI), vol. 3071, pp. 136–154. Springer, Heidelberg (2004)
24. Milojičić, D., Douglis, F., Wheeler, R.: Mobility: processes, computers, and agents. ACM Press/Addison-Wesley Publishing Co., New York (1999)
25. Gärtner, F.C., Pleisch, S.: Failure detection sequencers: Necessary and sufficient information about failures to solve predicate detection. In: Malkhi, D. (ed.) DISC 2002. LNCS, vol. 2508, pp. 280–294. Springer, Heidelberg (2002)
26. Charron-Bost, B., Delporte-Gallet, C., Fauconnier, H.: Local and temporal predicates in distributed systems. ACM Trans. Program. Lang. Syst. 17(1), 157–179 (1995)
27. Gärtner, F.C., Pleisch, S. (Im)Possibilities of predicate detection in crash-affected systems. In: Datta, A.K., Herman, T. (eds.) WSS 2001. LNCS, vol. 2194, pp. 98–113. Springer, Heidelberg (2001)
28. Fetzer, C., Schmid, U., Süsskraut, M.: On the possibility of consensus in asynchronous systems with finite average response times. In: Proceedings of the 25th IEEE International Conference on Distributed Computing Systems (ICDCS 2005), pp. 271–280 (2005)
29. Fischer, M.J.: The consensus problem in unreliable distributed systems (a brief survey). In: Karpinski, M. (ed.) FCT 1983. LNCS, vol. 158, pp. 127–140. Springer, Heidelberg (1983)
30. Ilarri, S., Mena, E., Illarramendi, A.: Using cooperative mobile agents to monitor distributed and dynamic environments. Information Sciences 178(9), 2105–2127 (2008)
31. Chandra, T.D., Toueg, S.: Unreliable failure detectors for reliable distributed systems. Journal of the ACM 43(2), 225–267 (1996)
32. Lamport, L.: Time, clocks, and the ordering of events in a distributed system. Communications of the ACM 21(7), 558–565 (1978)
33. Guerraoui, R., Olivera, R., Schiper, A.: Stubborn communication channels. Technical Report 98-278, Département d'Informatique, École Polytechnique Fédérale de Lausanne (1998)
34. Brewington, B., Gray, R., Moizumi, K., Kotz, D., Cybenko, G., Rus, D.: Mobile agents in distributed information retrieval. In: Intelligent Information Agents, pp. 355–395. Springer, Heidelberg (1999)

Personalized Ontology-Based Recommender Systems for Multimedia Objects

Krzysztof Juszczyszyn, Przemysław Kazienko, and Katarzyna Musiał

Wrocław University of Technology, Wyb.Wyspiańskiego 27, 50-370 Wrocław, Poland
{krzysztof,kazienko,katarzyna.musial}@pwr.wroc.pl

Abstract. A framework for recommendation of multimedia objects based on processing of individual ontologies is proposed in the chapter. The recommendation process takes into account similarities calculated both between objects' and users' ontologies, which reflect the social and semantic features existing in the system. The ontologies, which are close to the current context, provide a list of suggestions presented to the user. Each user in the system possesses its own Personal Agent that performs all necessary online tasks. Personal Agents cooperate each other and enrich lists of possible recommendations. The system was developed for the use in the *Flickr* multimedia sharing system.

1 Introduction

Multimedia sharing systems like Flickr or YouTube, as typical examples of Web 2.0 applications, play ever increasing role in the recent society. They enable their users to upload, download, manage, and browse multimedia contents such as photos, videos, animations called commonly multimedia objects (MOs). In most multimedia sharing systems, users simultaneously interact, collaborate and influence one another forming, in this way, a kind of social community. Hence, users can not only tag multimedia objects they have published but also comment the items added by others, include them to their favorites, etc. Additionally, users have the opportunity to set up new, direct relationships with other system users as well as establish groups of collective interests and directly enumerate their friends or acquaintances.

A typical multimedia sharing system accumulates vast amount of data about published multimedia objects, relationships between them, tags and various types of user activities. However, the information hidden in this data is poorly structuralized and do not provide any comprehensive view onto the relationships between MOs nor the system users. Besides, users often utilize tags and descriptions containing colloquial and improperly used terms, which are not really comprehensible for the others. Moreover, they tend to assign to the MOs they publish only a few relevant tags or no tags at all, so that many multimedia objects do not possess any appropriate and verified descriptions nor tags. As a result, there are many MOs in the system that contain similar multimedia content and completely divergent tags and textual description. The process of autonomous creation and assignment of tags to multimedia objects performed by their authors is often called *folksonomy* [7, 24].

A. Håkansson et al. (Eds.): Agent and Multi-agent Technology, SCI 289, pp. 275–292.
springerlink.com

This diverse and vast amount of data about both multimedia objects and user activities gives the opportunity to complex analysis. It can also be exploited to create complex ontologies that would provide the comprehensive view onto both the multimedia objects existing within the system, the relationships between them as well as the users operations connected with these multimedia objects. Next, the knowledge built into these ontologies can be utilized by the recommender system to suggest to the active user the items, which are the most suitable for them.

2 Related Work

Nowadays, recommender systems became more and more popular and often constitute the integral part of the recent web sites. They help people to make decision, what items to buy, which news to read [25], which movie to watch or even who they can invite to their social network [14]. On the other hand, especially in e-commerce, these kinds of system provide the powerful tool to maintain the loyalty of the customers and increase the sales [12]. Recommender systems are especially useful in environments with vast amount of information since they cope with selection of a small subset of items that appears to fit to the users' preferences [1, 17, 23, 26].

Overall, the recommender systems are usually divided into three main categories: collaborative filtering, content-based filtering, and hybrid recommendation [1]. The collaborative filtering technique relies on opinions about items delivered by users. The system recommends products or people that have been positively evaluated by other people, whose ratings and tastes are similar to the preferences of the user who will receive recommendation [1, 8, 25]. In the content-based filtering the items that are recommended to the user are similar to the items that user had picked and rate high in the past [22]. The hybrid method combines two previously enumerated approaches [9, 12, 25].

However, in some cases, especially in Web 2.0 applications, we can use available knowledge about application domain in order to generate some recommendations [27].

An ontology is a conceptualisation of a domain into a machine-readable format typically in the form of a structure consisting of concepts, attributes, relationships, and axioms [6]. The problem of finding similarities between ontology elements and also between entire ontologies as complex structures plays important and growing role in shaping online user communities and managing the content of Web portals [11] such as Flickr.

There exist numerous researches that applied the idea of ontology in the recommendation process. The first examples were Quickstep and Foxtrot systems proposed in [19], where the collections of research papers were classified using ontological classes. The advantage of such methods is that they facilitate to take into account the relations between commonly used search terms and their possible role in the application domain under consideration. The similar approach, using ontologically grounded user profiles, was proven to be successful in [3]. Another example where the view-based search method developed within the information retrieval community was combined with the ontology-based annotations and search was presented in [10, 16]. As a result the authors worked out the ontology and vie-based image retrieval and recommendation browser Ontogator.

Ontologies were also utilized to address the cold-start problem as well as interest acquisition problem [18]. It was performed by integrating the Quickstep recommender system, AKT ontology and OntoCoPI. Furthermore, the ontologies were applied in the recommender system for business collaboration [21]. The goal of the system was to facilitate knowledge workers to find the most relevant resources that enable to complete the given task. This process was performed based on continuous monitoring of worker's context and anticipating the person's potential requirements.

3 Problem Description

The typical multimedia sharing system accumulates vast amount of data about the published multimedia objects, relationships between them, tags and various types of user activities. However, the information hidden in this data is poorly structuralized and do not provide any comprehensive view onto the relationships between MOs nor the system users. Besides, due to open and flexible profile of the multimedia sharing system users often utilize the language, which is not really comprehensible for the others. It regards especially tags and descriptions containing colloquial and improperly used terms. Moreover, users often assign to the MOs they want to publish only few relevant tags or no tags at all. Hence, many multimedia objects do not possess any appropriate and verified descriptions or tags. As a result, there are many MOs that contain similar multimedia content and completely divergent tags and textual description.

On the other hand, the ontology concept appears to be suitable for this kind of data but acquiring such knowledge and keeping it up to date is not a trivial task.

Typical recommender framework process data gathered by the system and generate some suggestions to users. However, the users have no influence on the recommendation process. Sometimes, in the systems based on demographic filtering, users can only provide information about themselves: their interests, gender, age and other demographic data. This kind of data is usually inserted once at registration and never updated. The character of multimedia sharing systems requires a new method of recommendation, in which users would be able to change relationships between multimedia objects generated by the system. However, these relationships result from many different either semantic or social links. The former include common tags and similar descriptions, whereas the latter are consequences of relations between system users and can be derived from lists of favourites, groups, contact lists and comments to the same MOs. The exaggerated independence of the recommender system can be weakened by introduction of the automatically created individual ontologies that could be manually changed by the users.

4 Ontology-Based Recommendation in Multimedia Sharing Systems

4.1 General Concept

The main goal of the recommendation process in multimedia sharing system such as Flickr is to provide users the MOs that they would be interested in and in consequence they will watch and comment them or even add to their list of favourites. The suggestions generated by the recommender system respect the current context, i.e. the user and the individual MOs recently viewed by this given user. This context is monitored by *Personal Agents* (PA) created and assigned to every current user at the log-in procedure.

The general concept of the recommendation process for user u_x while viewing multimedia object a_i is presented in Fig.1. It consists of four main steps: user context monitoring (user u_x and the viewed MO a_i), ontology-based similarities calculation, integration of knowledge provided by ontologies and finally creation and presentation of the recommendation list for the current user u_x.

The online tasks particularly related to the current user are performed by user's Personal Agent that communicates with other Central Agents (CA) responsible especially for arduous offline processing, see Sec. 4.2.

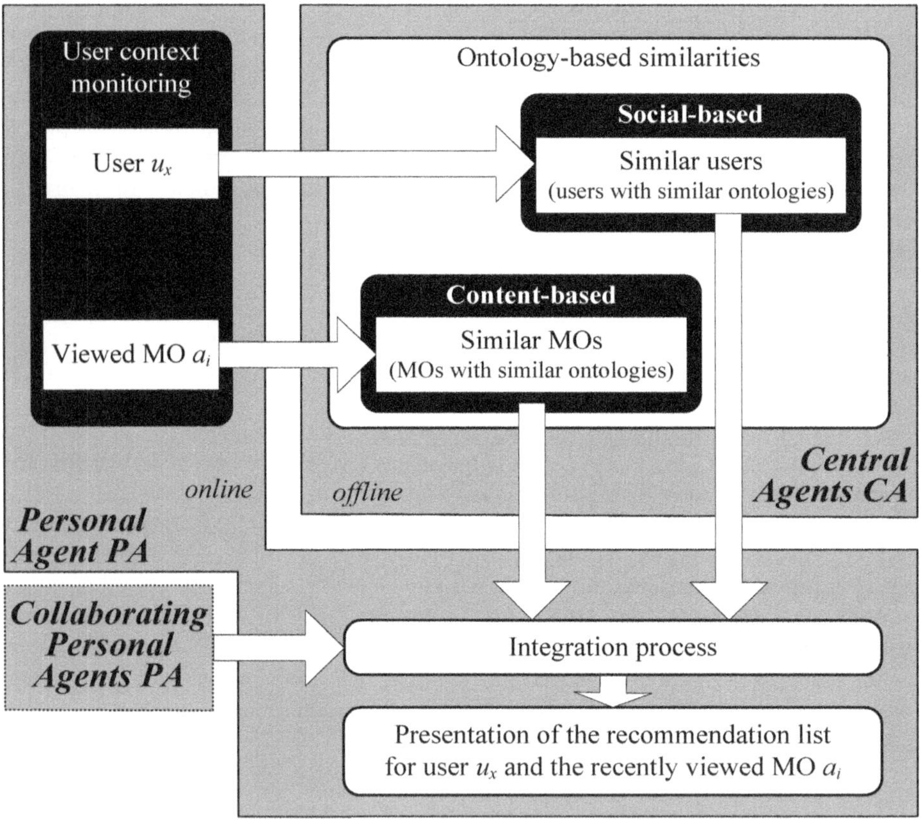

Fig. 1. The general concept of ontology-based recommendation process in the multimedia sharing system

The first step of the recommendation process, user context monitoring, is incorporated into basic function of the web server. It provides the data about current user u_x

and the MO a_i that is just requested by this user to the rest of the recommender engine. Both user u_x and MO a_i possess their own individual ontologies (see Sec. 4.3) that serve as the input to the second element of the framework, i.e. ontology-based similarities calculation (see Sec. 4.5). In this phase, user u_x is compared to all other users that are registered in the system by means of similarity function between their individual ontologies. As a result, we obtain the ranking list of users – potential authors of MOs ordered by their similarity to user u_x.

The similar procedure with is launched for MO a_i, although only k MOs closest to the given MO a_i are selected. The number k limits the length of the MO list and is used to reduce the number of necessary calculations performed online during the next step - integration. In the integration process, the list of k-nearest MOs is verified according to the closest of their authors. It means that for each of k MOs from the list its author is identified in the list of users with ontologies similar to person u_x.

The list of candidates for recommendation (MOs) is provided not only by means of ontology similarity but it is also expanded by other Personal Agents that can deliver their own lists of MOs relevant to the current context.

The basic concept of final recommendation list creation is to confront the similarity between MOs and suggestions provided by collaborating agents with the similarity between their authors and the current user u_x: the weight of the close MO is enriched by the weight of its author if it is on the list of the users close to u_x. Note that the weights reflect the level of similarity, i.e. the greater weight, the greater similarity. This regards also pairs of MOs. Finally, the top N multimedia objects are chosen from the enhanced list and presented to user u_x as recommendations.

4.2 System Architecture

The system is developed based on two kinds of agents (specialized modules): Personal Agents PA and Central Agents CA, Fig. 2. Personal Agents are individual assistants of the current user. They are responsible for the management of individual ontologies of both users themselves and MOs they authored. Ontologies can be changed automatically by Personal Agents due to activities of the user within multimedia sharing system, e.g. each addition of new MOs results in some new items in the personal ontologies of its author. Some other activities include new opinions, updates of the list of favourites, tags used, etc.

An additional task the Personal Agent performs is monitoring of the current context. The context contains two components: the user (u_x) and the viewed multimedia object MO (a_i). Personal Agent preserves all contexts for the current user session. These historical contexts are used while filtering recommendation lists to enable rotation mechanism, see Sec. 4.6.

Besides, PA is responsible for some crucial online actions related to the recommendation process, especially integration, filtering and ranking of recommendation lists and finally presentation of some top items to the user.

Additionally, each Personal Agent PAx collaborates with other Personal Agents PAy being online. Hence, Personal Agent PAx generates and spreads a new request to other PAy at each change of the current context. Agents PAy respond (or not) and provide their own suggestions of most appropriate multimedia objects relevant to the current context of PAx. The requested agent Pa$_y$ may not respond due to its overload. Agent PA$_y$ waits only some time for responses from others to avoid the dead-lock.

There are also some other agents in the system namely Central Agents. They maintain the databases necessary for all processing, i.e. ontologies, user preferences, etc. Their other essential duty is to perform all offline tasks, particularly periodical recalculation of ontology similarities, separately for users and multimedia objects. The multimedia sharing system contains some regular agents, which collaborate with the Personal Agent to create the proper page layout for the web browser.

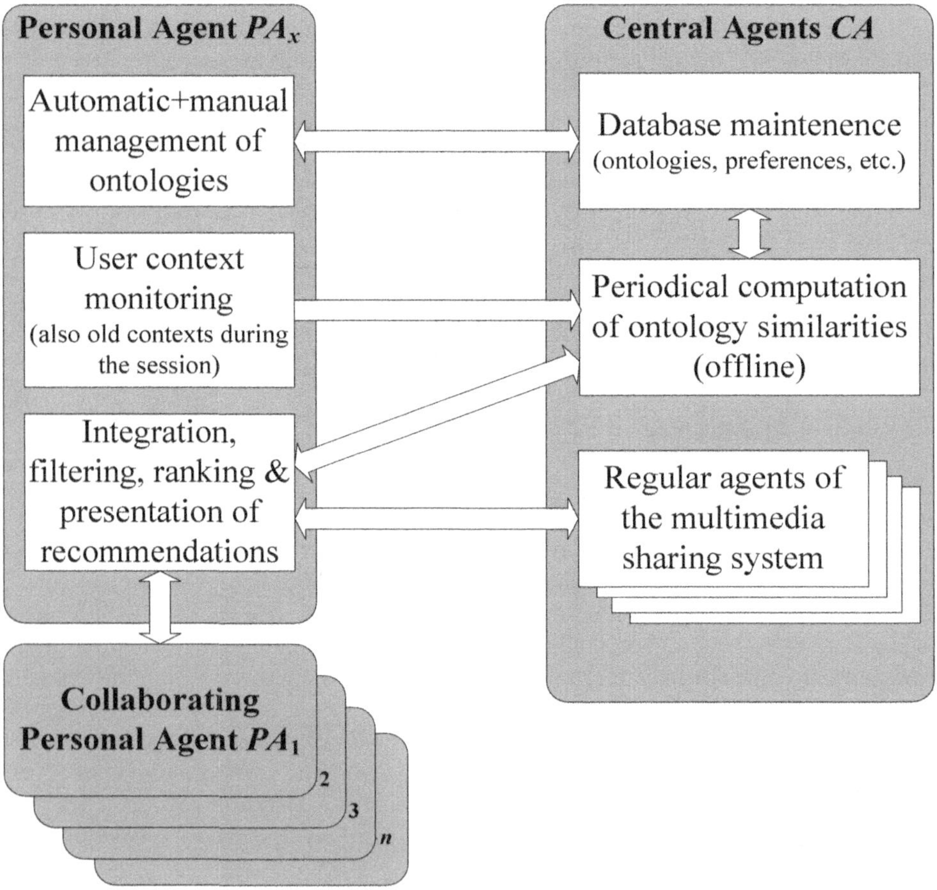

Fig. 2. Architecture of the system

4.3 Individual Ontologies

Domain Knowledge consists of the two types of ontologies, representing knowledge about users and multimedia objects. Both define the domain concepts and the basic relations met in our system. Their basic structure is predefined (Fig. 3a and Fig. 4a), however the individual set of concepts for each user and multimedia object may differ depending on the users' actions and their history. The examples that are presented in Fig. 3b and Fig. 4b provide better insight into how the individual ontologies can look like. Nevertheless, it should be emphasized that these ontologies used all extracted concepts whereas in a real system, every individual ontology can contain different types of concepts. Moreover, most of the concepts are optional, so for example the user does not have to possess favourite MOs or be a member of any group. The same situation appears regarding the multimedia object that e.g. does not have to belong to any group and in consequence the concept "groups" is not relevant for this particular object.

The individual user ontology represents knowledge about users and covers five main aspects of user activity (Fig. 3):

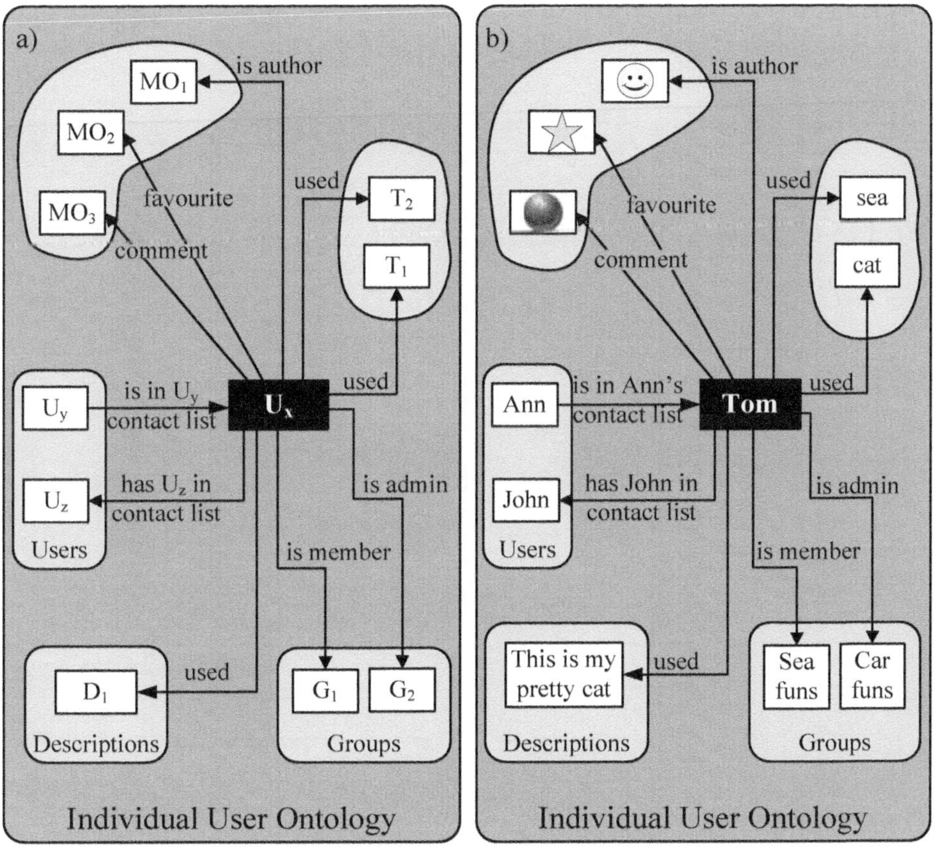

Fig. 3. The individual user ontology

- authored, favourite and commented MOs. Note that, these three activities can overlap. It means, that one user can e.g. both adds the given MO to the favourites and in the same time comments it,

- the tags used by the user to annotate MOs. These tags also called keywords usually reflect the content of the MO,
- the descriptions made by users in order to provide more information about their objects,
- users included in contact list as well as the fact of being included in someone's contact list,
- the fact of being a member or an administrator of user group.

On the other hand, the concepts within the individual multimedia object ontology reflect the knowledge about MOs uploaded to the system (Fig. 4):

- the users who authored, favourite or commented given object.
- description attached to MO by the author,
- the tags which describe a MO,
- the groups to which this MO belongs to.

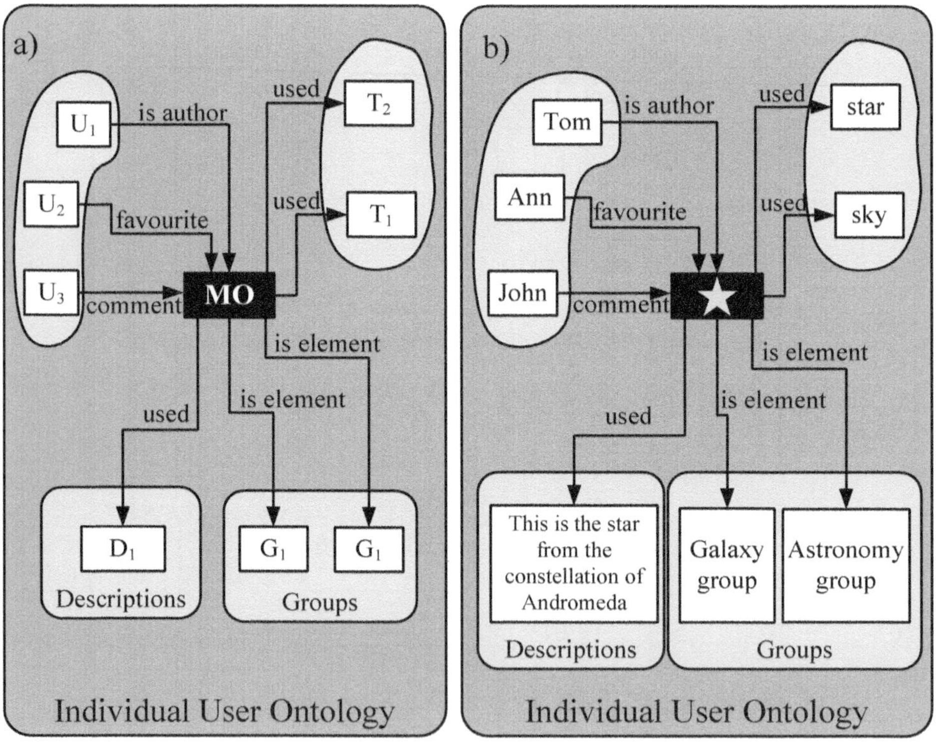

Fig. 4. The individual multimedia object ontology

The ontologies are created in the moment when the user or multimedia object appears in the system for the first time. When the MO is added by the user then the individual MO's ontology is created based on such information as tags, description and author of the photo. On the other hand, when new user registers to the system then the empty ontology for this person is created.

Of course the ontologies must be revised continuously. The changes will come both from the users who like to update their ontologies and from the system. Such consideration provokes the formation of two layers of ontology: the system and the user ontology. The former one will be managed by the system itself and the latter one will be changed by the user of multimedia sharing system. Moreover, each person can maintain only own ontology and the ontologies of this users' photos. However, the particular user cannot directly influence other users' ontologies. The ontology that will be presented in the system is the product of these two enumerated ontologies. The user ontology is stored in the system as the set of facts that denote the elements the user has changed, i.e. were added, modified or deleted.

In order to facilitate the process of updating the individual user ontology the appropriate mechanisms that support the users in their activities ought to be developed. One of them is to guarantee the user the access to the dictionaries such as WordNet, which enable to introduce the unified tags as the keywords for their photos. Furthermore some visualization tools ought to be available in order to support users in their actions regarding their ontologies.

4.4 Ontology Similarity Measure

It is assumed that in order to compute ontology-based similarities between users and Multimedia Objects (see Fig.1), an *ontology similarity measure* will be developed and applied to individual user's and MO's ontologies comparison. It should be noted that in majority of researches, addressing problems of ontology similarity, merging and alignment of ontologies, only similarity between the elements of ontological structures (typically: concepts and relations) is considered [2]. There are only few works which deal with comparing ontologies as whole knowledge structures [4, 15].

In our approach we use a *Taxonomic Precision* (*TP*), a similarity measure based on the notion of *semantic cotopy* (see def. 2) recently presented and analysed in [5]. The reason to choose this measure was its ability to compare ontologies as whole structures and along multiple dimensions. The legal values of *TP* are from the range [0;1].

As stated in [5] this definition of taxonomic precision may be influenced by the lexical term layer in the case of significant differences in domain models (ambiguous use of terms). However, in our approach, most of the concepts used in individual ontologies (both users' and MOs') come from global sets (user, group and object names, also – to some extend – tags), so this issue is not expected to appear. In our approach terms like tags, user names etc. are directly identified with concepts.

Moreover, we do not distinguish between relations in our ontologies, when applying similarity measures we treat them as taxonomies with root concepts *user* and MO.

Now we introduce the basic definitions needed to formulate the notion of *Taxonomic Precision*.

Definition 1. The ontology O is a structure $O := (C, root, \leq_X)$ where C is a set of concept identifiers and *root* is a designated root concept for the partial order $\leq_X$ on C.

Definition 2. *Semantic Cotopy* $sc(c,O)$ of a concept c from ontology O is a set containing c and all super- and subconcepts of c in O, excluding root concept $root(O)$.

Note: the above modification from standard definition (exclusion of the root concept) comes from the specific features of our system. In our case when comparing ontologies, the root concepts will always be different.

Definition 3. *Taxonomic Precision* of a concept c and the two ontologies O_1 and O_2 such that $c \in O_1$ and $c \in O_2$ is defined as:

$$tp(c, O_1, O_2) = \frac{|sc(c, O_1) \cap sc(c, O_2)|}{|sc(c, O_1)|} \quad (1)$$

Definition 4. *Global Taxonomic Precision* TP(O_1,O_2)of the two ontologies O_1 and O_2 is defined as:

$$TP(O_1, O_2) = \frac{1}{|C_1|} \sum_{c \in C_1} \begin{cases} tp(c, O_1, O_2) & if \quad c \in C_2 \\ 0 & if \quad c \notin C_2 \end{cases}, \quad (2)$$

where:
C_1, C_2 – the sets of concepts of O_1 and O_2 respectively.

4.5 Ontology Similarity Assessment

In order to decide if the given user or multimedia object is similar to another one their individual ontologies need to be processed. Intuitively, people *similar* to the given individual will be users who are utilizing the same sets of tags, are dealing with the same or alike MOs, are participating in the same groups, etc. The same concerns MOs. Because individual ontologies of the users do not represent information about the features of processed MOs or users in the contact list (the same concerns MOs' ontologies) we postulate their extension by adding relevant subconcepts. This action is temporary and performed only for the purpose of computing similarities.

The similarity assessment algorithm called OSA (Ontology Similarity Algorithm) for the two users' or MOs' ontologies (from here on denoted as O_1 and O_2) looks as follows:

The ontology similarity algorithm – OSA

Input:

- Ontologies O_1 and O_2 to be compared.

Note: we assume that they are both of the same type, i.e. either user's or MO's individual ontologies, as defined in sec 4.2.

Output:

- The value of similarity between O_1 and O_2 from the range [0;1].

1. begin
2. /* create extensions O_1^* and O_2^* of O_1 and O_2 */
3. $O_1^* = O_1$ $O_2^* = O_2$
4. for (each *user* concept $C_i \leq_X$ *root* in O_1^*) do
5. begin
6. find ontology O_i such that $root(O_i) = C_i$
7. attach all subconcepts of C_i from O_i as subconcepts of C_i in O_1^*
8. end
9. for (each MO concept $C_j \leq_X$ *root* in O_1^*) do
10. begin
11. find ontology O_j such that $root(O_j) = C_i$
12. attach all subconcepts of C_j from O_j as subconcepts of C_j in O_1^*
13. end
14. repeat steps 4-13 for O_2^*
15. calculate $TP(O_1^*, O_2^*)$ according to Def. 4.
16. return $TP(O_1^*, O_2^*)$
17. end

In order to compute the similarities between the O_1 and O_2, they will be extended by attaching concepts from individual ontologies of users and MOs met in O_1 and O_2. The motivation is to take into account their characteristic features, which could be omitted otherwise. For example, two different MOs in ontologies of two users are not signs of their similarity, but if they are tagged in the same way, by the same users and have similar descriptions – it should have positive influence on similarity between these users.

4.6 Recommendation Process

Based on the gathered information from the individual users and multimedia objects ontologies we have built the recommendation framework that enables users to view, comment, and add to the list of favourites the MOs that they will be keenly interested in. Moreover, if one finds the recommended MO interesting then this person can find the author of the photo and set up a new relationship with this user. The main goal of the system is to provide the most relevant recommendations to users. Moreover, by combining the several, different sources of data, the method facilities a bootstrap user to find interesting content in the multimedia sharing system and fulfil their expectations.

The overall view of the recommendation framework for the multimedia sharing system is presented in Fig. 5.

Before the recommendation process for the given person is launched the individual ontologies for all people as well as for all MOs are created. Note that these ontologies can be changed by both the system itself as well as each user can maintain their own ontology as well as ontologies of MOs added by them to the system (see Sec. 4.3)

The first step of the recommendation process is to capture the user context, i.e. that both the user who browse the MOs and the MO selected by this person to browse are identified. In order to facilitate the further explanation of the process let's assume that the user u_x is watching MO a_i.

The individual ontologies for the users and the MOs serve as the input data for the whole process and are utilized in the ontology-based similarities calculation phase. Note that these ontologies are periodically recalculated in order to assure their validity. Based on individual multimedia object ontologies the k-nearest MOs, which are close to MO a_i, are selected and list L_{MO} of these objects is established. List L_{MO} contains the weights that reflect the level of similarity between MO a_i and k-nearest MOs. This element can be seen as the content-based filtering, whereas the method of list L_U creation is also called social-based filtering.

List L_U is obtained by comparing the given user u_x's ontology with all other users' ontologies registered in the system. In this list, the weights that reflect the level of similarity between user u_x and all other users are stored.

Additionally, other user being currently online can enrich the list of considered MOs. The Personal Agent of the current user sends a query with its current context (MO a_i). Some collaborating Personal Agents of other users respond providing some short lists of MOs relevant to a_i. For example, imagine that Personal Agent PA_x for user u_x is in the context (u_x,a_i) and asks another PA_y for advice for MO a_i. In response, PA_y performs the entire recommendation process (except asking other PAs) for the context (u_y,a_i). Finally, PA_y returns to PA_x a short list L^R_{MO} of suggested MOs, Fig. 5.

It may also happen that the requested PA is too busy and refuses responding. The Personal Agent PA_x, which initializes the query, applies the appropriate tiny time-out to avoid too long anticipation for responses.

All suggestions (all lists L^R_{MO}) received from the collaborating PAs are aggregated into one coherent list L^C_{MO}. This list L^C_{MO} is, apart from L_{MO}, the second source of objects considered for recommendation. All three lists L^C_{MO}, L_{MO} and L_U have some weights assigned. These weights reflect with what confidence the certain MO should be recommended for the current user.

Afterwards, the user context filtering is performed. During the recommendation process some of the multimedia objects must be rejected from the lists of candidates (L_{MO} and L^C_{MO}) in order to avoid the situation in which the user have already seen the particular multimedia object. The MOs that ought to be omitted and in consequence rejected from the lists L_{MO} and L^C_{MO} are: the objects owned by user u_x as well as the objects that have already been commented or added to favourites by user u_x. Moreover, the weight of MOs that have been already viewed by user u_x should be made smaller. The level of reduction basically depends on two elements. Firstly, how often was the particular MO viewed in the past and secondly, when was this object browsed for the last time by the given user. Obviously, the MOs that were viewed several times and in the nearest past will have much lower weight than these that were viewed once and it was few months ago. As a result, we obtain two filtered lists L'_{MO} and $L^{C'}_{MO}$.

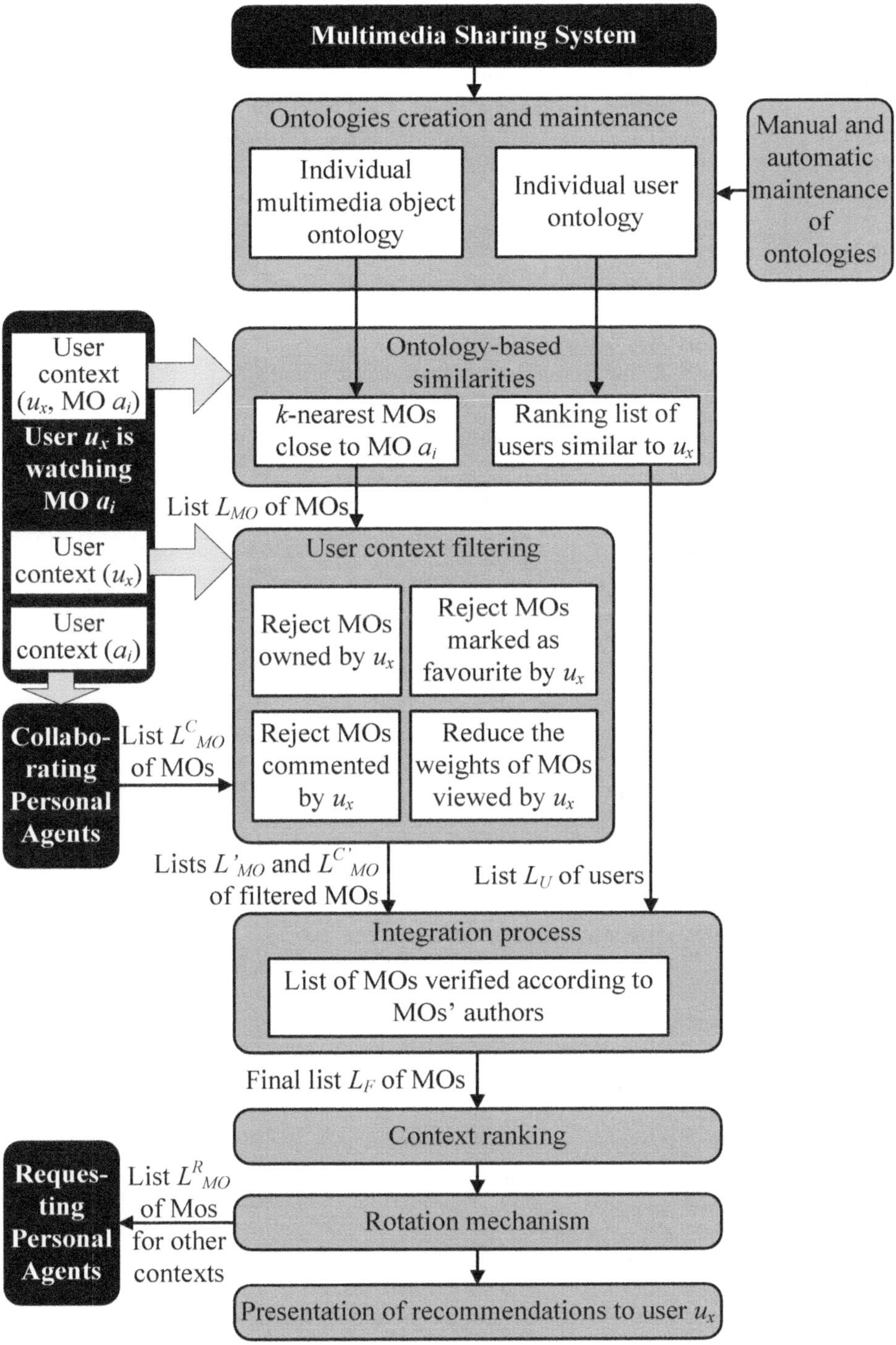

Fig. 5. Ontology-based Recommendation Process in Multimedia Sharing Systems

The next step is the integration of lists L'_{MO}, $L^{C'}_{MO}$ and L_U. The main idea here is to create the final recommendation list L_F and top N elements from this list are presented to the user u_x. List L_F is obtained by verifying the list aggregating both L'_{MO} and $L^{C'}_{MO}$ according to MOs' authors. It means that for each MO from list L'_{MO} and $L^{C'}_{MO}$ its author's weight from list L_U is considered and these weights are summed up. Note that for all of these weights the importance coefficients are assigned for the current context (u_x,a_i), as follows:

$$w^{final}(a_j, a_i, u_x) = \alpha \cdot w^{MO}(a_j, a_i) + \beta \cdot w^{C}(a_j, a_i) + \gamma \cdot w^{A}(u_x, u_j), \tag{3}$$

where:
$w^{final}(a_j, a_i, u_x)$ – the final weight for the MO a_j from the list L'_{MO} or $L^{C'}_{MO}$ of items most similar to MO a_i just viewed by u_x,
$w^{MO}(a_j, a_i)$ – the weight for the MO a_j from the list L'_{MO} of items most similar to MO a_i from the range [0;1],
$w^{C}(a_j, a_i)$ – the weight for the MO a_j from the list $L^{C'}_{MO}$ of items provided by collaborating Personal Agents; only a_j which are most similar to MO a_i are considered; if more than one agent PA suggests the same MO a_j with some weights then $w^{C}(a_j, a_i)$ is the arithmetic mean of these component weights; $w^{C}(a_j, a_i)$ is from the range [0;1],
$w^{A}(u_x, u_j)$ – the weight for the author u_j of MO a_j from the list L_U of users most similar to u_i from the range [0;1],
α, β, γ – importance coefficients with values from the range [0;1]. They are used to simulate and adjust the influence of the weights from lists L'_{MO} or $L^{C'}_{MO}$ and L_U, respectively. For example, if α and β are low and γ is high then the author's weight is more significant than MO weights.

Since values of both components are from the range [0;1], the value of final weight belongs to the range [0;3].

If MO a_j is not included in one of the list L'_{MO} or $L^{C'}_{MO}$ then its weight from that list equals 0, i.e. either $w^{MO}(a_j, a_i)$=0 or $w^{C}(a_j, a_i)$=0.

After the integration process the list L_F is sorted and finally top N selected MOs from L'_{MO} are suggested to person u_x. The rotary mechanism is used, to prevent the same MOs to be recommended to user u_x all the time [13].

The recommender system described above can be split into several processes (Fig. 6). Such division provides deeper insight and enables better understanding of the whole framework. The processes within the ontology-based recommender systems can be grouped into two sets: the users and system elements. The former ones are these in which the users modify their ontologies as well as the MOs' ontologies that belong to them. The latter ones are performed directly by the system itself. The system is responsible for creation and maintenance of the both users' and multimedia objects' ontologies (see Sec. 4.3). All enumerated and described above processes support the idea of ontology maintenance.

The system performs also all processes strictly connected with the generation of the recommendations. It creates and update recommendation lists both for each MO and for every single user. Furthermore, the system (more precisely the Personal Agent) is responsible for user context-based list selection, i.e. the appropriate lists for the user u_x that is watching MO a_i must be picked from the whole sets. Finally, the Personal Agent creates the final recommendation list L_F and by selecting N top elements from this list, recommends them to the user.

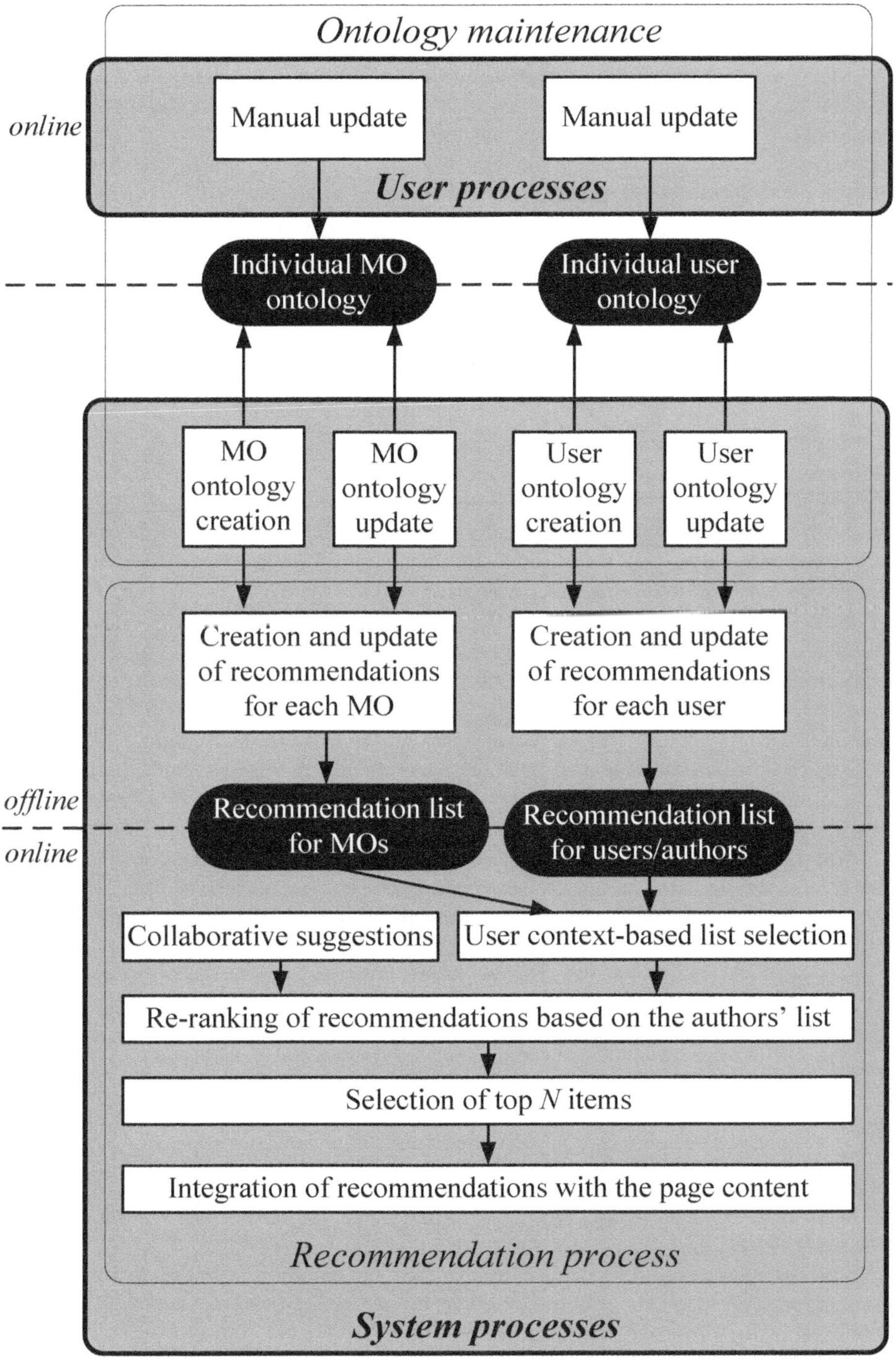

Fig. 6. Processes that exist within the recommender system

Moreover, the processes can be split into these performed offline and online (Fig. 6). The former set contains: the maintenance of the individual ontologies made by the system and periodical calculation of the ranking lists based on the ontologies similarities (performed by the Central Agents). The phase of periodical calculation of the ranking lists is the most time-consuming element of the whole recommendation framework. This is caused by the fact that the user is compared to all other users within the system as well as each MO is compared to all other MOs. In such a big and complex system as Flickr this can be a very significant problem. On the other hand, the online operations are maintenance of the individual ontologies made by the user, user context-based list selection, integration process during which the final recommendation list is created, and presentation of the suggestions to the user.

5 Discussion

During the development of the recommender process several issues, which need to be addressed, have appeared. Let us just enumerate few of them. One of the most important concerns is the high complexity of the performed calculations. In order to cope with that issue the processes that were established during the research (see Sec. 4.5) can be divided into two sets, i.e. these that can be executed offline and these that ought to be performed online as it was presented in Fig. 6.

Next thing that is connected with the efficiency issue is that the whole list of similar authors to the given one (L_U) is stored in the system. This list ought to be shortened, i.e. only m-nearest authors and their weights should be remembered and for the rest users one system weight should be assigned. This issue is especially significant in the multimedia sharing systems with large number of users and vast amount of data that describes these users, what in consequence leads to storing in the system the list of even few million of weights for each user. If the list is shortened only to m-nearest users then the integration process is faster because only m elements is searched through.

Another element that can be discussed is the integration process in which lists of MOs are verified according to MOs' authors. In the proposed method the three weights, two for the MO, which is in either list L_{MO} or L^C_{MO} and the third one for the author of this object from L_U, are aggregated, Eq. 3. Nevertheless, another approach can be applied in this phase. For example, weights can be multiplied by one another. However, note that in this case the outcome will be much more diversified as well as the weights cannot equal zero. This problem can be addressed by establishing the minimal non-zero value of the weight or by adding small number ε to each weight.

The descriptions which are represented in MOs' and users' ontologies are written in the natural language that must be properly processed. One of the methods is to abstract the most important words as well as to create the proper stop list of such words as personal pronouns, articles, etc. In our case we plan to use a chosen text comparing tool; the concepts *Description* from different ontologies will be considered the same if the tool returns text similarity value above given threshold.

The thing that should be also emphasized is that the cold-start problem, which appears in recommender systems based on collaborative or content-based filtering, do not exist in the proposed recommendation process. This is caused by the fact that the proposed recommender framework presents the hybrid approach and if new user registers to the system then only the similarity between the MO that this user is being viewed and other object from the system is calculated. On the other hand, if new MO is added then the system automatically creates the individual ontology for this object.

6 Conclusions

The proposed approach utilizes the technique that compares ontologies as whole structures to assess similarity between multidimensional profiles of users and multimedia objects in the recommender system. The ontologies provide the comprehensive view of the information gathered in the multimedia sharing system. As a result, we can execute the recommendation process, which takes into account many distinct features of system users and multimedia objects created and annotated by them.

The presented idea appears to be useful especially in large publishing systems like Flickr. Note, that there are many ways of further developments of the proposed scheme. They lay in more sophisticated mechanisms of ontology extension before similarity computation (see Sec. 4.5), providing users with advanced visual interfaces and conversational modules, which will use the underlying ontological structures. Another possible development direction is the recommendation of entire user profiles, i.e. other users themselves, instead of multimedia objects [20].

Acknowledgments. The work was supported by The Polish Ministry of Science and Higher Education, grant no. N516 037 31/3708.

References

1. Adomavicius, G., Tuzhilin, A.: Toward the Next Generation of Recommender Systems: A Survey of the State-of-the-Art and Possible Extensions. IEEE Transactions on Knowledge and Data Engineering 17(6), 734–749 (2005)
2. Andrea, M., Egenhofer, M.: Determining Semantic Similarity Among Entity Classes from Different Ontologies. IEEE Transactions on Knowledge and Data Engineering 15, 442–456 (2003)
3. Castells, P., Fersandez, M., Vallet, D., Mylonas, P., Avrithis, Y.: Self-Tuning Personalized Information Retrieval in a n Ontology-based Framework. In: Meersman, R., Tari, Z., Herrero, P. (eds.) OTM-WS 2005. LNCS, vol. 3762, pp. 977–986. Springer, Heidelberg (2005)
4. Croitoru, M., Hu, B., Dashmapatra, S., Lewis, P., Dupplaw, D., Xiao, L.: A Conceptual Graph Based Approach to Ontology Similarity Measure. In: Priss, U., Polovina, S., Hill, R. (eds.) ICCS 2007. LNCS (LNAI), vol. 4604, pp. 154–164. Springer, Heidelberg (2007)
5. Dellschaft, K., Staab, S.: On How to Perform a Gold Standard Based Evaluation of Ontology Learning. In: Cruz, I., Decker, S., Allemang, D., Preist, C., Schwabe, D., Mika, P., Uschold, M., Aroyo, L.M. (eds.) ISWC 2006. LNCS, vol. 4273, pp. 228–241. Springer, Heidelberg (2006)
6. Guarino, N., Giaretta, P.: Ontologies and Knowledge bases: towards a terminological Clarification. In: Towards Very Large Knowledge Bases: Knowledge Building and Knowledge Sharing, pp. 25–32. IOS Press, Amsterdam (1995)
7. Gruber, T.: Ontology of Folksonomy: A Mash-Up of Apples and Oranges. International Journal on Semantic Web & Information Systems 3(1), 1–11 (2005); also The First on-line Conference on Metadata and Semantics Research (MTSR 2005)
8. Ha, S.H.: Helping Online Customers Decide through Web Personalization. IEEE Intelligent Systems 17(6), 34–43 (2002)

9. Hsief, S.M., Huang, S.J., Hsu, C.C., Chang, H.C.: Personal Documents Recommendation System Based on Data Mining Techniques. In: IEEE/WIC/ACM International Conference on Web Intelligence, pp. 51–57. IEEE Computer Society, Los Alamitos (2004)
10. Hyvonen, E., Saarela, S., Vilijanen, K.: Ontogator: Combining View- and Ontology-Based Search with Semantic Browsing. In: XML Finland 2003, Kuopio (2003)
11. Juszczyszyn, K.: Virtual Communities and the Alignment of Web Ontologies. In: Dasgupta, S. (ed.), pp. 497–499. Idea Group Reference, Hershey (2006)
12. Kazienko, P., Kołodziejski, P.: Personalized Integration of Recommendation Methods for E-commerce. International Journal of Computer Science & Applications 3(3), 12–26 (2006)
13. Kazienko, P., Adamski, M.: AdROSA - Adaptive Personalization of Web Advertising. Information Sciences 177(11), 2269–2295 (2007)
14. Kazienko, P., Musiał, K.: Recommendation Framework for Online Social Networks. In: The 4th Atlantic Web Intelligence Conference, AWIC 2006, Beer-Sheva, Israel, June 5-7. Studies in Computational Intelligence, pp. 111–120. Springer, Heidelberg (2006)
15. Maedche, A., Staab, S.: Measuring Similarity between Ontologies. In: Gómez-Pérez, A., Benjamins, V.R. (eds.) EKAW 2002. LNCS (LNAI), vol. 2473, pp. 251–263. Springer, Heidelberg (2002)
16. Mäkelä, E., Hyvönen, E., Saarela, S.: Ontogator - A Semantic View-Based Search Engine Service for Web Applications. In: Cruz, I., Decker, S., Allemang, D., Preist, C., Schwabe, D., Mika, P., Uschold, M., Aroyo, L.M. (eds.) ISWC 2006. LNCS, vol. 4273, pp. 847–860. Springer, Heidelberg (2006)
17. McDonald, D.W.: Ubiquitous recommendation systems. IEEE Computer 36(10), 111–112 (2003)
18. Middleton, S.E., Alani, H., Shadbolt, N.R., De Roure, D.C.: Exploiting Synergy Between Ontologies and Recommender Systems. In: The 11th International World Wide Web Conference WWW, pp. 41–50 (2002)
19. Middleton, S.E., Shadbolt, N., De Roure, D.: Ontological User Profiling in Recommender Systems. ACM Transactions on Information Systems 22(1), 54–88 (2004)
20. Musiał, K., Kazienko, P., Kajdanowicz, T.: Social Recommendations within the Multimedia Sharing Systems. In: Lytras, M.D., Carroll, J.M., Damiani, E., Tennyson, R.D. (eds.) WSKS 2008. LNCS (LNAI), vol. 5288, pp. 364–372. Springer, Heidelberg (2008)
21. Ning, K., Gong, R., Decker, S., Chen, Y., O'Sullivan, D.A.: Context-Aware Resource Recommendation System for Business Collaboration. In: 9th IEEE International Conference on E-Commerce Technology and the 4th IEEE International Conference on Enterprise Computing, CEC/EEE 2007, July 23-26, pp. 457–460. IEEE Computer Society, Tokyo (2007)
22. Pazzani, M., Billsus, D.: Learning and revising user profiles: The identification of interesting web sites. Machine Learning 27, 313–331 (1997)
23. Perguini, S., Goncalves, M.A., Fox, E.A.: Recommender systems research: A Connection-Centric Survey. Journal of Intelligent Information Systems 23(2), 107–143 (2004)
24. Specia, L., Motta, E.: Integrating Folksonomies with the Semantic Web. In: Franconi, E., Kifer, M., May, W. (eds.) ESWC 2007. LNCS, vol. 4519, pp. 624–639. Springer, Heidelberg (2007)
25. Terveen, L., Hill, W., Amento, B., McDonald, D., Creter, J.: PHOAKS: A system for sharing recommendations. Communications of the ACM 40(3), 59–62 (1997)
26. Weng, L.T., Xu, Y., Li, Y.: A Framework for E-commerce Oriented Recommendation Systems. In: The 2005 International Conference on Active Media Technology AMT 2005, pp. 309–314. IEEE Press, Los Alamitos (2005)
27. Ziegler, C., Lindner, W.: SemanticWeb Recommender Systems. In: Lindner, W., Mesiti, M., Türker, C., Tzitzikas, Y., Vakali, A.I. (eds.) EDBT 2004. LNCS, vol. 3268, pp. 78–89. Springer, Heidelberg (2004)

Toward Emotional E-Commerce

Nicolae Jascanu, Veronica Jascanu, and Severin Bumbaru

Department of Computer Science, University "Dunarea de Jos" of Galati, Romania
{nicolae.jascanu,veronica.jascanu,severin.bumbaru}@ugal.ro

Abstract. We are emotional beings. We feel emotions all the time. They play a central role in our lives. They influence our actions, our decisions, our relations and our work. In the retail area of electronic commerce, emotions play a fundamental role. When we choose or buy something, our choice has a profound emotional dimension. Furthermore, we like to bargain for anything. We communicate our opinions, we are easily influenced by others opinions and we could make impulsive acts. We do all this because of emotions. The emotions are not bad, they just makes us what we are.

1 Introduction

We like to buy things and we like to bargain for them. These two fundamental truths are pushing the development of electronic commerce. We are living in an era when the offer and the demand are huge. In the last 30 years, the landscape of electronic commerce radically changed. There are a vast number of products and services available online and the websites are able to serve millions of people around the globe.

The huge demand and the multitude of choices stimulate the creation of new tools to support the customer-supplier interactions. The Web1.0 era proved that simple commerce websites that just lists products would not resist. What it really matters are the tools that support real life commerce acts, word-of-mouth advertising, bargaining, emotional implication, social shopping etc. The act of commerce is a profoundly social one. The society, culture, civilization or lifestyle shapes our needs and desires. In our society, the always-wired citizen becomes a very powerful factor of influence. From personal websites to forums, blogs and citizenship journalism, we have the tools to express ideas, attitudes, feelings or impressions. Our opinions do matter in this wired society.

The new era of Web2.0 commerce means customer satisfaction, social interaction and opinion proliferation. The oldest and the biggest e-commerce website, Amazon, is a living proof of this change. It is not only about the book you want to buy. That book is the center of a whole universe of recommendations, discussions, opinions and related products. Regarding recommendations, Amazon is one of the best websites around. Another approach comes from eBay, which creates a community of opinions around buyers, sellers and items. Even if the site maintains the same auction principles of selling, it managed to insert in the business model the social community.

A. Håkansson et al. (Eds.): Agent and Multi-agent Technology, SCI 289, pp. 293–321.
springerlink.com

These big players are still conservative. Many websites like Kaboodle, ThisNext, Etsy or StyleHive are leading a new way of doing commerce. These website are communities of shoppers that recommends and discuss products. This is their main profile. When you think at the Amazon, you see the classic e-commerce but when you think at the Kaboodle, you see the shopping community. The concept of collective intelligence is at the base of these shopping communities. James Surowiecki [1] says in his book "under the right circumstances, groups are remarkably intelligent, and are often smarter than the smartest people in them". A community that encourage its members to express the opinions and relies on these for the decision making process could have better results than a bunch of experts. The community recommends the products or services that suit us. Our relations and interactions with the social network shape our profile. The Web2.0 is about social interaction and tools that relieve the collective intelligence [2], [3].

All this philosophies and tools are here for us to understand and to find faster what we need or desire. Even so, the quantity of recommended products or services is huge. What we need is an automated system that further filters these recommendations and negotiates individual prices and offers. We need a multi-agent system that represents equally each entity of the ecosystem. In the case of electronic commerce, the entities are the customer, the supplier and the community.

Always there is a feedback. We buy things and after that, we express our satisfaction about them. Then we buy again or not because of our satisfaction. We could describe the evaluation of the service quality as a cognitive process where customer considers the goodness or badness of the commodity by comparing the service qualities with some predetermined standard [4], [5], [6], [7]. The real problem is the quantification of the satisfaction. It is obvious that the satisfaction contains an affective dimension. The same product is preferred or not by different people. Cognitive and affective responses can be seen as distinct, and having a separate influence on satisfaction formation [8], [9], [10].

Oliver [11], Oliver and Westbrook [12] defined affect as a mediator between cognitive evaluations and satisfaction. In his theory, Oliver argues that it is possible, regarding to a commodity, to experience both negative and positive affects at the same time [13]. We strongly base our theory on these assumptions. We perceive emotionally each characteristic in a different way. The satisfaction comes from the combination of positive and negative emotions of all service attributes. The customer's satisfaction is a whole universe of different opinions, not simply all positive or all negative emotions [14]. There are many aspects and facets of emotion involved in customer-supplier interactions, but ignoring the customer's emotions in the process of acquiring products or services it's a huge mistake for a supplier [15], [16].

In this work, we propose a multi-agent system for electronic commerce that integrates emotional models for each one of the three agents: the customer, the supplier and the community. We are using a continuous approach as a model for emotions based on the seminal work of James Russell. What we consider as a novelty, is the mapping of the emotional model to the electronic commerce field. We translate the consumer and supplier behavior in term of emotional models. More, we give an algorithm able to process negotiation configurations. By including in negotiation an emotional community agent to support the customer and the supplier, this work lays out the foundations for emotional based argumentation.

The remainder of the chapter has four sections. Section 2 is a very short introduction to affective computing. We motivate why we choose to use the circumplex model of affect. Sections 3, 4 and 5 describe the customer, supplier and community agents.

2 Affective Computing

For a long time, emotions were a research area only for philosophers. The seminal work in neuroscience of Damasio [17] proved that the emotions are fundamentals for intelligent behavior or social abilities. The last decade studies advances our understandings on emotions and their roles. Bates [18] introduces the notion of believable agents that mimics the human behavior, creating the illusion of intelligence. However, agents need emotions not only to be believable but also to be truly intelligent. With the introduction of the notion of Affective Computing by Picard [19] the scientific community started to really be interested in emotions.

From the multitude of the emotional models, we choose to use a continuous dimensional model: the circumplex model of affect proposed by Russell [20], [21]. We made this decision because its dimensions are easily interpretable from the electronic commerce and consumer behavior and satisfaction perspective. More, we have found out that the model is very intuitive and easy to use [22].

At the most fundamental level, a circumplex model means that some emotions are similar to each other yet measurably different from other emotions. Structural theories of emotion presume that emotions are not all unrelated and discrete but have certain underlying similarities and differences. By using the circumplex model, a claim is made that two affect dimensions can capture the majority of emotional experience. A circumplex suggests that emotions can be described in a circular fashion in two-dimensional space and that emotions do not simply aggregate together in several groupings. A circumplex implies that a very high or very low value on one dimension has a moderate value on the other dimension. A circumplex model of emotion suggests a clear structure for the effects emotion will have on behavior and thus has large heuristic value [23]. The circumplex model of affect provides a theoretical framework that can potentially advance our understanding of emotional life.

3 The Customer Agent

It is not easy to know what we like or what we want. Considering the vast amount of easy accessible choices offered through internet, this becomes cumbersome. There are so many similar products and services that the selection process is difficult. Even for exactly the same merchandise, we have to decide between suppliers, after-sales services, delivery costs, warranties etc. Moreover, for humans, choosing is far from being an algorithmic process. It depends on personal experiences, taste, education, emotional state, current needs, desires, hopes etc. The choosing process is a trade-off between rational and emotional issues. Let us say, we want to have a vacation in a Caribbean resort.

For a long time we wished for a vacation in Caribbean islands. Even if it is expensive and we have little available time, this year we will take our vacation. Of course, money and time are not the only issues. The resort should be near the beach, very quiet and have positive reviews regarding the existing services. We are willing to pay a little more for our dream, but if the price is too high, we will refuse the offer.

It is hard to translate in words our trade-offs between all of this issues. Maybe we will accept a higher price and a not so quiet resort, but a twenty minutes time to the beach will make the offer inacceptable. What will happen if the time to the beach is ten minutes or five minutes? Considering that our dream was a resort situated on the beach, five minutes are still acceptable but for ten minutes, we should think twice.

It is obvious that the price is a very important characteristic but it is not the only one. It is hard to order the characteristics over importance, because each value has a different importance. If the price is very good, not necessarily small, we could relax the importance of the other characteristics. Considering just one characteristic, we have different degrees of importance between its values. For the time to beach characteristic, zero to three minutes has the same importance, but from five to ten or twenty minutes the importance raises exponentially no matter what the value of price is.

Deciding which value is more important is not easy. When the product or service has more than two or three characteristics, it becomes difficult for a human to explain why it prefers one combination of values to other. Our intuition is that in the process of choosing are involved rational and emotional components, past experiences, intuition, desires and hopes. Our work is based on the following proposition.

Proposition 1. The result of choice process is an emotional state.

Considering that the result is an emotional state, the knowledge acquisition problem is a little bit simplified. We could use several instruments and theories about emotions from psychology in order to capture and interpret that emotional state. We choose to use the circumplex theory, for emotional knowledge acquisition and representation. The circumplex model of affect is a continuous model, so the customer does not need to express the emotional state explicitly by choosing a word or a category. The customer should translate the emotional state in terms of pleasure and arousal, which is far simpler.

Each product or service has its own characteristics. For every characteristic, several key values influence the customer emotional state. After deciding what the characteristics are, the customer will mark on circumplex the key values for each one. We will name from now on these key values as emotional references. For example, the time to beach characteristic could have emotional references at three, six, ten, twenty and twenty-five minutes.

Proposition 2. When the customer marks an emotional reference on the circumplex, it considers that the other characteristics have at least good values.

The second proposition states that the customer should focus only on the value of one characteristic. For example, the customer will mark on circumplex his emotional state when the time to beach is ten minutes considering that the price and number of days are good. It is obvious that, if the time to beach is big, like twenty or twenty-five minutes, the other characteristics should be very good for the customer to accept the offer. We think that this process of focusing just on a single characteristic at each time mimics brain mechanisms when we face a complex decision.

Let us see how such a configuration for an excursion to a Caribbean resort may look like. The customer agent will negotiate over three characteristics: the price, number of days and time to beach. The following figure presents the configuration of the customer agent. For the price characteristic, we will mark six emotional references for the following amounts of money: C=800$, V=900$, B=1000$, D=1500$, E=1700$ and F=1800$. If we could find a price of 800$ or 900$ we will feel very excited because it is a rare opportunity. Of course, the price of 1000$ is very good too and we know how hard is to find such a price. The price of 1500$ is a kind of a normal price for such an excursion, so we are not so happy about it, but it could be worse. A worse price for us is 1700$, and we have negative feelings about it. More, we will feel angry because there is a possibility to accept such a price if there are no other alternatives. Concerning the price of 1800$, we have strong negative feelings but we are not angry because we know that we will never accept such a price. So, 1800$ is the bottom line for the negotiation. As a remark, there is a kind of a linear change in the emotional state from value to value for the price.

For the number of day's characteristic, we mark five emotional references: G=9 days, O=7 days, H=6 days, I=4 days and J=3 days. There are not such big differences between the values, but the emotional state changes radically. Nine days are exceptional but we are satisfied with six days too. Finally, we will accept even three days if the price and time to beach are exceptionally good, but will feel angry about this compromise. For the time to beach characteristic, three minutes or less are extraordinary. From ten minutes or more, our emotional state changes radically. Maybe we will accept twenty-five minutes from beach, but for sure, the other characteristics should be very good.

The circumplex seems to be a good choice and an easy to use instrument if we want to capture our emotional state. Now we will try to interpret such configurations from the commerce point of view.

Between two products or services, it is commonsense to choose the one that makes a greater pleasure or makes you feel better. Between twenty-five minutes and three minutes time to beach, for sure we will choose the last one. The pleasure degree is not the only one that influences our decision. The arousal level has also a strong influence in the final decision.

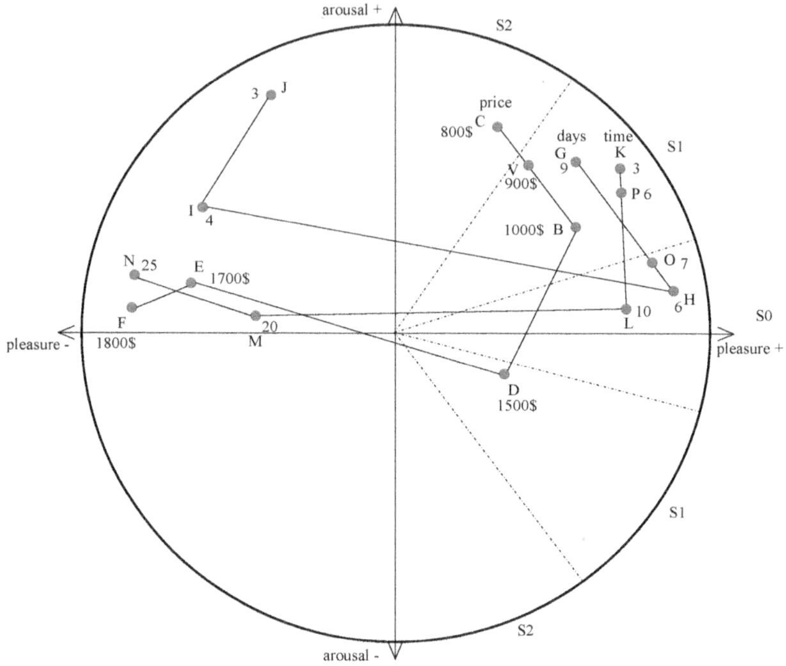

Fig. 1. This is the configuration of the customer agent for a Caribbean resort negotiation. There are three characteristics defined: price, number of days and time to beach. For each one, several emotional references associate a value with a specific emotional state.

Proposition 3. There is an order between emotional references from the commerce point of view.

We should state very clear that the order of the emotional references is not the same as the order of the references values. For example, the value of 500$ for the price characteristic could be perceived as a negative emotional state because it is cheap, but 800$ is extraordinary. The customer agent will place the price of 500$ at the end of the list of emotional references.

Our purpose is to create a method that takes such an emotional configuration and transforms it into an algorithm able to decide between any two configurations of real values. In other words, we need an algorithm that calculates a cost between two successive values of the same characteristic. A first step is to order the emotional references. We call this step the quantitative inference.

The second step consists in finding the cost between the current emotional reference and the next one. The cost depends on many factors. Practically we move from one emotion to other and somehow we must capture this in the algorithm. We call this step the qualitative inference.

3.1 Quantitative Inference

The quantitative inference algorithm will order the emotional references for each characteristic. The following propositions are commonsense if we look at the circumplex from the commerce point of view.

Proposition 4. Between two emotional references having the same arousal value, it will be selected the one with a greater pleasure value.

Proposition 5. Between two emotional references having the same pleasure value, it will be selected the one with a greater arousal value.

Proposition 6. Between two emotional references situated at the same distance from the circumplex center, it will be selected the one with a greater arousal value.

We define these three propositions for the first quadrant. For the other three quadrants, we will define similar propositions.

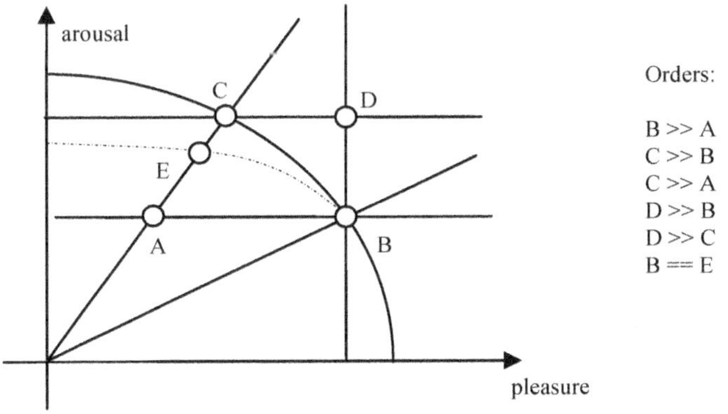

Fig. 2. These are the orderings for the emotional references corresponding to the first quadrant

We will use the symbol >> for preference ordering. From proposition 4, the emotional reference B is preferred over emotional reference A. From proposition 5, D is preferred over B. From proposition 6, C is preferred over B. From C >> B and B >> A, we have C >> A. Between A and C there is an emotional reference E which have the same preference level as the emotional reference B.

It is hard to tell where exactly the emotional reference E is situated. This is the reason why we are not using a rule engine for the quantitative inference. In fact, there is enough for us to know that such a reference exists. At last, we need to have an order between emotional references, so it does not really matter how we compute the order as long as we respect the propositions.

We could use for ordering the z-axis of a three dimensional surface. The corresponding height will order the emotional references. We aim for a surface like the one in the following figure.

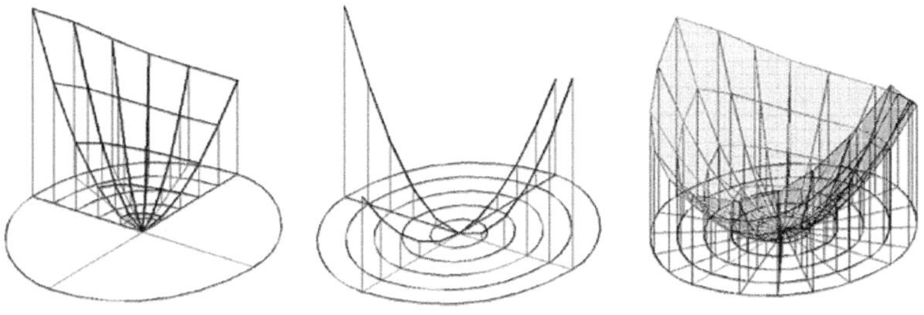

Fig. 3. Graphical representation of the three dimensional surface: first quadrant, main heights and entire surface

We use the following equation for the general surface H.

$$H = \left[\left(\frac{2\beta}{\pi} \right)^b (z_M - z_m) + z_m \right] \frac{r^a}{R^a} . \tag{1}$$

In the equation 1, we have the followings: $z_M > z_m$; $0 \le r \le R$; β is quadrant dependant; a and b are integers like two or three; z_m represents the height at maximum positive or negative pleasure and z_M the height at maximum positive or negative arousal. The chosen values for each parameter are not so important because the role of the surface is to set an order between emotional references.

For the first quadrant we have to replace $\beta = \theta_1; \ 0 \le \theta_1 \le \frac{\pi}{2}$, for the second quadrant $\beta = \pi - \theta_2; \ \frac{\pi}{2} \le \theta_2 \le \pi$, for the third $\beta = \theta_3 - \pi; \ \pi \le \theta_3 \le \frac{3\pi}{2}$ and finally for the fourth quadrant $\beta = 2\pi - \theta_4; \ \frac{3\pi}{2} \le \theta_4 \le 2\pi$.

Now, we could order the emotional references of each characteristic. For the price characteristic, we have the following order: C, V, B, D, E and F. For the number of days the order is G, O, H, I and J. For the time to beach characteristic, the order is K, P, L, M and N. In the following figures, we have a graphical representation of the heights for each emotional reference.

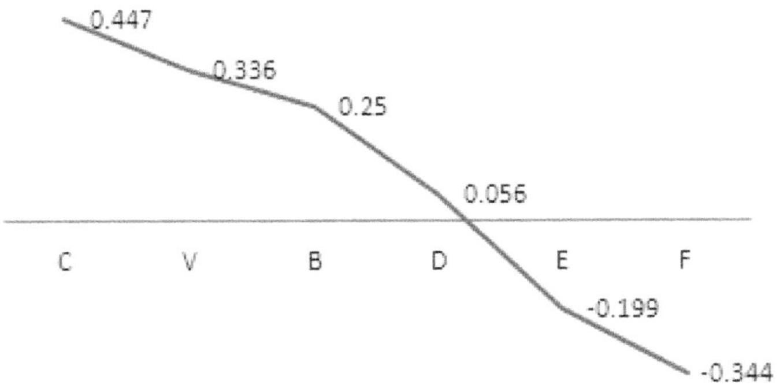

Fig. 4. The ordering and the corresponding heights for the price characteristic

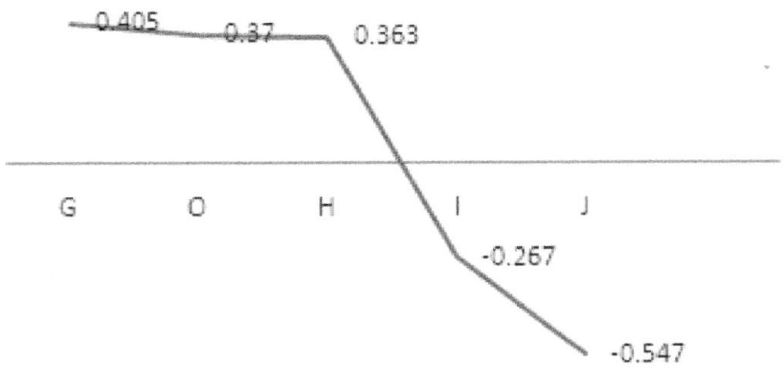

Fig. 5. The ordering and the corresponding heights for the number of days characteristic

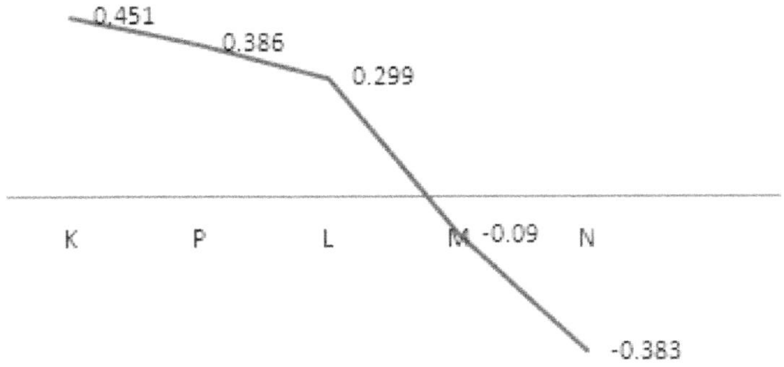

Fig. 6. The ordering and the corresponding heights for the time to beach characteristic

We have to make two observations: it is just a coincidence that the order of emotional characteristics resembles the order of real values and for the negative part of the circumplex, we consider the height of the surface as a negative height.

3.2 Qualitative Inference

We are all different. Every one of us has different needs and desires, motivations, dreams and abilities, way of thinking, social status and culture. Different people emotionally perceive the same value of a characteristic in different ways. For some people a price of 5000$ is a life standard but for others this price is inconceivable for a vacation. The same is true even for the cost of shifting from one emotional reference to other. For some people, shifting from a big plasma TV in the room to an ordinary TV is not such a big deal, but for others this is catastrophic.

The first proposition states that there is a link between circumplex as a model of affect and the consumer behavior. We will use the circumplex as a tool to understand someone's needs and necessities, what it is important or not for him or her and in what degree.

In this work, we give one possible interpretation for the circumplex model of affect from the commerce point of view. It is not so important how simple or complex is the algorithm as long as it offers a value for the cost of shifting between two emotional references. It is more important for the costs to reflect correctly the customer way of thinking.

The most interesting part of the circumplex from the commerce and negotiation point of view is the positive part. Here we have all kind of subtleties that have a direct influence over the course of negotiation. The negative part of the circumplex is less interesting because nobody wants products or services that have a negative emotional impact. While the inference engine tries to capture all sorts of fine details for the positive part, it treats in a sort of coarse-grained style the negative part. In the following figure, we divided the positive part of the circumplex into three symmetric emotional zones: S0, S1 and S2.

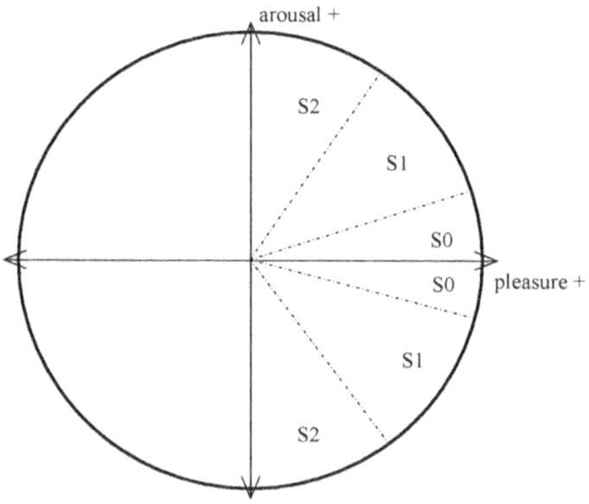

Fig. 7. Emotional zones defined for the positive pleasure part of circumplex: S0, S1 and S2

As a first step, the algorithm identifies the so-called negotiation profile. During a real encounter between a customer and a supplier, the supplier tries to understand what kind of person is the customer, adjusting the offer adequately. Knowing the customer is essential for the customer's business. If the customer is an adventurous teenager, the supplier will insist on the excitements of the offer. On the other hand, if the customer just wants a comfortable and stress-free vacation that offers pleasure and not so much excitement, the supplier will emphasize the facilities of the resort.

The negotiation profile reflects the customer's kind. We have divided the positive circumplex into three emotional zones: S0, S1 and S2. As we can see, these zones are symmetric. The zones represent the impact of emotional state on the shifting costs. From the negotiation and commerce point of view, the emotional states of elated or serene have the same impact on shifting costs. These emotions are very intense and practically we will threat them in the same way. The emotional S0 zone represents the stability, comfort and life quality. The references situated inside this zone makes pleasure but also are important to the customer. For example, the room should have an air conditioning system. It is not vital but it is important and the customer will enjoy more the staying. During negotiation, the inference engine will try to keep those references as long as it is possible. The happiness, the excitement, the relaxation are mapped inside S1 and S2 zones. These emotions are very important for the customer, but not essentials.

Therefore, we have three zones. Which one is more important? From strictly the emotional point of view, we do not know. However, from the commerce point of view the importance of these zones must be different. From these differences, we will understand the customer kind and build the negotiation profile. The profile deeply affects the shifting costs from one emotional reference to other.

We define the negotiation profile only for the positive part of the circumplex. From the distribution of the emotional references in the positive part, we will get the negotiation profile. The negotiation profile has five categories: pleasant, intense, extreme, common and uncommon. We will have a pleasant profile when almost all emotional references are inside the S0 emotional zone. The profile is intense when almost all references are inside the S1 and S2 zones. The profile is extreme if the references are almost exclusively inside the S2 zone. We will have a common profile if the emotional references are inside S0 and S1 zones and finally the profile is uncommon if the references are inside S0 and S2 zones.

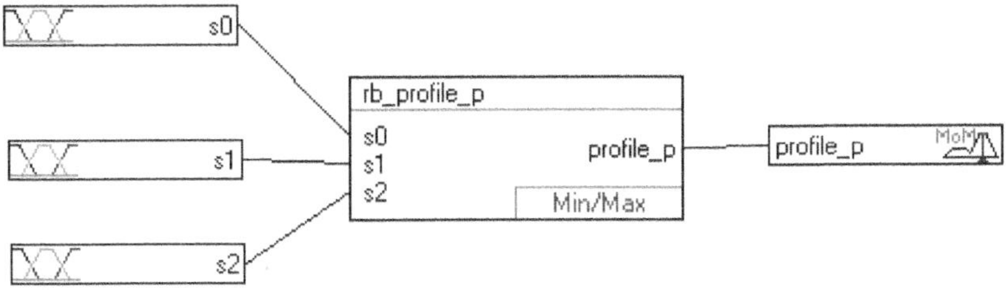

Fig. 8. The fuzzy system used to get the negotiation profile. The input variables represent the participation percent of each emotional zone.

In order to find out what negotiation profile we may have, we define a fuzzy system with three input variables like the one in the above figure. These input variables represent the participation percent of each emotional zone. For the input variables, we have defined a linguistic variable with three terms: small, average and big. The terms represent percents.

Each emotional reference has an amount of influence over the emotional zones. The amount of influence depends directly on the position of the reference inside the zone and the height of the three dimensional surface.

$$I_{re}^{Sx} = \mu^{Sx} H_{re} \quad , x \in \{0,1,2\} \ . \tag{2}$$

If the emotional reference is situated on the angle bisector of the zone, the value of μ^{Sx} is 1 for that zone and 0 for the adjacent zones. If the emotional reference is situated on the boundary between two zones, the value of μ^{Sx} is 0.5 for both zones. In the following table, we have the values of influence for the price, number of days and time to beach characteristics.

Table 1. The values of influence and the surface height for every emotional reference from the positive part of the circumplex

price		µS0	µS1	µS2	H
	C	0	0.3	0.7	0.447
	V	0	0.6	0.4	0.336
	B	0.3	0.7	0	0.25
	D	0.45	0.55	0	0.056
days		µS0	µS1	µS2	H
	G	0	0.8	0.2	0.405
	O	0.51	0.49	0	0.37
	H	0.7	0.3	0	0.363
time		µS0	µS1	µS2	H
	K	0.1	0.9	0	0.451
	P	0.2	0.8	0	0.386
	L	0.8	0.2	0.00	0.299

We use the following formula to find the participation percent of each zone.

$$Sx\% = \sum I_{re}^{Sx} \quad , x \in \{0,1,2\} \ . \tag{3}$$

The values of the input variables for the fuzzy system are the following table.

Table 2. The values of the input variables for the negotiation profile

profile_p		s0	s1	s2	total
		0.9	1.93	0.53	3.363
	%	26.9	57.4	15.71	100

In our case, the negotiation profile is common which means the most of emotional references are inside S0 and S1 zones. The fuzzy system uses the following set of rules to find the profile.

Table 3. The set of rules for the negotiation profile

IF			THEN
S0	S1	S2	profile_p
small	small	average	intense
small	small	big	extreme
small	average	small	intense
small	average	average	intense
small	average	big	extreme
small	big	small	intense
small	big	average	intense
average	small	small	pleasant
average	small	average	uncommon
average	small	big	uncommon
average	average	small	common
average	average	average	common
average	big	small	common
big	small	small	pleasant
big	small	average	uncommon
big	average	small	common

Now that we have the algorithm for the negotiation profile, we could move on. In order to find the cost of shifting from one emotional reference to other, we still need to define the following parameters: segment profile, segment angle, segment distance, one-to-fourth quadrant traversal and positive-to-negative traversal.

We define the segment as the distance between two consecutive emotional references. The segment profile is a very important parameter. The edges of the segment could be situated in the same zone or be in different emotional zones. Following the same ideas from the negotiation profile, the segment profile could be pleasant, intense, extreme, common or uncommon. In this case, the algorithm treats a particular case when we have only two emotional references, the edges of the segment. The segment profile is important because it offers a local view while the negotiation profile offers a global one. It is as the customer is the adventurous kind and the segment's profile is pleasant. How important is the segment in this case? What will happen if the segment's profile is also extreme? It is the segment more important in the last situation or not?

The segment angle is the angle between segment and the horizontal axis. When we shift from one emotional reference to other vertically, there is an arousal impact. If the shift is horizontal, there is a pleasure impact. Therefore, the angle is important and we should treat it differently. For example, if the negotiation profile and segment profile are both pleasant we talk about values affecting stability, comfort and life quality. In this case, the pleasure is more important than arousal so a horizontal segment or a segment with a small angle has a greater impact than a vertical or a big angle.

The segment distance is also important. Usually, a short segment has a small shifting cost while a long segment has a bigger shifting cost. The fuzzy linguistic variable distance has five terms: very small, small, average, big and very big.

The one-to-fourth quadrant traversal means a change in the arousal type. A serene state does not mean something bad. For us, it has the same importance as the excited state. Maybe you are looking just for a serene state, so the negotiation should insist on this. The traversal is not something like shifting from good to bad but from good to other kind of good. The traversal parameter is not very important. It just emphasizes the shifting cost.

The positive-to-negative traversal means a transition from good values to undesirable ones. The traversal also emphasizes the shifting cost and has a deeper impact than the one-to-fourth quadrant traversal.

All the above parameters are input variables in the fuzzy system built to calculate the shifting cost of a segment for the positive part of the circumplex. The system has five rule blocks with a total of 91 rules. The system is in its early alpha stages but it seems that the shifting costs mimic the customer's behavior during negotiation. We work now to validate systematically the results of the system.

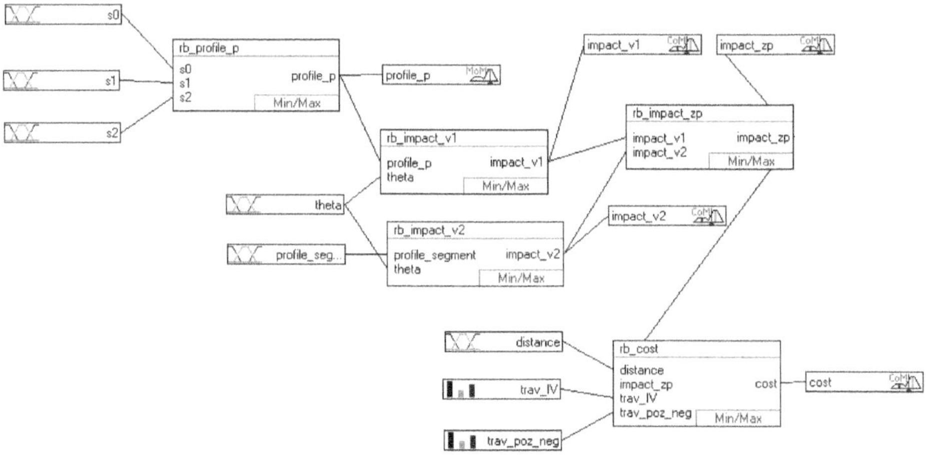

Fig. 9. Fuzzy system for positive qualitative analysis

We calculate the segment shifting cost for the negative part of the circumplex with a simpler fuzzy system. Because we are in the negative part, the values are anyway undesirable. The best strategy for the negotiation engine is to avoid as long as possible the negative part.

The input variables are segment distance, segment angle and two-to-three quadrant traversal. All these parameters mean the same things as the ones from the positive part.

The rule block has only a few rules. We list the rules in the following table.

Table 4. The rule set for the negative shifting cost

IF			THEN
distance	theta	trav_IIIII	cost
very small			very small
small	small		very small
small	average		small
small	big		average
average	small		small
average	average	false	average
average	average	true	big
average	big	false	big
average	big	true	very big
big			big
very big			very big

Table 5. The shifting costs for every segment of each characteristic

price		cost		cost
	CV	0.3177	EF	0.2369
	VB	0.3884		
	CB	0.5212		
	BD	0.8177		
	DE	1		
days		cost		cost
	GO	0.4409	IJ	0.6208
	OH	0.2891		
	GH	0.5496		
	HI	1		
time		cost		cost
	KP	0.2108	MN	0.4072
	PL	0.4105		
	KL	0.5		
	LM	1		

In the above table, we have the costs for each segment.

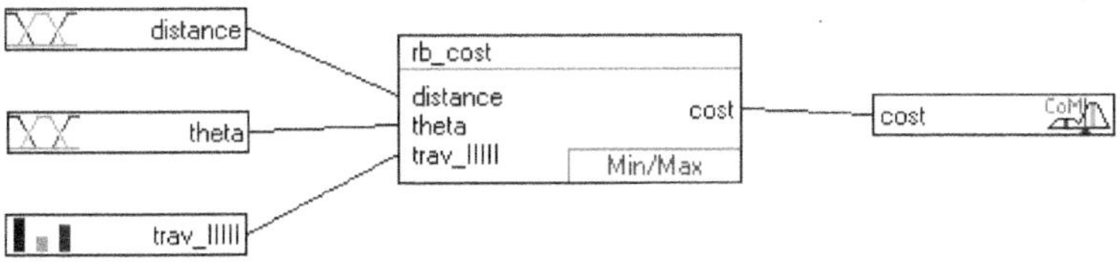

Fig. 10. Fuzzy system for negative qualitative analysis

The above figure represents the simplified fuzzy system used to find the shifting costs for the negative part of the circumplex.

From the shifting costs table we remark that $CV + VB \neq CB$, $GO + OH \neq GH$ and $KP + PL \neq KL$. These situations are normal because each segment intersects different emotional zones and has different dimensions. In the following figure are the shifting costs.

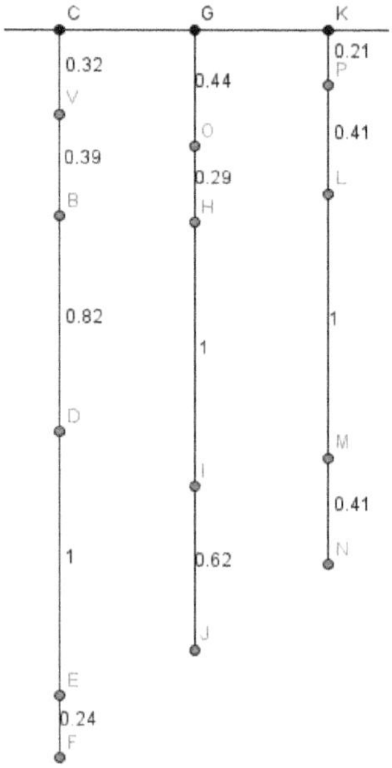

Fig. 11. The characteristics, segments and the associated costs

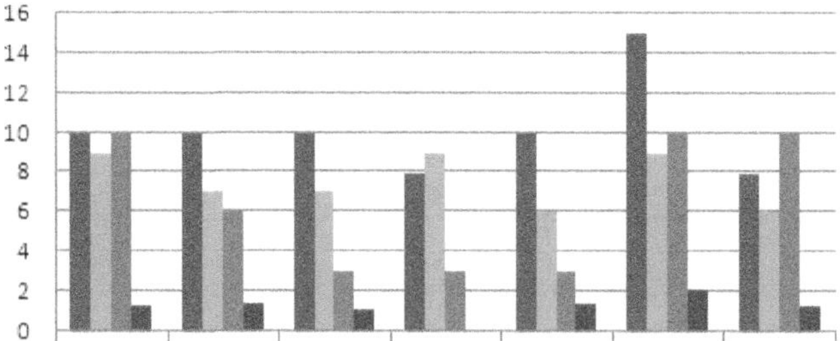

	BGL	BOP	BOK	CGK	BHK	DGL	CHL
pricex100	10	10	10	8	10	15	8
days	9	7	7	9	6	9	6
time	10	6	3	3	3	10	10
cost	1.33	1.36	1.15	0	1.44	2.15	1.35

Fig. 12. Several configurations with the associated costs

The best combination of emotional references is C, G and K, which have a zero total cost, and the worst is F, J and N with a total cost of 7.15. In the above figure, we present several possible combinations and associated cost of each one. Let us take for example BGL = 1000$, 9 days, 10 min and DGL = 1500$, 9 days, 10 min. It is obvious that the BGL combination is preferred because of the smallest price. Between BGL = 1000$, 9 days, 10 min and BHK = 1000$, 6 days, 3 min we prefer BGL because 6 days are a too short vacation at a price of 1000$. The 3 minutes time to beach is very appealing, but a 6 days vacation is too short for the price of 1000$. Interesting comparisons are between BGL, BOP and CHL. It is very difficult even for a human to trade off between so close values for price, days or time.

As an observation, if someone does not agree with the order of combinations, we must remember that this is the emotional configuration. The best person to analyze the results is the person who configured the circumplex. If the person is satisfied with the inference results than the model is valid.

3.3 Configuration Formalism

From a customer point of view, configuring the circumplex for negotiation should be a trivial task. Using a graphical interface, the customer defines the negotiation characteristics and the emotional references of each one. The real problem is how we formalize all this for the inference engine to work. In this section, we define such formalism. This is a work in progress but for now, it is suitable for many situations.

Proposition 7. A negotiation configuration is a 3-tuple

$$Cfg^{neg} = \left(ctg(Q) \wedge keys(\overline{K_1 K_i}), meta, chr_j \right) . \tag{4}$$

where:

1. $ctg(Q) \wedge keys(\overline{K_1 K_i})$ represents the category and the keywords that define the negotiation. In this way, both the customer and the supplier agents know about the subject of the negotiation. For our example with the Caribbean resort, the category could be */travel/hotels & resorts/* with the keywords *vacation, beach, Caribbean, islands.* Using this mechanism, the customer agent could search for and find supplier agents able to negotiate on the desired subject.
2. *meta* is intended to be used for future enhancements like the user profile, temporal restrictions etc. In this work, we use the value of *null* for this parameter.
3. chr_j represents the negotiation characteristics and is a set with the following parameters:

$$chr_j = \left(name, domain, unit, ref_1, ..., ref_n \right) . \tag{5}$$

where:

a. *name* is the characteristic's name like the ones from our example: price, number of days or time to beach.

b. *domain* represents the domain of values for each characteristic. The domain could be continuous or discrete. We define here only three domains: *number, text* and *date-time*. The *number* domain is a continuous one and represents ascending ordered numbers from left to right. The *text* domain is a discrete one. For example, for the color characteristic we could have the values of red, green, or blue. The *date-time* is also a continuous domain ordered from left to right. There is the possibility to define own domain of values. For example, if the characteristic is the LCD's diagonal size, we could define a domain only for several specific sizes. If the domain is ordered, the order is ascending from left to right and we will use right parentheses: [80, 94, 102, 106, 132]. If the domain is not ordered, we will use accolades: {94, 80, 102}.

c. *unit* represents the measurement unit. For the price characteristic we could have the *$* unit, for the number of days it will be *day* and for time to beach will be *min*. As future research, it should be defined a universal repository of measurements units to match the customer agent units with the ones from the supplier agent.

d. ref_k represents the emotional reference and it has the following form:

$$ref_k = (x, y, [\![...]\!]) \ . \tag{6}$$

where x and y are the circumplex coordinates with values between [-1,1]. The $[\![...]\!]$ represents the set of emotional reference real values. For example, when we negotiate a vacation, the *departure date* characteristic has a *delighted* emotional reference with real values for the following dates: between 9 and 14, on 22 or 24 and not on 28 or 30. The reference set of values is ordered from left to right no matter from many intervals is composed.

Proposition 8. We will use the following symbols to define the set of emotional references real values:

1. $\{\}$ the reference has any values,
2. $\{z\}$ the reference's value is z,
3. $\{z...t\}$ the reference's values are between z and t inclusive, where $z < t$,
4. $\{z\} \mapsto$ the reference's values are greater than or equal to z,
5. $\leftarrow\!| \{z\}$ the reference's values are smaller than or equal to z,
6. $||$ the reference has no value,
7. $| z |$ the reference has any values but not z,
8. $| z...t |$ the reference's values are not between z and t inclusive, $z < t$
9. $z \mapsto$ the reference's values are strictly greater than z,
10. $\leftarrow\!| z$ the reference's values are strictly smaller than z,
11. $\neg [\![...]\!]$ the complementary values of the set $[\![\ ]\!]$

Using the defined symbols, we have the following equivalencies:

$$\begin{aligned}
&\neg[\![\{\}]\!] \Rightarrow [\![||]\!] \\
&\neg[\![\{z\}]\!] \Rightarrow [\![| z |]\!] \\
&\neg[\![\{z...t\}]\!] \Rightarrow [\![| z...t |]\!] \\
&\neg[\![\{z\} \mapsto]\!] \Rightarrow [\![\leftarrow\!| z]\!] \\
&\neg[\![\leftarrow\!| \{z\}]\!] \Rightarrow [\![z \mapsto]\!] \\
&\neg[\![||]\!] \Rightarrow [\![\{\}]\!] \\
&\neg[\![| z |]\!] \Rightarrow [\![\{z\}]\!] \\
&\neg[\![| z...t |]\!] \Rightarrow [\![\{z...t\}]\!] \\
&\neg[\![z \mapsto]\!] \Rightarrow [\![\leftarrow\!| \{z\}]\!] \\
&\neg[\![\leftarrow\!| z]\!] \Rightarrow [\![\{z\} \mapsto]\!]
\end{aligned} \tag{7}$$

Now we are able to specify the real values of the emotional characteristics. For example, if the values are between 300 and 800 we will have the following equivalencies:

$$[300...800] \Leftrightarrow [\![\{300\} \mapsto, \leftarrow\!| \{800\}]\!] \Leftrightarrow [\![\{300...800\}]\!]$$

For the *departure date* characteristic, the *delighted* emotional reference will look like this:

$$(\textit{departure date, date-time,,}$$
$$(0.5, 0.5, [\![\{9\} \mapsto, \leftarrow\!| \{14\}, \{22\}, \{24\}, | 28 |, | 30 |]\!]))$$

The complementary values of the *delighted* emotional reference will look like this:

$$(\textit{departure date, date-time,,}$$
$$(0.5, 0.5, [\![\leftarrow\!| 9, 14 \mapsto, | 22 |, | 24 |, \{28\}, \{30\}]\!]))$$

We will give another example. If we wish to buy an air conditioning unit of 9000 or 12000 BTU, we could specify this by defining a domain and let the inference engine to negotiate for these the values.

$$(\textit{air conditioning,[9000,12000,18000,24000],BTU,(0,0,}[\![\{9000\}, \{12000\}]\!]))$$

Proposition 9. The negotiation configuration may contain the following three types of characteristics.

1. *not-negotiable* - the not-negotiable characteristics have only one reference. The reference's values are conditions for negotiation to take place. If the supplier does not fulfill all the not-negotiable characteristics, the negotiation does not take place. For example, we need an air conditioning unit in our room. If the offer does not specify such a device, the offer is simple rejected.

We will represent the not-negotiable references in the middle of the circumplex like a neutral emotional state.

2. *desirable* – these kind of characteristics have only one emotional reference. The values of the reference are desirable but not necessary. For example, the customer will be happy if there is air conditioning but this is not necessary. Because we speak about a preference and not an obligation, it means that we accept also the case when the reference's values are not satisfied. It is like the characteristic have two references: the one that we defined and an opposite one. Will we lose much if the reference is not satisfied? The inference engine generates a new reference in the middle of the circumplex. The values of the generated reference are complementary to those values of the defined emotional reference. Between these two references, we have a segment so we could find the shifting cost.

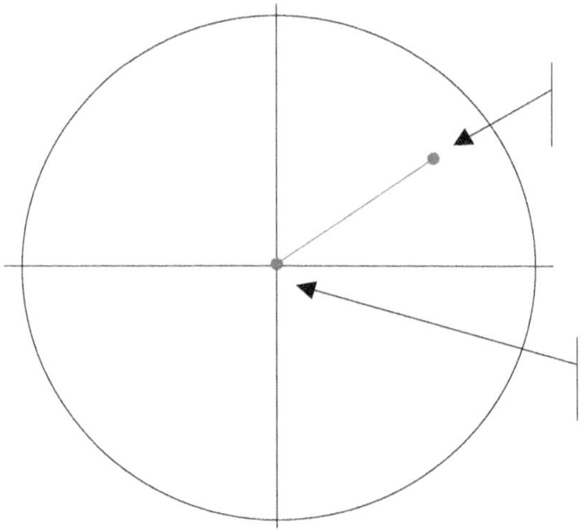

Fig. 13. The desirable characteristic for the existence of air conditioning unit

The configuration of the inference engine for this characteristic looks like this:

$$(\mathit{air\ conditioning},[9000,12000,18000,24000],\mathit{BTU},$$
$$A{:}(0.48,0.57,[\![\{\,\}]\!]),B:(0.0,0.0,[\![\,||\,]\!]))$$

3. *negotiable* – are those characteristics that have at least two emotional references. From our example, the price, numbers of days and time to beach characteristics are all negotiable.

The negotiation configuration is the starting point for the effective negotiation with the supplier agent. The emotional references are only major milestones for the negotiation. It is obvious that the supplier's offer has more values than the customer configuration has. For the nonexistent values, the customer's inference engine will try to interpolate new emotional references.

Proposition 10. The characteristic's emotional references may use a value only once.

In other words, a value has only one position in circumplex for the same characteristic. If this is not true, we could have an odd situation like having for the price of 1000$ an elated emotional state and an angry one. At this time, we do not treat these situations.

In order to accept interpolated values, the emotional references from the segment edges should have values that not overlap. We use one of the following two symbols to indicate that the segment accepts interpolated values: $\leftarrow\!\mid$ for smaller values or $\mid\!\rightarrow$ for greater values. In the following figure, we have two consecutive emotional references. The A reference has values between 800 and 900 and the B reference between 300 and 500.

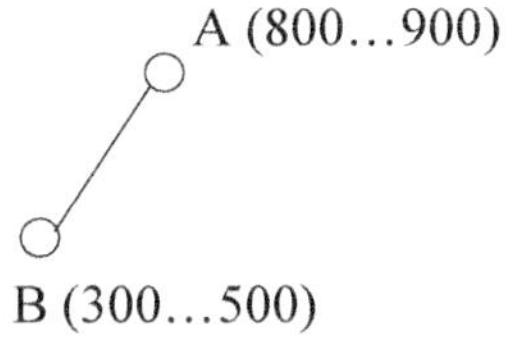

Fig. 14. A segment with the A-B order for the emotional references

The configuration for the segment looks like this:

$$A : [\![\leftarrow\!\mid \{800...900\}]\!]$$
$$B : [\![\{300...500\} \mid\!\rightarrow]\!]$$

The position of the interpolated reference with value of 700 is on the segment, between A and B. Using simple proportion the exact position is calculated. Of course, the position depends on the nature of the domain, but we will not explain all the combinations here.

Let us have a more complicated situation.

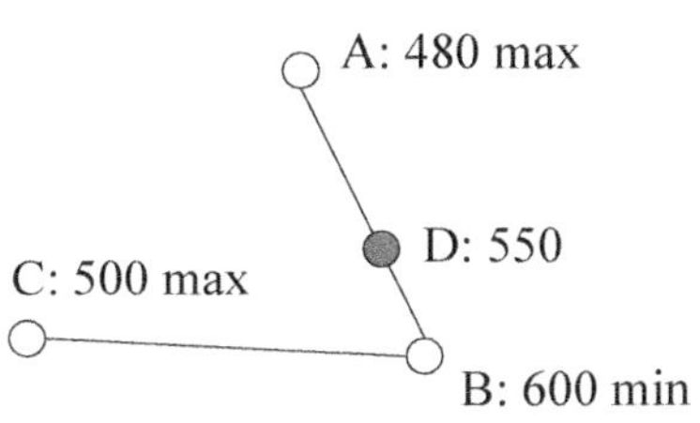

Fig. 15. Three ordered emotional references A-B-C

The configuration looks like this:

$$A : [\![\leftarrow\!| \{460...480\} |\!\rightarrow]\!]$$
$$D : [\![\leftarrow\!| \{550\} |\!\rightarrow]\!]$$
$$B : [\![\leftarrow\!| \{600...700\} |\!\rightarrow]\!]$$
$$C : [\![\leftarrow\!| \{300...350\}, \{400\}, | 420 |, \{500\} |\!\rightarrow]\!]$$

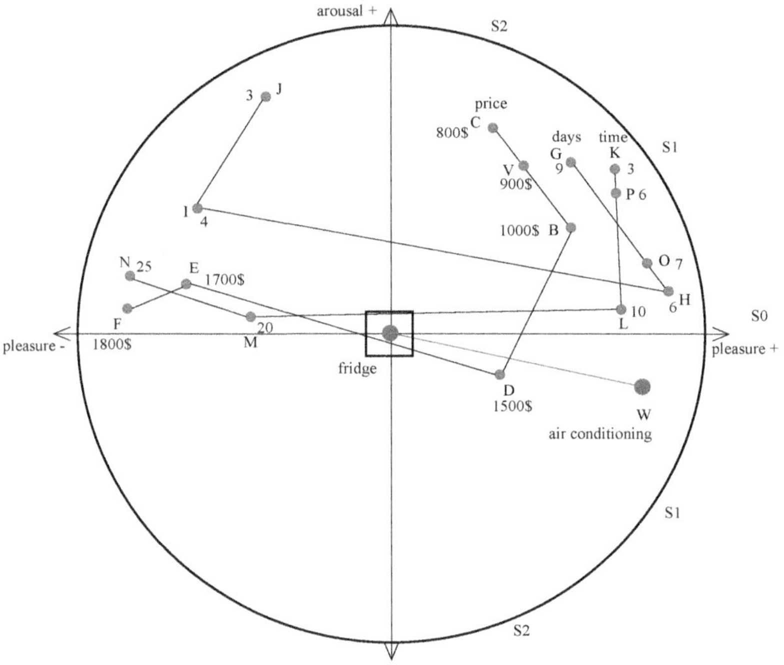

Fig. 16. A more complete negotiation configuration for the vacation to Caribbean islands

We must observe that the order of references does not depend on the real values of each one. The emotional reference D is an interpolated reference and has the value of 550. The inference engine will place the reference D on the first segment that respects the conditions. In our case, the segment is A-B. As an observation, the possible values of the segment B-C are already between the values of the segment A-B. For this reason, between B and C there will not be any interpolated references. Therefore, the values of 501 or 502 will be between A and B references. It is strange, but accordingly with proposition 10 it is correct. In addition, we must not forget that we work with emotional references. It is perfectly fine for a value to be in one place and the next value in another place. Because the customer did not explicitly configured the position for all possible emotional references, the inference engines tries to find the most believable position.

Now we will consider a more complete example to see if we could use the formalism in real scenarios.

We configure the customer agent to negotiate over a vacation to a Caribbean resort. The configuration will have the following three characteristics:

a. the not-negotiable characteristic *fridge*,
b. the desirable characteristic *air conditioning*,
c. the negotiable characteristics of *price*, *number of days* and *time to beach*.

The rectangle from the center of the circumplex represents the fridge characteristic. For the air conditioning system, the inference engine creates a new reference in the center of the circumplex.

Using the formalism from this chapter, the negotiation configurations for the vacation will be the following one.

The configuration of the negotiation

```
(/travel/hotels & resorts/ ∧ vacation beach Caribbean
islands, null,
      (fridge,,,(0,0,⟦{}⟧)),
      (air conditioning,[9000,12000,18000,24000],BTU,
            (0.91,-0.15,⟦{}⟧),
            (0,0,⟦||⟧)),
      (price, number,$,(0.36,0.68,⟦←| {800} |→⟧),
            (0.45,0.54,⟦←| {850} |→⟧),
            (0.61,0.32,⟦←| {1000} |→⟧),
            (0.36,-0.14,⟦←| {1500} |→⟧),
            (-0.59,0.14,⟦←| {1700} |→⟧),
            (-0.84,0.06,⟦←| {1800} |→⟧)),
      (number of days, number,,(0.61,0.54,⟦←| {9} |→⟧),
            (0.81,0.25,⟦←| {7} |→⟧),
            (0.87,0.15,⟦←| {6} |→⟧),
            (-0.57,0.35,⟦←| {4} |→⟧),
            (-0.41,0.74,⟦←| {3} |→⟧)),
      (time to beach, number, min,
            (0.76,0.48,⟦←| {3} |→⟧),
            (0.77,0.39,⟦←| {6} |→⟧),
            (0.79,0.09,⟦←| {10} |→⟧),
            (-0.42,0.05,⟦←| {20} |→⟧),
            (-0.82,0.23,⟦←| {25} |→⟧))
)
```

4 The Supplier Agent

It is not easy to convince a supplier to change its infrastructure and to adopt a new system. That is why the supplier agent should be no more than a lightweight service. We could imagine the agent as an unstructured table. From the supplier infrastructure, we extract and filter the information for the agent.

Each record of the table represents a negotiation configuration. Each configuration has several characteristics. The characteristic of selling priority is one necessary characteristic for each configuration. We choose to use the term priority over profit because there are situations when the profit is not easy to be calculated or makes no sense. By using priorities the systems becomes very flexible. Depending on the sales, we could adjust the priority for a product or services.

Proposition 11. Apart from the priority characteristic, the negotiation configuration could have other three types of characteristics:

1. *simple* – these are characteristics with values. The definition of the domain and the formalism used are identical with the one from the customer agent.
2. *non-negotiable* – this are mandatory characteristics for a customer. The configuration of the customer agent must have this characteristic. The values are not important, but the characteristic must exist. The supplier also uses the non-negotiable characteristics to force age, behavior or any other kind of restriction it may want. The customer should respect all conditions before the negotiation process starts.
3. *negotiable* – let us consider the following scenario: during high season, the supplier rents its resort only for groups of ten to fifteen persons at a fixed price. At the end of the season, it accepts also smaller groups of five to ten persons at a smaller fixed price. During off-season, the supplier is open to negotiations and accepts even groups of two or three persons. The inference engine using the circumplex model of affect could implement such a situation in a similar way as the one from the customer agent. The supplier defines what is its emotional states regarding, for example, price and number of persons and the inference engine will negotiate using also this emotional characteristics.

Proposition 12. A negotiation configuration is a 2-tuple

$$Cfg^{neg} = \left(ctg(Q) \wedge keys(\overline{K_1 K_i}), chr_j \right) . \quad (8)$$

where:

1. $ctg(Q) \wedge keys(\overline{K_1 K_i})$ represents the category and the keywords that define the negotiation. We use the category and the keywords in the same way as the ones from the customer agent.

2. chr_j represents the negotiation characteristics and is a set with the following parameters:

$$chr_j = op\left(name, domain, unit, ref_1, ..., ref_n\right) . \quad (9)$$

where:

a. *op* is an symbol that identifies the type of characteristic. We use the symbol * for non-negotiable characteristics and the symbol % for the negotiable ones.
b. *name, domain, unit, refs* parameters have the significance as the ones of the customer agent. We have a priority characteristic if we use the symbol ! instead of the *name* parameter. A higher value means a higher selling priority.

For a high season, a supplier of Caribbean island vacations could have a negotiation configuration with number of persons, price per person, number of days, smoker, time to beach, air conditioning unit and fridge characteristics.

A common negotiation configuration for a supplier

```
(/travel/hotels & resorts/ ∧ vacation beach Caribbean
islands,
     ((
        (!,number,,(0,0,⟦{10}⟧)),
        *(no. persons, number,,(0,0,⟦15 ↦⟧)),
        *(price, number,,(0,0,⟦{280}⟧)),
        *(no. days, number,,(0,0,⟦{5...9}⟧)),
         (time to beach, number, min,(0,0,⟦{4...6}⟧)),
        *(smoker,,,(0,0,⟦||⟧)),
         (air conditioning,,,(0,0,⟦{}⟧)),
         (fridge,,,(0,0,⟦||⟧))))
```

For the off-season period, the supplier will define in addition an emotional configuration for number of persons and price per person characteristics.

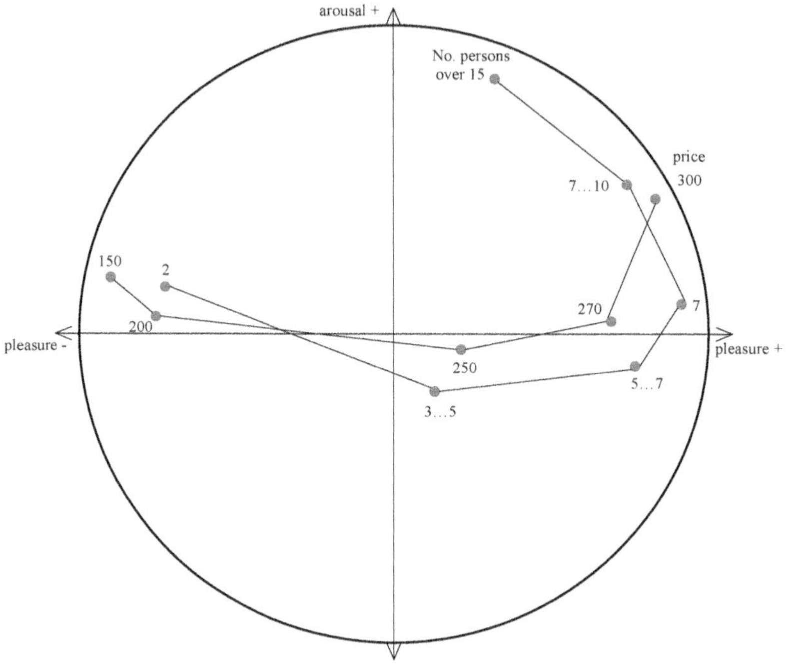

Fig. 17. The supplier could insert into the negotiation configuration such emotional configurations

The configuration of the inference engine will be update with the following entry:

```
(!,number,,(0,0,⟦{15}⟧)),
%(no. persons, number,,
        (0.4,0.8,⟦←| {15} |→⟧),
        (0.75,0.44,⟦←| 7 |→,{10} |→⟧),
        (0.93,0.09,⟦←| {7} |→⟧),
        (0.78,-0.11,⟦←| {5} |→,7 |→⟧),
        (0.18,-0.14,⟦←| {3} |→,5 |→⟧),
        (-0.7,0.18,⟦{2} |→⟧)),
%(price per person, number,$,
        (0.87,0.42,⟦←| {300} |→⟧),
        (0.74,0.03,⟦←| {270} |→⟧),
        (0.36,-0.05,⟦←| {250} |→⟧),
        (-0.72,0.03,⟦←| {200} |→⟧),
        (-0.91,0.1,⟦←| {150} |→⟧)),
```

```
*(no. days, number,,(0,0,⟦{3}↦⟧)),
 (distance, number, min,(0,0,⟦{4...6}⟧)),
*(smoker,,,(0,0,⟦||⟧)),
 (air conditioning,,,(0,0,⟦{}⟧)),
 (fridge,,,(0,0,⟦||⟧))
```

5 The Community Agent

Each one of us, as a buyer, tries to find as much information as it is possible about the desired product or service. We look at the supplier's ads, we search the net, the forums, the blogs and we ask our acquaintances for supplementary information. No matter how we search for information, we need it in order to take informed decisions.

Let us say that we look for supplementary information regarding our vacation to Caribbean islands. Many travel websites lists the opinions of the tourists about their visited places. The opinion is stored as a textual description along with a rate. The textual description is the most helpful source of information for us. Here, the tourists describe at their best the experiences. They tell what they liked or not, what was funny, interesting or boring. They reveal how they were treated, what was the quality of services, what was bad or what was wrong. On the other side, the rating system is very inflexible. There are several very general fields and the tourist tries to convert his or her experience to a five star grade. Then, the system aggregates all this ratings showing the final rates for each field. The idea of showing an overview is very good but the rating system cannot capture the subtleties of the felt experience. On the other hand, reading hundreds of textual descriptions is very confusing.

The community agent is a repository of emotional opinions. Instead of writing a textual description with your feeling about a place or event, you could place an emotional reference on the circumplex. We describe the emotional references using a set of keywords. Therefore, the keywords are associated with a feeling. Imagine that you are able to express what you felt at that moment or soon after. After a vacation, you easily expressed tenths of emotional references. All this references could easily replace the textual description. Moreover, the references are already rated in a very intuitive and flexible system. The system is able to aggregate all the information even there are many opinions. An overview for a vacation to the Caribbean islands will show what is exciting, boring or relaxing.

We have developed on the Android mobile platform, an application able to capture emotional references. We display the search results in a condensed and intuitive format. As a parallel study, we have integrated the application with the Twitter service. In this way, our followers know about our feelings in real time.

The community agent is a very valuable source of information for both the customer and supplier agents. The strongest feature of the community agent is the process able format of the information. During the negotiation process, each agent consults the community agent. The customer agent seeks out for possible problems regarding the offer while the supplier agent understands better how its offer is perceived. We have the basis for a very powerful argumentation framework for the negotiation.

The community agent could assist the customer during configuration or the marketing tools of the supplier could use it to understand their products or services.

6 Conclusions

In this work, we have introduced the main actors of the multi-agent system. There are many more things to do from now on. We wish to validate extensively the model and further develop the capabilities of each one of the agents. Further research will study various negotiation protocols. Our wish is to develop an argumentation framework that deals with the defined emotional aspects. Our intuition is that this is a natural evolution step for our system in order to offer a more human-like experience for the retail area of electronic commerce.

References

1. Surowiecki, J.: The Wisdom of Crowds (2005)
2. O'Reilly, T.: What Is Web 2.0. In: Design Patterns and Business Models for the Next Generation of Software (2005)
3. O'Reilly, T.: Web 2.0: Compact Definition (2005)
4. Militaru, D.: Consumer behavior in Electronic Commerce Environments and Fashion Effect. In: ICEC, Minneapolis, USA (2007)
5. Oliver, R.L.: Satisfaction: A behavioral perspective on the consumer. McGraw-Hill, New York (1997)
6. Yi, Y.: A critical review of consumer satisfaction. In: Zeithaml, V.A. (ed.) Review of marketing, Chicago, pp. 68–123 (1990)
7. Karat, J.: Beyond task completion: evaluation of affective components of use. In: Handbook of human-computer interaction, vol. 59, pp. 1152–1164. Springer, Heidelberg (2002)
8. Desmet, P.M.A., Tax, S.J.E.T., Overbreeke, C.J.: Designing products with added emotional value: development and application of an approach for research through design. The Design Journal 4, 32–47 (2000)
9. Desmet, P.M.A., Hekkert, P.: The basis of product emotions. In: Pleasure with products: beyond usability, London (2002)
10. Schwartz, N.: Feelings as information: Informational and motivational functions of affective states. In: Handbook of motivation and cognition: Foundations of social behavior, pp. 527–561. Guilford Press, New York (1990)
11. Oliver, R.L.: Cognitive, affective, and attribute bases of the satisfaction response. Journal of Consumer Research 20 (1993)
12. Oliver, R.L., Westbrook, R.A.: Profiles of consumer emotions and satisfaction in ownership and usage. Journal of Consumer Satisfaction, Dissatisfaction and Complaining Behavior 6, 12–27 (1993)
13. Liljander, V., Strandvik, T.: Emotions in service satisfaction. International Journal of Service Industry Management 8, 148–169 (1997)
14. Stauss, B.: The qualitative satisfaction model. In: Advancing Service Quality: A Global Perspective, New York, pp. 35–44 (1996)
15. Pullman, M., Gross, M.: Ability to experience design elements to elicit emotions and loyalty behaviors. Decision Sciences 35, 551–578 (2004)

16. Vilnai, I., Rafaeli, A.: Aesthetics and professionalism of virtual servicescapes. Journal of Service Research 8, 245—259 (2006)
17. Damasio, A.R.: Descartes' Error: Emotion, Reason and the Human Brain. Putnam Pub. Group (1994)
18. Bates, J.: The role of emotion in believable agents. Communications of the ACM 7, 122–125 (1994)
19. Picard, R.W.: Affective Computing. MIT Press, Cambridge (1997)
20. Russell, J.A.: A circumplex model of affect. Journal of Personality and Social Psychology 39, 1161–1178 (1980)
21. Russell, J.A.: Core affect and the psychological construction of emotion. Psychological Review 110, 145–172 (2003)
22. Sundström, P., Ståhl, A., Höök, K.: In Situ Informants Exploring an Emotional Mobile Meassaging System in Their Everyday Practice. In: IJHCS on Evaluating Affective Interfaces, vol. 65, pp. 388–403 (2007)
23. Russell, J.A., Fehr, B.: Relativity in the perception of emotion in facial expression. Journal of Experimental Psychology 116, 223–237 (1987)

Web-Based Affective Human-Agent Interaction Generation

Xia Mao and Zheng Li

School of Electronic and Information Engineering
Beihang University, Beijing, China, 100191
{moukyoucn,buaa_david}@yahoo.com.cn

Abstract. Virtual agents, as a promising technology for human-computer interaction, have become focus of research community in resent years. They serve as communicative fellows in a variety of applications. Employing virtual agents to realize human-computer communication on the web is a promising way to make the interaction attractive. In order to make use of intelligent interaction in the web by virtual agents, an important issue is that we should have a scripting language, which is easy to be used by authors. In this chapter, we discuss our research on the Multimodal Interaction Markup Language (MIML), which is a powerful and easy-to-use XML-based language. Different from the related languages in existence, MIML can script not only the presentations of virtual agents, but also their affective capability. We will describe the architecture of MIML, the facial expression recognition, speech emotion recognition, emotional speech synthesis ActiveX controllers and illustrate one scenario that instantiates the affective web-based human-agent interaction scripted by MIML. With the MIML we designed, web-based affective interaction can be described and generated easily.

1 Introduction

The Human-Computer Interaction has moved ahead from Character-order User Interface (CUI) and Graphical User Interface (GUI) towards Perception User Interface (PUI). The multimodal interface as a new domain of PUI is attracting more and more attention of research community. In multimodal interface, people can communicate with computer through natural language, gestures (pointing at something), eye gaze (looking down or up to a particular object), body posture (leaning back or forward) and facial expression (smile, cry or nose wrinkling) etc. However, computer does not have face and body and can not interact with people through the methods mentioned above. As a result, virtual agents who can provide effective means of imitating human skills such as presenting information or engaging in a conversation received a considerable amount of interest [1-2]. They have shown their potential to allow users to interact with computer in a natural and intuitive manner through human communicative means. In order to let the virtual agents become a truly friendly and intelligent partner in our information space, we must endow the virtual agents with more perceptive intelligence. Emotion is proven to play an important role in human intelligence. It is fundamental to human experience, influencing cognition, perception, and everyday

A. Håkansson et al. (Eds.): Agent and Multi-agent Technology, SCI 289, pp. 323–345.
springerlink.com

tasks such as learning, communication, and even rational decision-making [3-5]. Toda proposed that emotion could provide machine with autonomy and was the ultimate origin of intelligence [6]. Minsky brought forward machine could never have intelligent behavior without emotion [7]. So, in order to realize natural, harmonious and believable intelligent human-agent interaction, we must integrate emotion with virtual agents, that is to say, they ought to have capabilities to express their own emotion and recognize user's emotion states. At the same time, the tools that can facilitate the scripting of the virtual agents with emotion intelligence are needed. It is not easy for non-expert users to write multimodal interface with virtual agents, because of the complexity of describing various behaviors if virtual agents based on a particular agent system with low-level description language. For the purpose to resolve the problem and make people write multimodal interface as easily and efficiently as people building website using HTML, many researchers have devoted themselves into the creation of a medium-level description language commonly applicable to many virtual agents.

In this chapter, we will describe the Multimodal Interaction Markup Language (MIML) that derives from the Multimodal Presentation Markup Language (MPML), a language specifically designed for non-expert users allowing them to direct the behavior of multiple virtual agents when creating web-based presentation [8-9]. The distinctness between MIML and MPML is that MPML can only script the presentation capability (verbal and non-verbal behavior) of virtual agents, while MIML can also describes their emotion detection capability. Two ActiveX controllers are attached to MIML, which is called real-time facial expression recognition and real-time speech emotion recognition respectively.

The remaining chapter is organized as follows. The next section reviews some of the related research on scripting the behaviors of virtual agents. Section 3 is dedicated to introducing the MIML system. We first give an overview of the architecture of the MIML system, and then describe the tags we created to control the behavior of virtual agents. The explanation of facial expression recognition, speech emotion recognition and emotional speech synthesis ActiveX controllers are also given. In Section 4, the usefulness of the MIML is illustrated by discussing one actual implementations of web-based intelligent tutoring system. Section 5 discusses how we evaluated MIML in the context of an experimental application. Finally, we summarize and conclude this chapter.

2 Related Work

In recent years, there is an emerging number of scripting and representation languages with both common and similar objectives, but which have taken different approaches to specify their objectives. We start here by analyzing several broadly used languages.

2.1 VHML

The VHML(Virtual Human Markup Language), which gathers several languages (such as emotion markup language, gesture markup language, facial animation markup language etc), provides tagging structures for facial and bodily animation, gesture, speech, emotion, as well as dialogue management and so on. For example: tags representing signals (head turns left, eyes blink) or tags representing emotion (happiness, sadness). The language is XML-based and covers different abstraction levels. Despite its expressive power, current applications of the language mainly focus on "Talking Heads" rather than full-body characters [10-11].

2.2 BEAT

The BEAT (Behavior Expression Animation Toolkit), another XML-based language designed for generating virtual agents from text input, emphasizes the reusability of the nonverbal behavior generators by separating the concept of behavior generation and behavior realization. It has been used for tagging both the agents' input and output. The input is an utterance that parsed into a tree structure, which is manipulated to include information about nonverbal signals and then specified again in XML. The toolkit is then able to generate appropriate and synchronized nonverbal behaviors and synthesized speech specified by the output language containing tags describing the type of animation to be performed and its duration. Yet, the behavior generators have access to a variety of information about the text to be spoken, at different levels of abstraction, and therefore do not provide a clean interface to communicative intent. The Spark system modified BEAT to work within an avatar-based chat system, using the behavior generators to automatically animate the delivery of text messages between chatting participants [12].

2.3 APML

The APML (Affective Presentation Markup Language) is an XML-based language that has been designed to represent the Poggi's taxonomy communicative functions which are defined as a pair (meaning, signal), where the meaning item corresponds to the communicative value of the signal item. The APML allows to specify the meaning of a communicative act which is then instantiated to a specific communicative signal, depending on the virtual agents' personality or culture. The APML serves as the Mind-body interface to represent the Body's input [13].

2.4 BML/FML

The researcher lay down a general structure for multi-modal behavior generation system, the SAIBA (Situation, Agent, Intention, Behavior, Animation) framework, and specify the macro-scale multi-modal generation consisting of processing stages on three different levels: (1) planning of a communicative intent, (2) planning of a multi-modal realization of this intent, and (3) realization of the planned behaviors. The interface between stages (1) and (2) describes communicative and expressive intent without any reference to physical behavior. The FML (Function Markup Language) was proposed for specifying such information. It is meant to provide a semantic description that accounts for the aspects that are relevant and influential in the planning of verbal and nonverbal behavior. The interface between stages (2) and (3) describes multi-modal behaviors as they are to be realized by the final stage of the generation pipeline. The BML (Behavior Markup Language) was brought forward for this purpose, which can provide a general, player-independent description of multi-modal behavior that can be used to control the virtual agents [14-18].

2.5 MPML/SCREAM

The MPML (Multimodal Presentation Markup Language) compliant with standard XML, is designed for non-expert users allowing them to direct the behavior of multiple virtual agents when creating web-based presentations. In order to facilitate the generation of different kinds of presentations, MPML provides appropriate tagging structures that enable authors to utilize features that are familiar from presentations given by human presenters in web-based presentation environments. Through MPML, we can script the virtual agents to communicate information by using multiple modalities, such as synthetic speech, facial displays, hand gestures, head movements and body postures. Furthermore, with an interface with the SCREAM (Scripting Emotion-based Agent Minds) , the virtual agents not only can simply respond with a pre-defined utterance dependent on the user's speech input, but also can produce dynamic response which is decided by its mental states. The SCREAM system is a scripting tool that enables users to create emotionally and socially appropriate responses of virtual agents. Users may design the mental make-up of a virtual agent by configuring a set of parameters and behaviors relevant to affective communication and obtain quantified emotional reactions that can be input to an animation engine. [8-9, 19-25]

2.6 TVML

TVML (Television program Making Language) is a text-based language that can describe all the necessary elements for an agents-based TV program production. TVML Player is the software that can interpret the script written in TVML and generate an agents-based TV program in real time. This language enables the user to create an agents-based TV program on a desktop workspace simply using text-based script [26].

Besides the TVML, there are also some representation languages that do not follow the XML style including PAR and STEP. The PAR (Parameterized Action Representation) system allows for a fine-grained specification of parameters to modify the execution of virtual agents' actions, such as duration, purpose, emotion and personality [27-28]. The STEP (Scripting Technology for Embodied Personal Language) employs distributed logic programming and action composition operations to specify motions of VRML-based virtual agents [29]. While those two approaches permit more sophisticated control of agents, they might counteract our purpose of creating a markup language that can be used by average users rather than experts.

From the analysis above, we conclude that all the scripting and representation languages can be broadly categorized into two kinds: script the virtual agents' body and script the virtual agents' minds. Nevertheless, neither of them can script the virtual agents' emotion detection capability.

3 The MIML System

This section is devoted to explaining the MIML system architecture. The MIML is an extension of MPML. The reason we replace "P" with "I" is mainly because MIML can describe not only the presentation (such as emotion expression and text-to-speech capability etc) of virtual agents, but also their emotion detection capability to realize the intelligent interaction. The use of the MsAgent [30-31] is assumed by default just like MPML. Due to the need of speech dialogue feature, it is necessary to incorporate voice commands and TTS (Text-To-Speech) engines. Other agent systems can be used with appropriate diver programs. An overview of the MIML system architecture is shown in Figure 1.

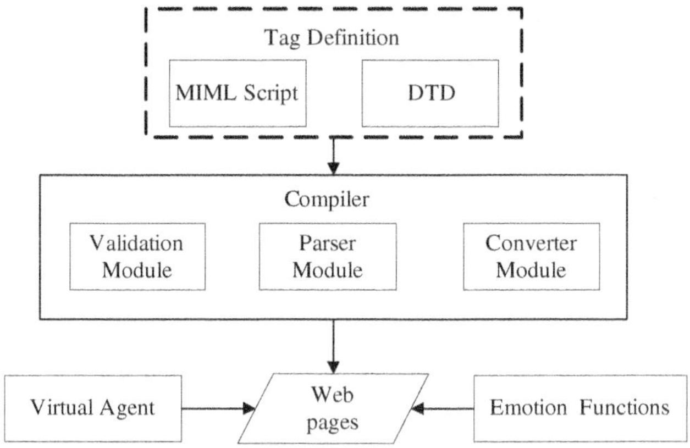

Fig. 1. Architecture of MIML

- The tags are defined to control the agents' behaviors. Document Type Definition (DTD) for the tags is created to define the grammar
- The validation module invokes the Document Type Definition (DTD) file to check the errors contained in the MIML script file
- The parser module employs the Simple APIs for XML (SAX) which is provided in MSXML.DLL to parser the MIML script
- The converter module transforms the parsed MIML script to Vbscript which is executable in web browser
- The virtual agents module uses the Microsoft Agent package (MsAgent) to provide the 2D animation sequences
- The emotion module endows the virtual agents with the emotion detection and expression capability. The facial expression recognition function and speech emotion recognition ActiveX controllers are designed so they can be called by the MIML tags on the web pages. We also enhance the emotion expression capability by the emotional speech synthesis controller.

3.1 Tags in MIML

MIML is an XML-based language that composes a tag tree described in Figure 2. We inherit some tags from MPML and all the tags are easy to learn and remember, as they follow the conventions of HTML. The root tag pair of an MIML script document is <miml> and <miml>, which contains all other tag pairs. The tag <head> specifies general information and the tag <body> refers to the sequence of events comprising the actions of virtual agents.

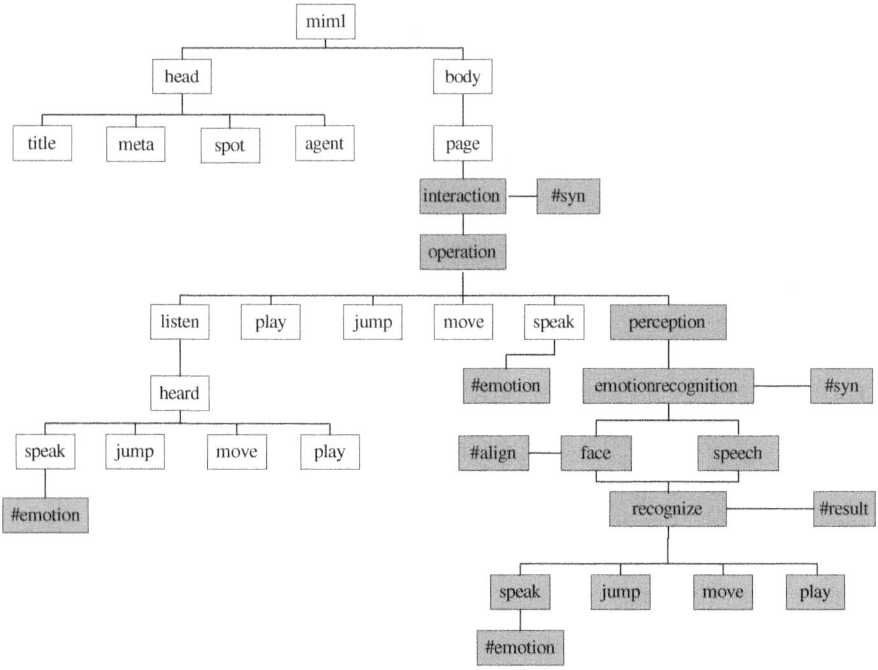

Fig. 2. Tag Structures of MIML

The tags in white boxes in Figure 2 are used to control the agents' basic behavior similar to MPML. The tags proposed to create web-based affective interaction are placed in colored boxes. Document Type Definition (DTD) for the tags is created to define the grammar and allows the author to make the virtual agents recognize the user's emotional state through facial expression and speech in the web-based interaction system. The virtual agents can make various responses (verbal and non-verbal behavior) according to the detective result. The tags are introduced as follows:

Title: This tag defines the name of the miml script document.

```
<title>
    Affective Computing
</title>
```

Meta: This tag defines the information of the author.

```
<meta name ="Zheng Li" />
```

The "name" attribute denotes the name of the MIML script author.

Spot: The tag defines the position of the virtual agents.

```
<spot id = "spot1" location= "500, 360"/>
```

The "id" attribute denotes the ID number of the location and can be omitted if unnecessary. The "location" attribute represents the position on the web page.

Agent: The tag defines the name of the virtual agent.

```
<agent id = “1” name= “genie”/>
```

The “id” attribute denotes the ID number of the agent and can be omitted if unnecessary. The “name” attribute represents the name of the agent.

Page: The tag defines the background of the interaction. The page is web page designed by the author and it appears behind the agent.

```
<page id = “1” ref = “main.html”/>
```

The “id” attribute denotes the ID number of the page and can be omitted if unnecessary. The “ref” attribute represents the name of the web page.

Interaction: The tag is the root tag pair to describe human-agent interaction which includes one sub-tag <operation>.

```
<interaction syn= “par”>
    <operation>
        ...
    </operation>
    <operation >
        ...
    </operation >
</interaction >
```

To enable interaction level and modality level synchronization, we employed synchronization mechanism like the one used in Synchronized Multimedia Integration Language (SMIL) [32] for the <operation> element. The “syn” attribute in the <interaction> tag controls synchronization of <operation> elements in it. If “syn” is “par” (parallel), all <operation> elements are executed in parallel. If “syn” is “alt” (alternative), one of the <operation> elements is executed alternatively, and if “syn” is “seq” (sequential), all <operation> elements are executed in document order.

Operation: This tag is the root tag that includes a set of human-agent exchanges.

```
<operation>
        <speak>
            Hello, nice to meet you, I am genie.
        </speak>
        <jump des=“affective.html”>
</operation>
```

Listen and Heard: These two tags enable the agent to accept speech input from a user.

```
<listen>
        <heard value = “back”>
            <jump des=“last.html”/>
        </heard>
        <heard value = “forward”>
            <jump des=“next.html”/>
```

```
        </heard>
</ listen >
```

The <listen> tag lets the agent enter into the listening mode. If a speech input text recognized by the speech recognizer matches with a value specified in a <heard> tag, then the actions described inside the <heard> tag are executed. If no match is found, the listening mode will be closed and the interaction will proceed to the next step. The Microsoft voice commands engine is used for these two tags.

Play: This tag defines which action the agent will play, such as "point right", "look left", "show", "think" etc. Most Microsoft agents typically have about 30-50 predefined actions [30-31].

```
<play act = "think" />
```

The "act" attribute represents the name of the action.

Jump: This tag defines which web page will jump to.

```
<jump des="next.html"/>
```

The "des" attribute represents the name of the web page that will jump to.

Move: The tag controls the movement of the virtual agent on the screen to the destination spot.

```
<move  spot= "spot1"/>
```

The "spot" attribute denotes the position of the screen. The value "spot1" is defined beforehand; other, a direct specification of the destination spot is possible using x and y argument instead, for example, {x="200" y="100"}.

Speak: The agent speaks one or more sentences through a TTS (text-to-speech) or emotional speech synthesis engine.

```
<speak  emotion = happiness>
    I am very happy.
  ...
</speak>
```

The "emotion" attribute represents which emotion the agent will speak with and can be omitted if unnecessary. If it is omitted, the TTS will be used directly. If not, the emotional speech synthesis engine will be called.

Perception: This tag is the root tag of the perception module.

```
<perception>
    <emotionrecognition syn= "par">
      ...
    </emotionrecognition>
</perception>
```

Emotionrecognition: This tag is the root tag of the emotion recognition module.

```
<emotionrecognition syn= “alt”>
    <face align = “right”>
       ...
    </face>
    <speech>
       ...
    </speech>
</emotionrecognition>
```

The “syn” attribute in the <emotionrecognition> tag controls synchronization of <face> element and <speech> element in it. If “syn” is “par” (parallel), the <face> and <speech> elements are executed in parallel. If “syn” is “alt” (alternative), one of the <face> and <speech> elements is executed alternatively, and if “syn” is “seq” (sequential), the <face> and <speech> elements are executed in document order.

Face: This tag is used to control the facial expression recognition.

```
<face align=right>
   ...
</face>
```

The “align” attribute represents the position of the facial expression recognition ActiveX controller on the web page.

Speech: This tag is used to control the emotion speech recognition.

```
<speech>
   ...
</speech>
```

Recognize: The tag <recognize> and “result” attribute attached to it are to be employed with the <face> and <speech> tags as parent. They are equivalent to the C “switch” and “case” instruction respectively. It compares the return value of emotion recognition function with the value of “result” attribute, and executes the script included in the <recognize> tag if they are identical.

```
<face align = “right”>
     < recognize  result =“happiness”>
         <speak emotion = “happiness”>
              I am happy.
         </speak>
     </ recognize >
     < recognize result =“surprise”>
          <speak emotion = “surprise” >
              I am surprise.
         </speak>
     </recognize>
</face>
```

3.2 MIML Compiler

MIML compiler is the software that can interpret the script written in MIML and generate an interaction system in real time. It is composed of validation module, parser module and converter module. Firstly, the validation module invokes the DTD to check the MIML script text file for syntactical errors. Then the parser module calls SAX (Simple APIS for XML) in MSXML.DLL to parse the MIML script. Finally the converter module generates Vbscript code that is executable in web browser to perform an interaction. The backgrounds of the MIML interaction are constituted of HTML pages. The Vbscript code will be embedded into the appointed HTML pages automatically by the converter. Currently, MIML assumes Microsoft Internet Explorer 6.0 (or higher) to run the interaction. Some of the rules for the converter are listed in Figure 3.

```
<perception>
   <emotionrecognition>
      <face align="right">
        <recognize result="happiness">
            <speak emotion="happiness">
                Oh,you are smiling.
            </speak>
        </recognize>
        <recognize result="anger">

            <speak emotion="sadness">
                Oh,you are angry.
                What happened?
            </speak>
              ...
        </recognize>
      </face>
   </emotionrecognition>
</perception>
```

```
Sub FaceReFunction
     a = Face.Camload()
     Set FaceReRequest1 = agent.Play("RestPose")
End Sub

Sub AlertUser(a)
     If a = "happy" Then
          agent.Play "Pleased"
          agent.Speak("Oh,you are smiling.")
     ElseIf a = "anger" Then
          agent.Play "Sad"
          agent.Speak("Oh,you are angry. What happened?")

     ...

     End If
End Sub

Sub Agent_RequestComplete (ByVal Request)
     Select Case Request
       Case FaceReRequest0
         ID = SetTimeout("FaceReFunction", 100)
       Case FaceReRequest1
         ID = SetTimeout("AlertUser(a)", 100)
     End Select
 End Sub
```

Fig. 3. Rules for Converter

3.3 Emotion Functions

In order to get an idea about the effectiveness of machine-based emotion recognition compared to humans, a review of research has been done by Huang, Chen, and Tao [33]. They investigated the performance of machine-based emotion recognition employing both video and audio information. Their work was based on human performance results reported by De Silva et al. [34]. Their research indicated that the machine performance was on average better than human performance with 75% accuracy. In addition, comparing detection of confusions indicated similarities between machine and human. These results are encouraging in the context of our research for endowing virtual agents with affective detection capability. The facial expression recognition function and speech emotion recognition function are attached to MIML. They are realized as ActiveX controller so they can be called by the MIML tags on the web pages. Meanwhile, the emotional speech synthesis controller is designed to enhance the emotion expression capability.

3.3.1 Facial Expression Recognition

An automatic facial expression analyzer was developed in our research group [35]. Three main components constitute the analyzer: face detection, facial feature extraction and facial expression classification. We adopt the concept of integrated image proposed by Viola [36] and the Adaboost method to train the face detector and the facial expression classifier. However, the performance of facial expression recognition could be influenced by noise or occlusion on the face. In many scenarios there is the opportunity for occlusion. The most common causes of occlusion can be pose variation, glass wearing, and hair or hand covering etc. The ability to handle occluded facial features is most important for achieving robust facial expression recognition. We recently improved our facial expression recognition system using a robust approach. In contrast to normal methods that do not deal with the occlusion regions separately, this new method is to detect and eliminate the facial occlusions. Thus, the procedure of facial occlusion removal is added to normal classification procedure. Our method of robust facial expression recognition consists of four stages: face reconstruction using Robust principal component analysis (RPCA), occlusion detection using saliency detection, occlusion region reconstruction, and AdaBoost classification, as shown in Figure 4.

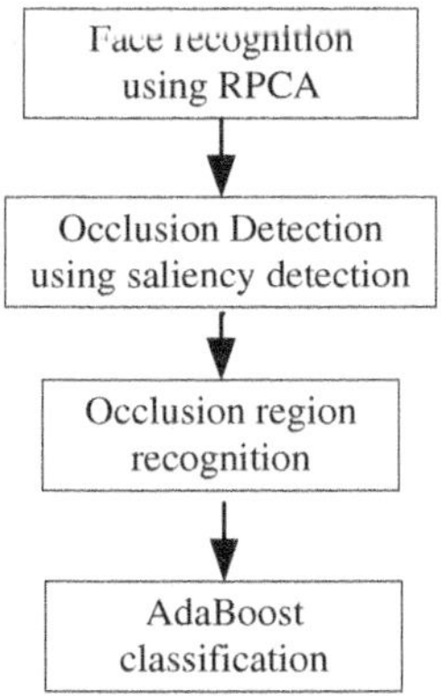

Fig. 4. Robust Facial Expression Recognition

Face Recognition Using RPCA

Robust principal component analysis (RPCA) is robust to outliers (i.e. artifacts due to occlusion, illumination, image noise etc.) in training data and can be used to construct low-dimensional linear-subspace representations from noisy data. When the face contains a small fraction of the subjects with glasses, or forehead overlaid by hair, or chin surrounded by hands, the pixels corresponding to those coverings are likely to be treated as outliers by RPCA. Hence, the reconstructed image of the original face will possibly not contain the occlusions.

Occlusion Detection Using Saliency Detection

To find the occlusion regions on the face, we adopt the method of saliency detection. Firstly, the original face image is transformed into gray level and normalized to $I(x, y)$ using histogram equalization. Then, $I(x, y)$ is reconstructed to $R(x, y)$ using RPCA. We can obtain the residual image $D(x, y)$ between the reconstructed image $R(x, y)$ and $I(x, y)$ by:

$$D(x, y) = |R(x, y) - I(x, y)|$$

Then, the residual image $D(x, y)$ is put to a saliency detector to find the local places with high complexity, which is hypothesized to be the occlusion on the face. The measure of the local saliency is defined as:

$$H_{D,Rx} = -\sum_i P_{D,R_X}(d_i) \log_2 P_{D,R_X}(d_i)$$

Where $P_{D,R_X}(d_i)$ is the probability of descriptor (or difference image) D taking the value d_i in the local region R_x. We apply the saliency detection in the residual image over a wide range of scale, and set the threshold value of $H_{D,Rx}$. The region with biggest $H_{D,Rx}$ value over the threshold is set to the occlusion region. If all regions have $H_{D,Rx}$ less than the threshold, it is presumed that no occlusion exists. Note that we just choose one occlusion region in one operation of saliency detection even if there are multiple regions with saliency value over the threshold.

Occlusion Region Reconstruction

Detailed information is most important to facial expression recognition. To avoid the wrong information introduced by face reconstruction in non-occluded region, we adopt the mechanism of occlusion region reconstruction rather than the total face reconstruction. To obtain the new face image $P(x, y)$, pixel values of the detected occlusion region will be replaced by the reconstructed face using RPCA. Thus, the wrong information in the occlusion region may be shielded while the other regions of the face retain the same. To further decrease the impact of occlusion for facial expression reconstruction, we perform occlusion region reconstruction for several iterations until the difference of the reconstructed face between two iterations is below a threshold. The new face image $P_t(x, y)$ in iteration t can be obtained by:

$$P_t(x, y) = \begin{cases} I(x, y) & (\mathrm{x}, \mathrm{y}) \notin R_{occlusion} \\ R_t(x, y) & (\mathrm{x}, \mathrm{y}) \in R_{occlusion} \end{cases}$$

Where $I(x, y)$ is the normalized image, $R_t(x, y)$ is the reconstructed image using RPCA in iteration t, and $R_{occlusion}$ defines the occlusion region. Note that

$$R_t(x,y)=\begin{cases} RPCA(I) & t=1 \\ RPCA(P_{t-1}) & t>1 \end{cases}$$

Where *RPCA* designates the RPCA procedure, and t is the iteration index.

AdaBoost Classification

We employ harr-like features for feature extraction and implement multiple one-against-rest two-class AdaBoost classifiers for robust facial expression recognition. In the algorithm, multiple two-class classifiers are constructed from weak features which are selected to discriminate one class from the others. It can solve the problem that weak features to discriminate multiple classes are hard to be selected in traditional multi-class AdaBoost algorithm. The proposed algorithms were trained and tested on Beihang University Facial Expression Database (BHUFE) [35]. This database consists of 57 university students in age from 19 to 27 years old and includes videos with hand and glass occlusion when displaying kinds of facial expressions. We also randomly add occlusions on the face to generate occluded faces. Five basic emotions can be recognized including anger, happiness, sadness, disgust and surprise. The average recognition rate is 81.5%.

3.3.2 Speech Emotion Recognition

Speech is a primary and efficient way to express emotions. With the development of human machine intelligent interaction, speech emotion recognition has attracted increasing attention in the artificial intelligence field. An automatic speech emotion recognition system based on a hybrid of Hidden Markov Models (HMMs) and Artificial Neural Network (ANN), which was focused on the speaker dependent emotion recognition, was developed in our research group [37]. In this chapter, aiming at speaker independent emotion recognition, a HMMs fusion system based on a new algorithm named weighted ranked voting is advised to improve recognition rate greatly. In addition, relative speech features are used to enhance the precision of emotion recognition system as well as to make it more robust.

Feature Extraction and Relative Feature Calculation

Study on emotion of speech indicates that pitch, energy, duration, formant, Mel prediction cepstrum coefficient (MPCC) and linear prediction cepstrum coefficient (LPCC) are effective absolute features to distinguish certain emotions. In the paper, for each frame, six basic features, including pitch, amplitude energy, box-dimension, zero cross ratio, energy-frequency-value, first formant frequency, as well as their first and second derivatives, are extracted. Besides, 10-order LPCC and 12-order MPCC are also be extracted. Although absolute features of speeches corresponding to same emotion have large differences among different speakers, the differences of feature change induced by emotion stimulation are small relatively. Hence, relative features which reflects feature change is more credible than absolute features for emotion recognition. Relative features used in the paper embody alterations of pitch, energy or other features. They are obtained by computing the change rate relative to natural speech. Features of the kind are robust to different speakers because its calculation is combined with normalization of the features of neutral speeches. For computing relative features, the reference features of neutral version of each text and each speaker should be obtained by calculating the statistics of some frame-based parameters. In this chapter, the statistic features used are means of dynamic features, including pitch, amplitude energy, energy-frequency-value, box-dimension, zero cross ratio, and first formant frequency as well as their first and second derivatives. Then, the six statistic features are used to normalize the corresponding dynamic features for each emotion speech, including training samples and test samples. Assuming $Mf_i, i = 1, 2, \cdots, 18$ are reference features of neutral version, $\vec{f_i}, i = 1, 2, \cdots 18$ are the corresponding dynamic feature vectors, the relative feature vectors $R\vec{f_i}$ can be obtained according to following formula:

$$R\vec{f}_i = (\vec{f}_i - Mf_i)/(Mf_i + 0.0000001)$$

where $\vec{f_i} = [f_{i1}, f_{i2}, \ldots\ldots f_{iL}]^T$, $R\vec{f_i} = [Rf_{i1}, Rf_{i2}, \cdots Rf_{iL}]^T$ and L indicates the length of feature vector.

Isolated HMMs

The HMMs are left-right discrete models. The most pervasive methods, Forward-Backward Procedure, Viterbi Algorithm and Baum Welch re-estimation are employed in this chapter. Baum Welch re-estimation based on likelihood training criterion is used to train the HMMs, each HMM modeling one emotion; Forward-Backward Procedure exports the likelihood probability; Viterbi Algorithm, focusing on the best path through the model, evaluates the likelihood of the best match between the given speech observations and the given HMMs, then achieves the "optimal" state sequences. The recognizing process based on HMMs is shown as Figure 5. A speech sample is analyzed and then represented by a feature vector, according to which the likelihood between the speech sample and each HMM is computed. Then the emotion state corresponding to maximum likelihood is selected as the output of the classifier through comparison.

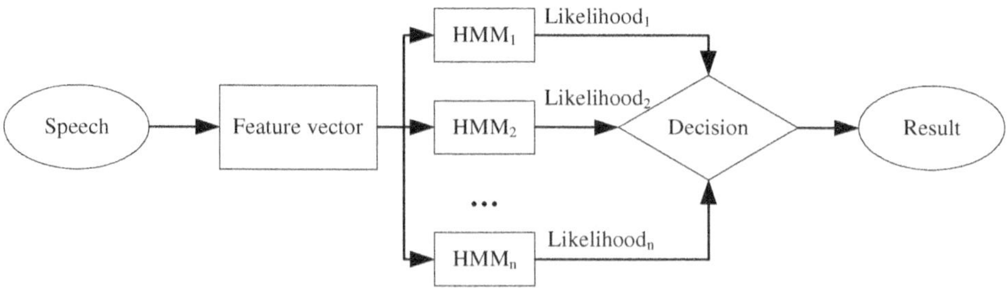

Fig. 5. Emotion Recognition by HMMs

HMMs Fusion System

For the complexity of speech emotion recognition, single classifier systems have limited performance. In recent years, classifier fusion proves to be effective and efficient. By taking advantage of complementary information provided by the constituent classifiers, classifier fusion offers improved performance. Classifier fusion can be done at two different levels, namely, score level and decision level. In score level fusion, raw outputs (scores or confidence levels) of the individual classifiers are combined in a certain way to reach a global decision. The combination can be performed either simply using the sum rule or averaged sum rule, or more sophisticatedly, using another classifier. Decision level fusion, on the other hand, arrives at the final classification decision by combining the decisions of individual classifiers. Voting is a well-known technique for decision-level fusion. It can mask errors from one or more classifiers and make the system more robust. Voting strategies include majority, weighted voting, plurality, instance runoff voting, threshold voting, and the more general weighted k-out-of-n systems. In this chapter, four HMMs classifiers, which have different feature vectors (see Table1), are used. HMMs classifier takes only the emotion which satisfies the model most as the recognition result. But the correct result often should be the emotion which satisfies the model secondly or thirdly. So a new algorithm named weighted ranked voting, which is a reformed version of ranked voting method provided by C De Borda [38], is proposed. Ranked voting method permits a voter to choose more than one candidate in proper order. Moreover, the improved algorithm also makes the voted emotions attached by different weights.

Table 1. Feature Vector of 4 HMMs Classifiers

Classifier	Feature vector
1	pitch, box- dimension, energy with their first and second derivatives; 10-order LPCC
2	energy-frequency-value, box- dimension, formant with their first and second derivatives; 12-order MFCC
3	pitch, zero cross ratio, formant with their first and second derivatives; 12-order MFCC
4	all features extracted in the paper

For a speech sample and a classifier, the voting weight of a certain emotion is determined according to the likelihood between the speech and the HMM model corresponding to the emotion. Firstly, the likelihood values between the speech sample and HMM models are calculated. Secondly, the emotion states are sorted according to likelihood. Then, the voting weights of the first three emotions are allocated according to the order. In the paper, the weight is determined as Table 2. Finally, the weights from four classifiers corresponding to each emotion are summed up and the emotion which has maximum value is selected as result.

Table 2. Weight Allocation for Voting

	First	Second	Third
Weight	1	0.6	0.3

The steps are listed as follows for each speech sample.

- step1: Initialize weight value as 0 for each emotion.
- step2: Sort emotions according to likelihood for each classifier.
- step3: Vote the first three emotion attached by weight according to Table 3 for each classifier.
- step4: Sum up the weights from four classifiers for each emotion and choose the emotion which has the biggest weight sum as the recognition result.

To evaluate the performance of the proposed classifier in this chapter, Database of Emotional Speech was set up to provide speech samples. This corpus contains utterances of five emotions, twenty texts and five actors, two males and three females. Each speaker repeats each text three times in each emotion, meaning that sixty utterances per emotion. For classifier evaluation, 1,140 samples of eight speakers, which have been assessed, are used. The evaluation was done in a "leave-one-speaker-out" manner. One feature vector, formed by six relative features combined with LPCC or MPCC, is used. The average recognition rate is 74.6% for five basic emotions including anger, happiness, sadness, disgust and surprise.

3.3.3 Emotional Speech Synthesis

Speech is the easiest way to convey intention, and it is one of the fundamental methods of conveying emotion, on a par with facial expression. In MPML, the agent can express emotion through TTS by modified the speech parameters. Due to the limitation of the TTS engine, only the speed, pitch average, pitch range and speech intensity are controlled. We enhance this by the emotional speech synthesis controller. The variety rule of prosodic features containing pitch frequency (F0), energy and velocity are concluded by analyzing emotional speech in our Emotional Speech Database. The autocorrelation function (ACF) method based on Linear Predictive Coding (LPC) and wavelet transform approach are employed to extract the F0 and tone respectively. Then prosodic features regulation is set up by utilization of Pitch Synchronous OverLap Add (PSOLA) and the original peace speeches are transformed into appointed emotional speech, including happy, anger, surprise and sad, based on the rules and regulation. Figure 6 illustrates the work flow of our approach.

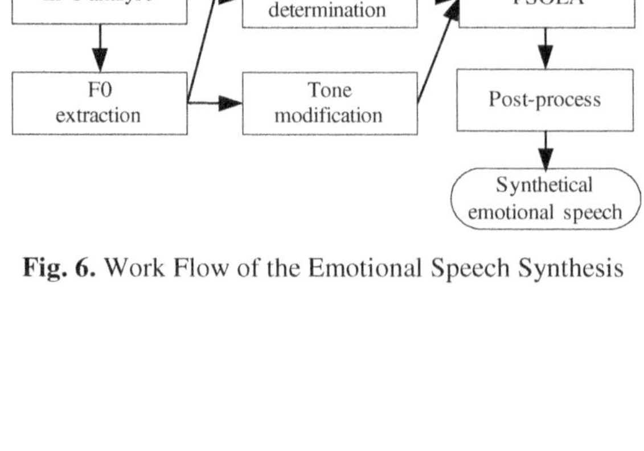

Fig. 6. Work Flow of the Emotional Speech Synthesis

- **Pre-process.** Include noise elimination, pre-emphasis and amplitude normalization
- **LPC analyze.** Partition the original speech into frame, take LPC analysis of each frame, get the LPC residual function and first order reflection coefficient
- **F0 extraction.** Get the F0 through the autocorrelation analysis of the residual function and the F0 profile curve of the original speech
- **Surd and sonant separation.** Do the surd and sonant separation according to first order reflection coefficient, signal energy and frequency extraction result
- **Glottal Closure Instances (GCI).** Determine the GCI according to the F0 extraction
- **Tone modification.** Extract the tone information using wavelet transform; modify the tone information according to the target-emotion and adopt inverse wavelet transform to get the F0 curve
- **PSOLA**. Use PSOLA technology to transform original speech into appointed emotional speech
- **Post-process.** De-emphasis, i.e. do the anti-operation of pre-emphasis in pre-process to restore speech effect

3.4 Feature

From the introduction aforesaid, we can conclude that the MIML is a medium-level description language allowing ordinary people to write web-based affective human-agent interaction easily. The feature of MIML can be summarized as:

- **Easy of use.** The tags defined in MIML follow the convention of HTML and the tags' names clearly indicate their meaning. Writing MIML script is similar to writing Web contents with HTML.
- **Extensibility.** The MIML serves as a core for integrating various components. New features or functions can be added into the MIML system easily by incorporating new tags.

4 Illustration

In this section, we will describe one web-based scenario that instantiates the intelligent tutoring system scripted by MIML. We believe that an important factor in the success of human one-on-one tutoring is the tutor's ability to identify and respond to the student's affective state. This characteristic may give the student the sensation that there is someone behind the program who follows his learning development and cares about him as a human tutor would. Using MIML, user can easily set up the intelligent web-based tutoring system. The script fragment of the tutoring system is demonstrated in Figure 7. Figure 8 is one of the web page results after compiling of the MIML script. In Figure 7, line 6 means the background web page is "main.html". In line 8-11, the agent "genie" expresses his welcome to the learner with happiness emotion, then, the web page will jump to "affective.html". In this web page, the agent will take the pedagogical action (line 17-24), meanwhile, one of the facial expression recognition function and speech emotion function is executed alternatively to detect the learner's emotion state (line 26-55). The agent is capable of adapting to these states wisely. If the result is unhappiness, the agent will make detailed explanation to the learner (line 30-36 and line 42-48). Otherwise, the learning process will continue (line 37-39 and line 49-51).

```
1 <miml>
2     <head>
3        <agent id="genie" character="genie"/>
4     </head>
5     <body>
6        <page id="1" ref="main.html">
7           <interaction>
8             <speak emotion="happiness">
9                Hello, I am genie, nice to meet you. Welcome to our
10               tutoring system! Let's talk about Affective Computing.
11            </speak>
12            <jump des="affective.html"/>
13          </interaction>
14       </page>
15       <page id="2" ref="affective.html">
16          <interaction syn="par">
17           <operation>
18            <speak emotion="happiness">
19               Affective Computing is computing that relates
20               to, arises from, or deliberately influences
21               emotion or other affective phenomena.
22               ...
23            </speak>
24           </operation>
25         <!-- Invoke the emotion recognition function -->
26           <operation>
27            <perception>
28             <emotionrecognition syn="alt">
29              <face align="right">
30                <recognize result="unhappiness">
31                  <speak>
32                    It seemed as if you didn't understand what I said, be patient,
33                    I will explain it again.
34                     <jump des="detail.html"/>
35                  </speak>
36                </recognize>
37                <recognize result="default">
38                   <jump des="continue.html"/>
39                </recognize>
40              </face>
41              <speech>
42                <recognize result="unhappiness">
43                  <speak>
44                    It seemed as if you didn't understand what I said, be patient,
45                    I will explain it again.
46                  </speak>
47                   <jump des="affective.html"/>
48                </recognize>
49                <recognize result="default">
50                   <jump des="continue.html"/>
51                </recognize>
52              </speech>
53             </emotionrecognition>
54            </perception>
55           </operation>
56         </interaction>
57       </page>
58    </body>
59 </miml>
```

Fig. 7. Script Fragment of Web-based Tutoring system

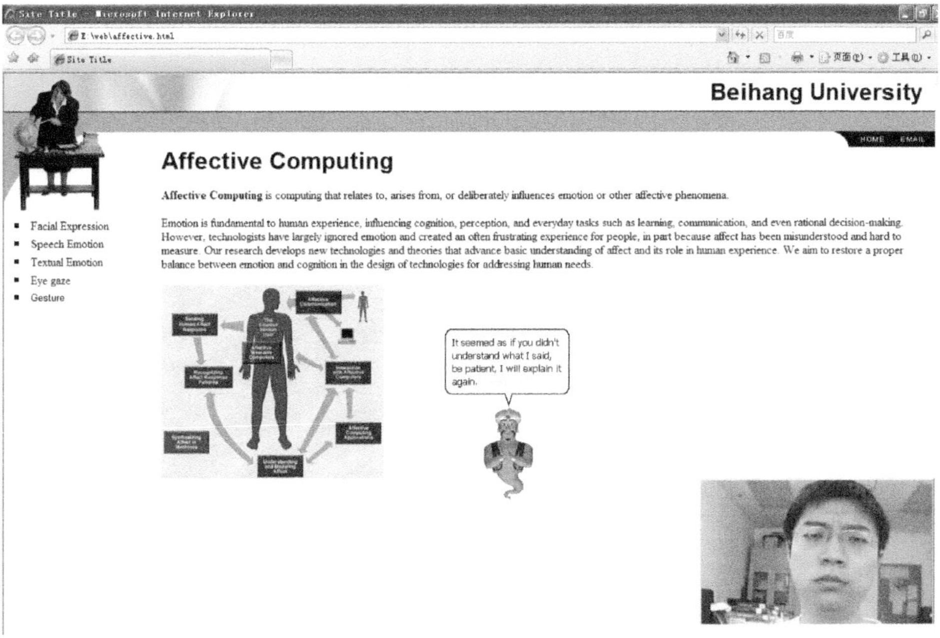

Fig. 8. Result of the Script Fragment

5 Evaluation and Results

A 60-person user study was conducted to quantitatively measure the "performance" of MIML. In this section, we first present the experiment design and the performance measurement results, followed by some informal comments.

5.1 Experiment Design and Performance Measurement Results

Sixty users were asked to use three kinds of agent scripting languages, including TVML, MPML and MIML, to describe the human-agent interaction scenario they want to get. Then they were required to evaluate the three languages against four aspects: usage, interactivity, intelligence, and adoption. For each aspect, users were asked to evaluate how much they agreed with a statement, on a scale from 1 (strongly disagree) to 7 (strongly agree). The quantitative results of the questionnaire are shown in Figure 9. The length of each bar indicates the mean score, which is also written at the right end of each bar.

The results of the performance measurement evaluation were generally positive. On three aspects, MIML well outperformed the other two languages. As for the item "easy of use", it is mainly because MIML has more tags than MPML, users need to learn more about how to use them. In general, the results were more or less expected and suggested that the MIML worked well enough to bring measurable benefit to the intelligent human-agent interaction.

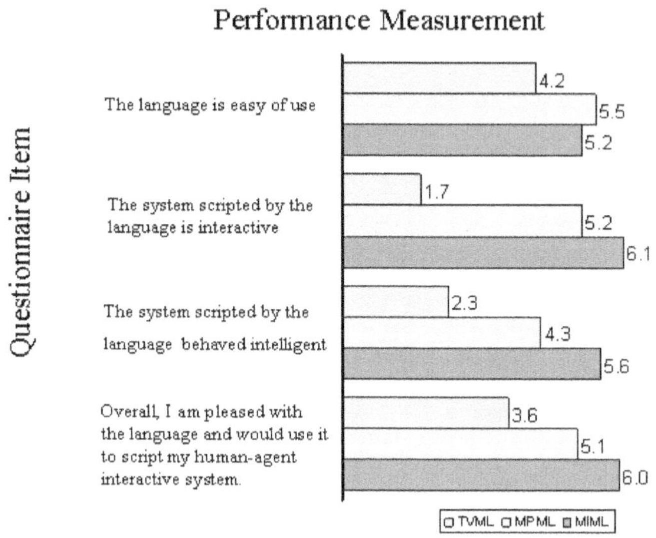

Fig. 9. User Testing Questionnaire Results

5.2 Informal Subject Comments

In the post-study interview the users gave both positive and negative comments about MIML and the positive comments were far more than negative ones as we expected. We only list main negative comments for improvement in the future.

Several subjects complained that the emotional recognition accuracy needs to be improved. Though we mentioned that the machine performance was on average better than human performance with 75% accuracy [33-34], we will devote ourselves to improving the recognition rate of the emotional functions.

Another subjects remarked that in despite of the agent's emotion detection capability can be scripted easily, the agent sometimes misinterpreted the user's intention and did not allow the user to "undo" incorrect system decisions. Our conceivable remedies include: (1) the user explicitly confirms a decision of the system, e.g. by simple verbal input ("Yes", "No"), and (2) the system detects the user's acknowledgement (head nodding) or negative signal (head shaking) through posture recognition, i.e. We will extend MIML with the "undo" tags and function.

We conclude this section with the statement of a subject who expressed excitement about the MIML with the words "Overall, I would say it is a very nice language. I can get the interaction scenario I want easily. Moreover, the agents can sense my affective states. I think it is intelligent and attractive."

6 Conclusion and Future Work

Recent years have seen many efforts to involve virtual agents as a crucial component of application fields including tutors in E-learning system, recommenders in E-commerce system, actor in entertainment system and partners in chatting system etc.

That results in the growing number of scripting languages for controlling the behaviors of the virtual agents. However, all the languages in existence can not script the intelligent perceptive capability, especially the emotional detection behavior. In this chapter, we have described the architecture of MIML, a markup language for scripting the web-based intelligent interaction of virtual agents which can not only simply respond with a pre-defined utterance dependent on the user's input, but also produce dynamic response which is decided by user's emotional state. The discussion of one actual web-based intelligent interaction implemented with MIML, the tutoring system, was intended to show how to mark up the interaction system that features virtual agent with intelligent behavior.

Since the current implementation is based on the recognition rate of the facial expression and speech emotion, we are planning to improve the recognition accuracy, and besides, we will add more tags and modules into the MIML architecture. The modules related to the emotional intelligent perception include textual emotion detection module. With this module, the virtual agents can detect user's emotion state through text input. Meanwhile, other intelligent perception modules, such as eye-tracking module, gesture and body detection module (nodding or shaking head), "undo" module are also planned to integrate into the MIML architecture. Eventually, we aim to make the MIML a powerful and easy-to-use virtual agent scripting tool for non-expert users to script the attractive, natural, harmonious and believable intelligent human-agent interaction.

Acknowledgments

This article is derived and extended from the paper [39] originally published by Springer. This work is supported by the High Technology Research and Development Program of China (863 Program, No.2006AA01Z135), the National Research Foundation for the Doctoral Program of Higher Education of China (No.20070006057) and National Nature Science Foundation of China (No.60873269). We would like to appreciate the Professor Mitsuru Ishizuka at University of Tokyo and the Associate Professor Helmut Prendinger at the National Institute of Informatics, who have provided many free resources on their home pages.

References

1. Allen, J., Byron, D., Dzikovska, M., Ferguson, G., Galescu, L., Stent, A.: Towards Conversational Human-Computer Interaction. AI Magazine 22(4), 27–38 (2001)
2. Cassell, J., Sullivan, J., Prevost, S., Churchill, E.: Embodied Conversational Agents. The MIT Press, Cambridge (2000)
3. Picard, R.: Affective Computing. The MIT Press, Cambridge (2000)
4. Preece, J., Rogers, Y., Sharp, H.: Interaction Design, Beyond Human-Computer Interaction. John Wiley&Sons, Inc., Chichester (2002)
5. MIT Media Lab, `http://affect.media.mit.edu/`
6. Toda, M.: Basic Structure of the Urge Operations, in the urge theory of emotion and cognition. SCCS Technical report, Chuyko University, Nagoya (1994)
7. Minsky, M.: The Society of Mind. Simon and Schuster, New York (1986)

8. Prendinger, H., Descamps, S., Ishizuka, M.: MPML: A Markup Language for Controlling the Behavior of Life-like Characters. Journal of Visual Languages and Computing 15(2), 183–203 (2004)
9. Prendinger, H., Ishizuka, M.: Life-Like Characters-Tools, Affective Functions and Applications. Cognitive Technologies Series. Springer, Heidelberg (2004)
10. Marriott, A., Stallo, J.: VHML - Uncertainties and problems, A discussion. In: Proc. AAMAS 2002 Workshop on ECA-Let's Specify and Evaluate Them, Bologna, Italy (2002)
11. VHML Home Page, `http://www.vhml.org/`
12. Cassell, J., Vilhjalmsson, H., Bickmore, T.: BEAT: The Behavior Expression Animation Toolkit. In: Proc. SIGGRAPH 2001, Los Angeles, USA, pp. 477–486 (2001)
13. DeCarolis, B., Caroglio, V., Bilvi, M., Pelachaud, C.: APML: a Mark-up Language for Believable Behavior Generation. In: Proc. AAMAS 2002 Workshop on ECA-Let's Specify and Evaluate Them, Bologna, Italy (2002)
14. Kopp, S., Wachsmuth, I.: Synthesizing Multimodal Utterances for Conversational Agents. Computer Animation and Virtual Worlds 15(1), 39–52 (2004)
15. Kopp, S., Krenn, B., Marsella, S., Marshall, A., Pelachaud, C., Pirker, H.: Towards a Common Framework for Multimodal Generation: the Behavior Markup Language. In: Gratch, J., Young, M., Aylett, R.S., Ballin, D., Olivier, P. (eds.) IVA 2006. LNCS (LNAI), vol. 4133, pp. 205–217. Springer, Heidelberg (2006)
16. Heylen, D., Kopp, S., Mareslla, S., Pelachaud, C., Vilhjalmsson, H.: The Next Step towards a Function Markup Language. In: Prendinger, H., Lester, J.C., Ishizuka, M. (eds.) IVA 2008. LNCS (LNAI), vol. 5208, pp. 270–280. Springer, Heidelberg (2008)
17. Heylen, D., Maat, M.: A Linguistic View on Functional Markup Language. In: Proc. AAMAS 2008 Workshop on Functional Markup Language, Estoril (2008)
18. Kreen, B., Sieber, G.: Functional Markup for behavior Planning. Theory and Practice. In: Proc. AAMAS 2008, Estoril (2008)
19. Prendinger, H., Ishizuka, M.: Scream: Scripting Emotion-based Agent Minds. In: Proceeding of AAMAS 2002 Workshop on ECA-Let's Specify and Evaluate Them, Italy (2002)
20. Prendinger, H., Descamps, S., Ishizuka, M.: Scripting Affective Communication with Virtual Characters in Web-based Interaction System. Applied Artificial Intelligence (2002)
21. Prendinger, H., Ishizuka, M.: Virtual Characters Tools, Affective Functions and Applications. Cognitive Technologies Series. Springer, Heidelberg (2004)
22. Prendinger, H., Ishizuka, M.: The Empathic Companion: a Character-based Interface that Addresses User's Affective States. Journal of Applied Artificial Intelligence 19(3-4), 267–285 (2005)
23. MPML Home Page, `http://www.miv.t.u-tokyo.ac.jp/MPML/mpml.html`
24. Nischt, M., Prendinger, H., Ishizuka, M.: MPML3D: A Reactive Framework for the Multimodal Presentation Markup Language. In: Gratch, J., Young, M., Aylett, R.S., Ballin, D., Olivier, P. (eds.) IVA 2006. LNCS (LNAI), vol. 4133, pp. 218–229. Springer, Heidelberg (2006)
25. Ullrich, S., Bruegmann, K., Prendinger, H., Ishizuka, M.: Extending MPML3D to Second Life. In: Prendinger, H., Lester, J.C., Ishizuka, M. (eds.) IVA 2008. LNCS (LNAI), vol. 5208, pp. 281–288. Springer, Heidelberg (2008)
26. TVML Home Page, `http://www.nhk.or.jp/strl/tvml/`
27. Badler, N.: Parameterized Action Representation for Virtual Human Agents. The MIT Press, Cambridge (2000)
28. PAR Home Page, `http://hms.upenn.edu/software/par/`

29. Huang, Z., Eliebs, A.: STEP: a Scripting Language for Embodied Agent. In: Proc. of PRCAI 2002 Workshop on Virtual Animated Agent - Tools, Affective Functions and Applications, Tokyo (2002)
30. Microsoft Agent Home Page, `http://www.microsoft.com/msagent`
31. Microsoft, Developing for Microsoft Agent. The Microsoft Press (1998)
32. SMIL Home Page, `http://www.w3.org/AudioVideo/`
33. Huang, T., Chen, L., Tao, H.: Bimodal Emotion Recognition by Man and Machine. In: ATR Workshop on Virtual Communication Environments (1998)
34. DeSilva, L., Miyasato, T., Nakatsu, R.: Facial Emotion Recognition Using Multimodal Information. In: Han, Y., Quing, S. (eds.) ICICS 1997. LNCS, vol. 1334, pp. 397–401. Springer, Heidelberg (1997)
35. Mao, X., Xue, Y.L.: Beihang University Facial Expression Database and Multiple Facial Expression Recognition. In: Proc. of ICMLC 2006, pp. 369–372 (2006)
36. Viola, P.: Rapid Object Detection Using a Boosted Cascade of Simple Features. In: Proc. CVPR 2001, pp. 511–518 (2001)
37. Mao, X., Zhang, B., Luo, Y.: Speech Emotion Recognition Based on a Hybrid of HMM/ANN. In: Proc. WSEAS 2007, pp. 181–184 (2007)
38. Moulin, H.: Axioms of cooperative decision making. Cambridge University Press, Cambridge (1988)
39. Mao, X., Li, Z., Bao, H.Y.: Describing and Generating Web-based Affective Human-agent Interaction. In: Lovrek, I., Howlett, R.J., Jain, L.C. (eds.) KES 2008, Part I. LNCS (LNAI), vol. 5177, pp. 625–632. Springer, Heidelberg (2008)

Author Index

GPSR Compliance

The European Union's (EU) General Product Safety Regulation (GPSR) is a set of rules that requires consumer products to be safe and our obligations to ensure this.

If you have any concerns about our products, you can contact us on ProductSafety@springernature.com

In case Publisher is established outside the EU, the EU authorized representative is:

Springer Nature Customer Service Center GmbH
Europaplatz 3
69115 Heidelberg, Germany

Batch number: 10755466

Printed by Printforce, the Netherlands